Technology Due Diligence:
Best Practices for Chief Information Officers, Venture Capitalists, and Technology Vendors

Stephen J. Andriole
Villanova University, USA

INFORMATION SCIENCE REFERENCE

Hershey · New York

Acquisitions Editor:	Kristin Klinger
Development Editor:	Kristin Roth
Senior Managing Editor:	Jennifer Neidig
Managing Editor:	Jamie Snavely
Assistant Managing Editor:	Carole Coulson
Copy Editor:	Jeannie Porter
Typesetter:	Amanda Appicello
Cover Design:	Lisa Tosheff
Printed at:	Yurchak Printing Inc.

Published in the United States of America by
Information Science Reference (an imprint of IGI Global)
701 E. Chocolate Avenue, Suite 200
Hershey PA 17033
Tel: 717-533-8845
Fax: 717-533-8661
E-mail: cust@igi-global.com
Web site: http://www.igi-global.com/reference

and in the United Kingdom by
Information Science Reference (an imprint of IGI Global)
3 Henrietta Street
Covent Garden
London WC2E 8LU
Tel: 44 20 7240 0856
Fax: 44 20 7379 0609
Web site: http://www.eurospanbookstore.com

Copyright © 2009 by IGI Global. All rights reserved. No part of this publication may be reproduced, stored or distributed in any form or by any means, electronic or mechanical, including photocopying, without written permission from the publisher.
Product or company names used in this set are for identification purposes only. Inclusion of the names of the products or companies does not indicate a claim of ownership by IGI Global of the trademark or registered trademark.

Library of Congress Cataloging-in-Publication Data

Andriole, Stephen J.

  Technology due diligence : best practices for chief information officers, venture capitalists and technology vendors / Stephen J. Andriole.

       p. cm.

  Includes bibliographical references and index.

  Summary: "This book develops a due diligence framework for anyone resolving technology decisions intended to help their business achieve positive results"--Provided by publisher.

  ISBN 978-1-60566-018-9 (hardcover) -- ISBN 978-1-60566-019-6 (ebook)

 1. Information technology--Management. 2. Chief information officers. 3. Decision making. I. Title.

  HD30.2.A536 2008

  658.1'62--dc22

                              2008009113

British Cataloguing in Publication Data
A Cataloguing in Publication record for this book is available from the British Library.

All work contributed to this encyclopedia set is new, previously-unpublished material. The views expressed in this encyclopedia set are those of the authors, but not necessarily of the publisher.

If a library purchased a print copy of this publication, please go to http://www.igi-global.com/agreement for information on activating the library's complimentary electronic access to this publication.

Table of Contents

Preface ix

Acknowledgment xviii

Section I:
Due Diligence Strategies and Tactics

Chapter I
The Due Diligence Process 1
Due Diligence Criteria 2
Criterion #1: The "Right" Technology 4
Criterion #2: Few or No Infrastructure Requirements 9
Criterion #3: Budget Cycle Alignment 10
Criterion #4: Quantitative Impact 13
Criterion #5: Changes to Processes and Culture 14
Criterion #6: Solutions 15
Criterion #7: Multiple Exits 16
Criterion #8: Horizontal and Vertical Strength 17
Criterion #9: Industry Awareness 19
Criterion #10: Partners and Allies 19
Criterion #11: "Politically Correct" Products and Services 20
Criterion #12: Recruitment and Retention 22
Criterion #13: Differentiation 22
Criterion #14: Experienced Management 23
Criterion #15: "Packaging" and Communications 25
Criteria Weighting 27

Organization and Execution	31
Organizing the Right Team	32
Finding and Leveraging Savvy Consultants	33
Scheduling	34
Developing Powerful Business Cases	35
The Business of Business Technology Cases	36
The Recommendation	39
Execution	40
Outcomes	40
References	45
Endnotes	46

Chapter II
Acquisition Targets 47

Applications Targets	47
Software Applications Investment Guidelines	56
Communications Targets	58
Communications Technology Investment Guidelines	64
Data Targets	66
Data Investment Guidelines	72
Infrastructure Targets	73
Infrastructure Investment Guidelines	76
Security Targets	79
Security Investment Guidelines	85
Advanced Technology Targets	86
Advanced Technology Investment Guidelines	91
Services Targets	92
Services Investment Guidelines	97
References	97
Endnote	98

Chapter III
Business Technology Trends Analysis 99

Business Technology Trends Analysis Methodology	100
Current and Future Business Models and Processes by Vertical Industry	108
Advanced Technology Trends	109
Five Technology Trends That Matter	111
The Combined Effect	131
Implications	132
Endnote	134

Section II:
Due Diligence Case Studies

Chapter IV
Venture Investing in Wireless Communications Technology:
The ThinAirApps Case **136**
Introduction to the Case.. 136
The ThinAirApps Opportunity.. 137
ThinAirApps Solutions 140
Product Line.. 141
Pricing... 141
Sales 142
Customers 142
Marketing.. 143
History and Accomplishments.. 143
Management Team 143
Risks 145
Financials 146
Forecast 146
Additional Funds.. 146
Investment Highlights 147
Due Diligence 148
Conclusion 156
Endnotes... 156

Chapter V
Enterprise Investing in Remote Access Technology:
The Prudential Fox Roach/Trident Case **157**
Introduction to the Case.. 157
Description of the Investment Opportunity.................................... 159
Due Diligence 165
Conclusion 174
Endnotes... 174

Chapter VI
Venture Investing in Voice-Over-IP (VOIP):
The NexTone Communications Case.. **175**
Introduction to the Case.. 175
Background... 177
Prior Financing 178
Management Team 178
Market... 179
Competition.. 179
Technology Drivers 181

Market Evolution 182
Intellectual Property (IP) 185
Due Diligence 185
Conclusion 188
Endnote 190

Chapter VII
Enterprise Investing in Radio Frequency Identification (RFID):
The Oracle Case 191
Introduction to the Case 191
Oracle's RFID Investment Strategy 193
Oracle's RFID Development Roadmap 195
RFID Hardware Considerations 196
RFID Standards 196
Integration 197
Manageability, Scalability, Security 197
Oracle RFID Partners 199
The Horizontal and Vertical Market Strategy 201
Due Diligence 202
Conclusion 208
Endnotes 209

Chapter VIII
Technology Product and Service Development in an Enterprise
Architecture Service Capability: The LiquidHub Case 210
Introduction to the Case 210
Background 210
The Challenge 211
The Approach 212
The LiquidHub Enterprise Services Transformation Roadmap 221
Due Diligence 222
Conclusion 228
Endnote 228

Chapter IX
Venture Investing in E-Mail Trust Solutions: The Postiva Case 229
Introduction to the Case 229
Background 229
The Market and the Opportunity 231
Productivity 233
Business-to-Business Integration and Messaging Optimization 234
The Postiva Solution 236
Revenue and Business Model 238

Competitive Advantages.. 242
Competition.. 243
Technology.. 245
Summary of Projected Financial Performance 246
Management: Officers, Directors, and Key Consultants 246
Due Diligence 251
Conclusion 256
Endnote 267

Chapter X
Investing in Knowledge-Based User-Computer Interaction:
The TechVestCo Case.. 258
Introduction to the Case.. 258
The Need for Easier to Use Software Applications......................... 258
The User Interface Workbench 259
Components of the Interactive Knowledge-Based Workbench 260
Due Diligence 263
Conclusion 268
Endnote 269

Chapter XI
Enterprise Investing in Wireless Technology:
The Villanova University Case... 270
Introduction to the Case.. 270
Background 270
Challenge 271
Solution 271
Results 273
Due Diligence 273
Conclusion 279
Endnote 279

Section III:
Due Diligence Tools and Techniques

Chapter XII
Tools of the Trade.. 281
Technology Trends Analysis Methodology..................................... 281
Off-the-Shelf Tools for the Due Diligence Analysis 285
Due Diligence Project Management... 285
A Due Diligence Template 287
Reference... 291
Endnote 291

Section IV:
Appendices

Appendix A
Technology Trends Analysis: Trends in Pervasive Computing 293
Pervasive Computing Technology Trends 298
A Pervasive Computing Action Plan 324

Appendix B
Technology Trends Analysis: Trends in Intelligent
Systems Technology 327
Intelligent Decision-Making and Transaction Support 330
Intelligent Systems Technology and the Range of Computable Problem 337
AI Tools and Techniques 338
Why You Need to Understand It 342
The Range of Applications 342
Case Study 344

Appendix C
Technology Trends Analysis: Trends in Business
Technology Integration 349
Do You Think? 352
Acceptable Addiction 353
Having Fun Yet? 354
Privacy 356
Who is Technology? 356
Now What? 358
The New Compact, or "Alignment" in the 21st Century 360
How to Think 367
Business Technology Trends 377
10 Take-Aways 379
Where is it All Going? 381
Endnote 391

About the Author 392

Index 394

Preface

Every day Chief Information Officers, venture capitalists, technology vendors—and everyone that buys and deploys information technology—confront strategic and tactical technology investment challenges. Should Starbucks continue to invest in technology-based initiatives to attract more customers? Should Wal-Mart continue to invest heavily in radio frequency identification (RFID) technology? Should a private equity fund spend money on a broadband communications company or a software development company? Should Rohm and Haas, a global specialty chemical company, implement a standard financial reporting system? Should Microsoft build a suite of large scale enterprise applications to challenge those offered by SAP and Oracle?

How does this all work? Who sets the agenda? How do options get evaluated? How do we select among competing alternatives? Where are the land mines? How do we exploit opportunities?

Due diligence is a term used for a number of concepts involving either the performance of an investigation of a business or person, or the performance of an act with a certain standard of care. It can be a legal obligation, but the term more commonly applies to voluntary investigations.[1]

Technology due diligence refers to the process by which alternative technologies and technology services are vetted. Some organizations and CIOs are disciplined in the way they assess alternative technologies and technology services, while others are not so organized. In a perfect world, every technology investment decision is made with complete information gathered by a team of experienced due diligence professionals. In the real world, the due diligence process is often rushed, plagued by the unavailability of information, and conducted by people who have limited experience—all the more reason for discipline. Most of the prescriptive research on due diligence applies to mergers and acquisitions (Gordon, 1996; Harvey & Lusch, 1995; Lajoux, 2000; Perry & Herd, 2004), portfolio management (Weill & Aral, 2006) and macro trends in business technology (Andriole, 2005). Some analyses have been applied to venture capital due diligence (Camp, 2002; McGrath, Gunther, Keil, & Tukiainen, 2006), and some in the much larger context of business technology alignment (Prahalad & Krishnan, 2002). Very few analysts have focused on technology due diligence. None have focused on technology due diligence from the three intersecting perspectives discussed here.

Due diligence is the process by which we screen and select options. Some see due diligence as a religion; others are agnostic. The approach described here is part quantitative, part qualitative, part analytical, and part intuitive—because due diligence itself is part art, part science, and part luck. The due diligence conducted around technology decisions is complex. When we get it right, we enable enormous impact, but when we get it wrong we can wreak havoc on the organizations, corporate cultures, and markets we are trying to serve.

Due diligence is organized around a set of constant criteria that can be applied to technology investment decisions of all kinds. The 15 criteria described and applied here were derived from both an analysis of the (admittedly sparse) literature and 15 years of experience conducting due diligence for the three constituencies represented in the book's analysis of the due diligence process and the case studies. In fact, the final 15 criteria were culled from earlier lists of 20 and 25 criteria. Over time, we realized that the final 15 criteria addressed all of the major risks and opportunities found in most investment situations. A formal due diligence process was then defined around the criteria.

This book develops a due diligence framework for anyone who must select among competing technologies or invest in technologies intended to help their business achieve results. Among the most obvious executives and managers in this group are Chief Information Officers (CIOs), Chief Technology Officers (CTOs), venture capitalists (VCs), and technology vendors who all have somewhat distinct technology acquisition and investment requirements. But there are many more Chief Executive Officers (CEOs), Chief Financial Officers (CFOs), Chief Operating Officers (COO), directors, and managers who wrestle with technology acquisition challenges every day. This book will help everyone who spends serious money on computing and communications technology.

When you have finished the book, you should have a good understanding of how to make good technology investments regardless whether you are sitting at Microsoft, Accenture, Wal-Mart, or Benchmark Capital.

The focus of the book is on technology due diligence that results in a technology investment. The investment targets include everything from software applications, communications, data, security, and technology services. The lenses used to vet investment opportunities and challenges are organized around the specific requirements that all technology investors—including especially CIOs, VCs, and technology vendors—need to satisfy to achieve their objectives.

I have participated in many due diligence projects over the years. TechVestCo, Inc, has developed technology offerings over the years that required due diligence. At CIGNA, an employee benefits company best known for its healthcare offerings, I was responsible for evaluating current and new technology to support the CIGNA infrastructure as well as the applications that contributed to the growth of the business. At Safeguard Scientifics, Inc., a public venture capital and operating company, we listened to hundreds of business plans a year, selecting anywhere from 15 to 30 in which to invest. I have consulted with strategists and developers in large and small companies as they have struggled to develop products and services they wanted to sell. I have also developed research and development (R&D) programs for government and industry. One of these assignments was at the Defense Advanced Research Projects Agency (DARPA) where I served as the Director of Cybernetics Technology. The game there—as it is everywhere—is prioritization: given a finite amount of money to spend, which projects should be funded, which should be queued, and which should be permanently shelved? At DARPA, we made trade-offs all the time about which research programs to fund, which ones to delay, and which ones to ignore altogether. As an example of just how important these decisions can be, imagine if DARPA had decided not to invest in the transmission control protocol/Internet protocol (TCP/IP)—the communications foundation of today's Internet and World Wide Web? More recently, I have continued to refine the criteria and the process at TechVestCo, Inc., and The Musser Group.

What have I learned? First, there are differences among the due diligence that occurs within a company to acquire or deploy technology vs. investors seeking a return on their technology investment, and companies trying to develop technology products and services to sell to existing and new customers. It is not that the evaluation criteria are different; in fact, they are amazingly similar. The differences can be found in how the criteria are analyzed and weighted.

There is also a difference among the targets of due diligence. Technology comes in multiple flavors including hardware, software, and services. The evaluation of each of these flavors requires a different perspective (in addition to the different perspectives that arise from the source and object of the technology investment). Finally, there are differences in the impact the technology is expected to have. Private equity venture capitalists want to make as much money as they can—as quickly as they can—for their general and limited partners. CIOs want to improve operating efficiencies while contributing to the growth and profitability of their companies. Technology vendors want to discover and build the next "killer app" so they can capture more market share.

Here are 15 criteria that apply to all technology investments—regardless of where you sit or the nature of the deal:

1. Products and services that are on the right technology/market trends trajectory
2. Products and services that have the right infrastructure story
3. Products and services that sell clearly into budget cycles and budget lines
4. Products and services whose impact is quantitative
5. Products and services that do not require fundamental changes in how people behave or major changes in organizational or corporate culture
6. Products and services that, whenever possible, represent total, end-to-end "solutions"
7. Products and services that have multiple exits
8. Products, services, and companies that have clear horizontal and vertical strategies
9. Products and services that have high industry awareness and recognition
10. Products, services, and companies that have the right technology development, marketing and channel alliances, and partnerships
11. Products and services that are "politically correct"
12. Companies that have serious people recruitment and retention strategies in place
13. Products, services, and companies that have compelling "differentiation" stories
14. Company executives that have wide and deep experience
15. Products and services companies that have persuasive products/services "packaging" and communications.

These 15 clusters of questions and issues constitute a framework that can be used to examine technology investments of all kinds. Chief Information Officers (CIOs) buy technologies all the time. How should they proceed? Private equity venture capitalists invest in technology companies all the time? How should they proceed? Technology companies must prioritize their research and product or service development budgets. How should they proceed?

The criteria clusters are, however, weighted differently depending on who is using them. We will build a general purpose framework agile enough to enable a variety of technology investment decisions and then apply it to a set of real cases.

Where did these criteria come from? A decade ago we began with an exhaustive set of criteria. In fact, there were well over 30 criteria we used in the early going. We then began to group the criteria according to areas like technology, differentiation, and management and then "test" them through a series of due diligence processes. We learned that 30 criteria were impractical and that the criteria overlapped. So we reduced the number of criteria in half based on the feedback we received from the application of thirty, then 20 and then 15. In other words, the 15 criteria were the result of a systematic analysis of multiple criteria packages and their relative contribution to multiple due diligence test cases.

We also interviewed due diligence professionals at venture firms to verify our application experiences. We found a near perfect correlation with the criteria lists that resulted from our experiments and what the interview data told us.

The objective of the criteria analyses was to develop a set of due diligence criteria that would serve us well over time—and support the due diligence that CIOs, vendors, and venture capitalists conduct all the time. There was a very practical aspect of our reduction procedure: we sought to identify the minimum number of truly diagnostic criteria that looked at the key aspects of a technology and its ecosystem—which we defined around the technology itself, its maturity, place in the marketplace, management, and marketing. In short, the objective was to identify a set of criteria that would "work" for as many analysts as possible.

The criteria were refined at TechVestCo, Inc., CIGNA, Inc., Safeguard Scientifics, Inc., and TL Ventures, LLC over the period from 1990 to 2005. Analysis of the diagnosticity of the criteria continued at Ascendigm, LLC and is ongoing at TechVestCo, Inc., and The Musser Group, LLC. Most of the case study-based evaluation of the criteria occurred from 1995 to 2005, resulting in the final list of 15.

ORGANIZATION OF THE BOOK

Section I of the book examines due diligence strategies and tactics.

Chapter I presents the criteria that should drive technology investments. The key aspect of this chapter is that due diligence criteria are generalizable to more than one kind of technology investment decision. In other words, if you master them, you can tweak them to the particular technology acquisition opportunity at hand. Chapter I also turns each due diligence exercise into a project that needs to be staffed, managed, and assessed. This is the proverbial due diligence process that lots of organizations believe they perform well but actually mismanage more often than anyone wants to admit. There are lots of "hard" and "soft" due diligence best practices. The process is often dangerously political and subjective. Chapter I also examines the range of expected outcomes, or "exits." What do investors want from their technology investments? Efficiency? Profit? Equity returns? Technology investment decisions should be framed around expected returns.

Chapter II examines the range of possible investments including software applications, data, communications, products, services, solutions, and "advanced technology." This last category is especially interesting to venture investors and corporate "bleeding edge" (early) technology adopters.

Chapter III looks at the special role that business technology trends analyses play in the technology creation, acquisition, and deployment process. I cannot emphasize more the importance of trends analysis. So many investments are out of sync with where the industry is going, where the major corporate R&D programs are going, and where corporate requirements are taking us. It is essential that investments be made in technology trends analyses prior to making investments in specific technologies or technology companies. Appendices A, B, and C at the end of the book present examples of formal trends analyses in three broad areas, pervasive computing, intelligent systems technology, and business technology integration. They are included here to provide you a feel for the critical role that technology trends analyses play in the due diligence process. Chapter III also discusses "5 Technology Trends that Matter" as an illustration of the trends analysis process and the value of using trends analyses as a foundation upon which the due diligence process rests.

Section I of the book describes "what," "how," and "why." Section II illustrates how the process can work extremely well—or fail miserably. Each of the cases in Section II are preceded with a situational context and followed by an analysis of how the due diligence criteria apply to the examples. All of the cases are real. These cases differentiate the book from others that have addressed due diligence methods and tools. The cases are "true": the descriptions of the investment opportunities are the actual descriptions presented to investors that had to decide which investment to make and which to avoid. The cases demonstrate how the 15 criteria can be used to conduct due diligence around the full range of investment opportunities discussed in the book. The cases adopt the three primary perspectives of technology investors: users of technology, venture investors in technology, and vendors looking to invest in technologies that will increase their market share.

Chapter IV examines how a venture capitalist decided to invest in a wireless communications technology company called ThinAirApps. This case explores the due diligence process that a team undertook that eventually resulted in an investment and an "exit," though the twists and turns of the story are interesting enough in and of themselves.

Chapter V looks at how a large real estate and mortgage broker decided to invest in a remote access technology for its 3,000+ agents. This case examines the criteria-based evaluation undertook by Prudential Fox Roach/Trident, the largest real estate/mortgage brokerage in the northeast.

Chapter VI switches gears back to venture capitalists and their due diligence process around a voice-over-IP (VOIP) investment. The significant thing about this process was the timing: the due diligence occurred before VOIP was as proven as it is today.

Chapter VII follows Oracle's decision to invest in RFID (radio frequency identification) technology and develop a series of data and services offerings around the new technology. The application here is primarily focused on transportation and, specifically, the tracking of passenger luggage with RFID tags. Oracle decided—through the application of the due diligence criteria—to make major investments in the technology and roll out substantial data and service offerings to a variety of clients.

Chapter VIII follows LiquidHub's decision to invest in a service offering wrapped around enterprise architecture. Should the company hire staff, develop engagement models, and market itself as an architecture consulting firm? Should it position itself as a service-oriented architecture (SOA) vendor? The due diligence criteria suggested that they do so—and they did.

Chapter IX looks at e-mail and a company's decision to dive deeply into the trusted e-mail management and performance area. The E-Privacy Group was way ahead of its time. The due diligence discussion focused on technology and services offerings in the messaging space.

Chapter X looks at the application of knowledge based expert systems to software design. In this case, the investment decision focused on the development of a knowledge-based system to design and test alternative user-computer interfaces. The decision was made to invest in the development of the platform and while the criteria suggested that the lights were green, the outcome turned bright red. The lesson? Good due diligence processes can lead to bad outcomes, or, put another way, discipline is not a guarantee.

Chapter XI changes the venue to a university—Villanova University—that used the criteria to decide upon a new communications infrastructure, specifically whether they should invest in pervasive wireless technology. This is a classic "go/no go" investment decision driven by the university's CIO. The decision was "go" and the outcome was good.

Section III of the book transitions from cases to prescriptive tools, and sends everyone off to make sound technology investments with a due diligence template. Chapter XII also identifies some specific tools and techniques as well as an overarching process methodology. Chapter XII is the book's "take-away." The three appendices illustrate what technology trends reports look like. They are integral to successful due diligence, since all technology investments are based on an understanding of where technology is going and the large context of our digital world. Figure 1 presents the book in a picture; Figure 2 includes the case studies.

Figure 1. Focus of the book

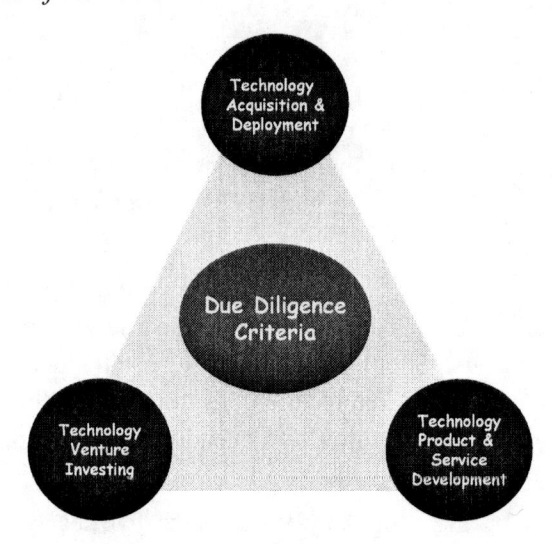

Figure 2. Focus of the book with cases

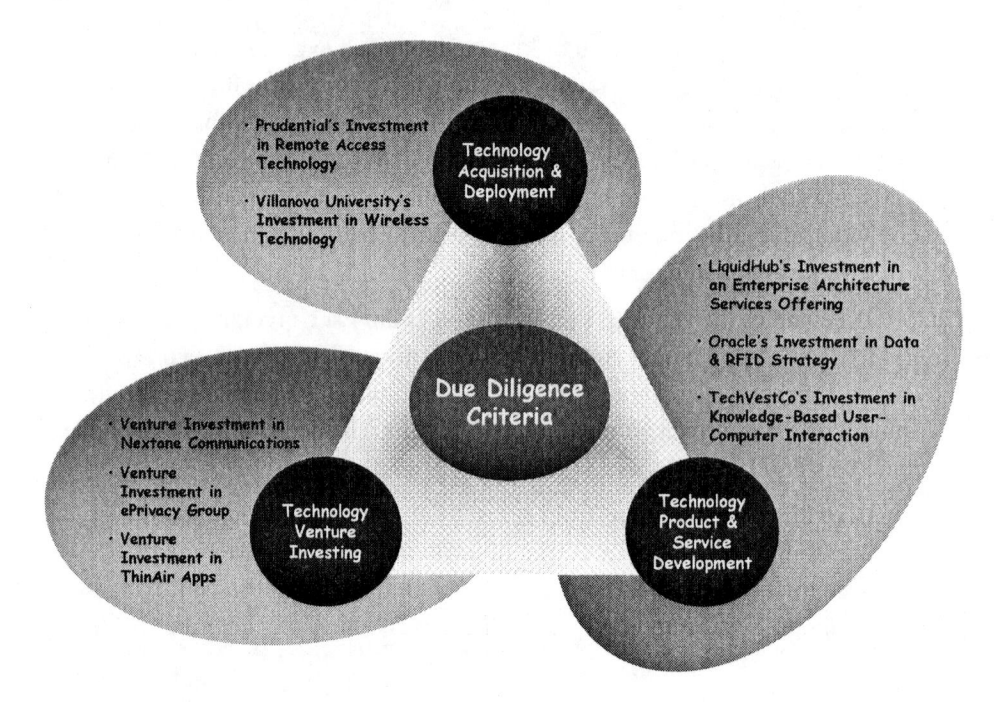

REFERENCES

Andriole, S.J. (2005). *The 2nd digital revolution*. IGI Global.

Andriole, S.J. (2007, March). The 7 habits of highly effective technology leaders. *Communications of the ACM, 50*(3), 67-72.

Bing, G. (1996). *Due diligence techniques and analysis*. Quorum/Greenwood.

Camp, J.J. (2002). *Venture capital due diligence*. John Wiley.

Harvey, M.G., & Lusch, R.F. (1995). Expanding the nature and scope of due diligence. *Journal of Business Venturing, 10*(1), 5-21.

Lajoux, A.R. (2000). *The art of M&A due diligence*. McGraw-Hill.

McGrath, R., Gunther, Keil, T., & Tukiainen, T. (2006). Extracting value from corporate venturing. *MIT Sloan Management Review, 48*(1), 50-56.

Perry, J.S., & Herd, T.J. (2004). Reducing M&A risk through improved due diligence. *Strategy & Leadership, 32*(2), 12-19.

Prahalad, C.K., & Krishnan, M.S. (2002). The dynamic synchronization of strategy and information technology. *MIT Sloan Management Review, 43*(4), 24-33

Weill, P., & Aral, S. (2006). Generating premium returns on your IT investments. *MIT Sloan Management Review, 47*(2), 39-48.

ENDNOTE

[1] Wikipedia

Acknowledgment

Lots of people contributed to this book—directly and indirectly. Over the years, I have learned a great deal about technology and technology management from lots of professionals, including especially George Heilmeier, Bob Fossum, Bob Young, Craig Fields, Clint Kelly, Nick Negroponte, Roger Schank, Bob Zito, Paul Weinberg, Mark Broome, Sam Palermo, John Waldron, John Pacy, Jerry Lepore, Mark Boxer, Len Adelman, Lee Ehrhart, Al Davis, Peter Freeman, Dick Fairley, Dick Lytle, Anne Wilms, Andrea Anania, Steve Fugale, John Carrow, Nora Swimm, Lee Yohannan, Jon Brasington, Rob Kelley, Joel Adler, Ralph Menzano, Jeff Worthington, Jeff Miller, Peter Whatnell, Vince Schiavone, Charlton Monsanto, Max Hopper, Bill Loftus, John Loftus, Frank Mayadas, Scott Snyder, Max Hughes, Don Caldwell and Paul Schoemaker, among others too numerous to mention here. I learned a great deal about technology due diligence from Dick Guttendorf, Rob Adams, Vince Schiavone, Raj Atluru, Chris Pacitti, Jim Ounsworth, Bob McParland, Brian Dooner, Dan McKinney, Mike Carter, Craig London, Tom Lynch, and, of course, Pete Musser, the godfather of venture investing in the Philadelphia region.

I would also like to thank the Labrecque family. I am very proud to hold the Thomas G. Labrecque chair of business at Villanova University. The chair that Mr. Labrecque endowed enabled me to pursue much of the research that led to this book. The support is very much appreciated and continues to enable the pursuit of applied research and development at the university. The Labrecque family continues to set the best possible example of the path from personal and professional success to understanding—and demonstrating—the importance of giving back. Again, I thank you.

My late wife, Denise, supported the entire process. I can still hear her asking, "how's the writing going?," not because she wanted to know when I would be free, but because she genuinely worried about the progress I was making. Thanks, Denise, for everything—again. You were always there and will always be there in my heart and mind. As always, my daughters, Katherine and Emily, supported me in the indirect way children do, as motivators and inspirers to do good things. Thanks, girls.

Hopefully everyone finds this book to be a useful experience. It is about as comprehensive as we could make it. We strongly believe in due diligence discipline and hopefully we have made a contribution to that discipline here.

Steve Andriole
Bryn Mawr, PA

Section I

Due Diligence Strategies and Tactics

Chapter I
The Due Diligence Process

As suggested in the preface, due diligence is a process designed to reduce uncertainty and increase the likelihood of productive investments. The focus here is on technology due diligence, or the process by which technology investment decisions are vetted to maximize impact and reduce risk. Research around technology due diligence is sparse. There are only a few analyses and case studies that look at the nuances of due diligence generally and technology due diligence specifically. A literature review reveals very few formal analyses of the overall process, though there are some useful sources, such as Gordon (1996), Harvey and Lusch (1995), Lajoux (2000), Perry and Herd (2004); a few on portfolio management, such as Weill and Aral (2006), and macro trends in business technology (Andriole, 2005). Some analyses have been applied to venture capital due diligence (McGrath, Gunther, Keil, & Tukiainen, 2006), and some in the much larger context of business technology alignment (Prahalad & Krishnan, 2002). As noted in the Preface, very few have focused on technology due diligence. None have focused on technology due diligence from the three intersecting perspectives discussed here.

The most relevant discussions for this book focus on merger and acquisition (M&A) due diligence, such as Cullinan, LeRoux, and Weddigen (2004), Breitzman and Thomas (2002), Bing (1996), Lajoux and Elson (2000), Howson (2003), and Perry and Herd (2004). Others focus on venture investing and the due diligence process that some venture capitalists apply (Camp, 2002; Zacharakis & Meyer, 1998). Still others focus on very specific aspects of due diligence—like patents (Panitch, 2000) and, as noted above, very few focus on technology due diligence (Marlin, 1998).

This chapter discusses technology due diligence. It first describes the criteria that can be used by Chief Information Officers (CIOs), Chief Technology Officers (CTOs), hardware and software vendors, and venture capitalists (VCs) to vet alternative technology decisions. It then turns to the processes by which due diligence projects can be organized.

DUE DILIGENCE CRITERIA

In a perfect world, every technology investment decision is made with complete information gathered by a perfect team of experienced due diligence professionals. In the real world, the due diligence process is often rushed, plagued by the unavailability of information, and conducted by people who have limited experience or—worse—have already decided that they love the technology and therefore "the deal." One of the core arguments here is that due diligence is part art, part science, and part luck. Recognizing this, however, by no means suggests that the process should not be structured by at least a checklist of things to consider before making technology investments. Ideally, the checklist is comprised of a set of criteria that are more or less quantifiable that lend themselves to situational weighting so that different investment perspectives can be accommodated.

Here are 15 criteria that apply to all technology investments—regardless of where you sit or the nature of the deal. The criteria frame the investment decision identifying the questions you need to ask about the investment opportunity before spending money. They include:

1. Products and services that are on the right technology/market trends trajectory
2. Products and services that have the right infrastructure story
3. Products and services that sell clearly into budget cycles and budget lines
4. Products and services whose impact is quantitative
5. Products and services that do not require fundamental changes in how people behave or major changes in organizational or corporate culture
6. Products and services that, whenever possible, represent total end-to-end "solutions"
7. Products and services that have multiple exits
8. Products, services, and companies that have clear horizontal and vertical strategies
9. Products and services that have high industry awareness recognition
10. Products, services and companies that have the right technology development, marketing, and channel alliances & partnerships

11. Products and services that are "politically correct"
12. Companies that have solid people recruitment and retention strategies in place
13. Products, services, and companies that have compelling "differentiation" stories
14. Company executives that have wide and deep experience
15. Products and services companies that have persuasive products/services "packaging" and communications

Depending on the investment perspective, some are more important than others. Some yield information more readily than others. Some are potentially dangerous, like when the due diligence team falls in love with the management team for the wrong reasons, and some are hard to quantify.

CIOs, CTOs, hardware and software vendors, VCs, and everyone that buys technology use some due diligence criteria as a means to vet ideas. Some are quite formal about the investment process while others prefer to fly by the seat of their pants. There is generally more discipline surrounding the creation and application of technology than we find in venture investing, especially seed and early stage investing (later stage venture investing tends to be somewhat more disciplined than seed and early stage investing). This is because CIOs (and others who spend significant amounts of money on technology) are expected to make the right decisions most of the time. They are consequently more careful about how they spend their firms' money, because if they are wrong too often they will get fired. Vendors are also careful since whole new product lines are expensive to develop, package, market, and sell: the last thing they want to do is invest in a new software application or communications technology that no one wants to use or one that the competition has released six months before the company's expected release date.

Private equity venture capitalists are allowed to fail far more often, since they live under a "high risk/high payoff" umbrella that actually protects them from the results of ill-informed investment decisions. Early stage technology VCs often seek returns on one-in-ten of their investments; and since most venture funds have a ten year life, they get to bat .100 for a decade without too many consequences—they still collect their management fees regardless of the overall performance of their funds (of course, if they fail to return the capital invested in the funds to their limited partners, they will probably have a tough time raising another fund—or not, depending on their personal and professional relationships). Many private equity VCs shoot from the hip relying on instinct and the track records of entrepreneurs much more than rigorous due diligence. At the same time, many of the 15 criteria certainly apply to the analyses of the business plans they evaluate, though the pressure to be right most of the time—the pressure felt by CIOs and vendors—is by no means heavy. The pressure to organize rigorous due diligence exercises is thus by definition less than it is for those playing with their own companies' money. This is one of the counter-intuitive aspects of technology investing. Many believe that due diligence rigor is the greatest in venture investing and much harder to find in technology investing at the user level. In fact, the reverse is often true.

The 15 criteria can help us organize decisions around where to invest, how to invest, and what return on investment (ROI) expectations are reasonable. We will discuss each of them generally and then with specific reference to the different lenses investors assume each time a technology investment is presented.

CRITERION #1: THE "RIGHT" TECHNOLOGY

This question assumes additional information. For example, the "right" technology assumes that the technology product or service is productive today—and likely to remain so. It assumes that the technology "works," and is capable of "scaling" (supporting a growing number of users). It assumes that the technology is secure. It assumes that the technology is part of a larger trend, such as the development of wider and deeper enterprise applications, like enterprise resource planning (ERP) applications sold by companies like SAP and Oracle. It assumes that the adoption of the technology will grow. It assumes that the foundation of the technology reflects larger digital progress, as suggested by Figure 1.

The evolution of technology is important to the identification of high impact technologies. Figure 1 describes the waves of technologies that have evolved since 1980. Note that we are now in a wave that assumes an enormous amount of efficient computing and communications infrastructure. In fact, since 2005 a new level of interoperability, reliability, scalability, and security has emerged that has—in effect—changed the technology concern game: before 2000 we worried a lot more about reliability and availability than we do now; today we focus more on the number of continuous transactions we can churn with an increasing outfacing architecture. Web 2.0 is the perfect example of how far we've come, but even here questions remain about the real impact Web 2.0 technologies will have on cost management (saving money) and revenue generation (making money; more on this later).

In the 1980s, information technology was relatively simple. There were applications and data bases that ran on large mainframe computers and technology management was in the hands of a few professionals. First generation (1G) automation and computer-to-computer connectivity was deployed: we were able to send and receive data across networks, though most of the "terminals" that people used were quite "dumb," with no local processing from the device that displayed the data. The 1990s saw the development of a new "architecture"; the way that applications and data were developed changed fundamentally. Instead of all or most of the computing power residing on mainframes, much of it was re-distributed to personal computers (PCs) connected to one another on local and wide area networks. All of this constituted first generation (1G) distributed computing, with "fat clients" (high capacity PCs), "skinny servers" (high performance computers located on networks and in early "server farms"), and "2 tier architectures" that enabled programmers to design applications to run independently on high capacity PCs or on mainframes—or both. By the mid-1990s things changed again. Enter the Internet and the World Wide Web (WWW)—and some new ("3 tier") architectures. The Internet was disintermediating book stores and travel agents, and programmers were designing applications that distributed computing power across computers on local and wide area networks and the Internet. Servers added capacity and some PCs actually got "skinnier" (less computational power) along with other devices like personal digital assistants (PDAs) that could talk to other devices over networks (including especially the Internet). Companies also started thinking much more seriously about digital supply chains around this time—just as security became increasingly important, given the dramatic rise in online transaction processing.

By 2000, computers were everywhere. The big change was that many devices had their own unique "IP address" (Internet protocol address). This means that individual computers, PDAs, servers, even home appliances were recognized over networks and can therefore perform certain operations as individuals working

Figure 1. Technology timeline

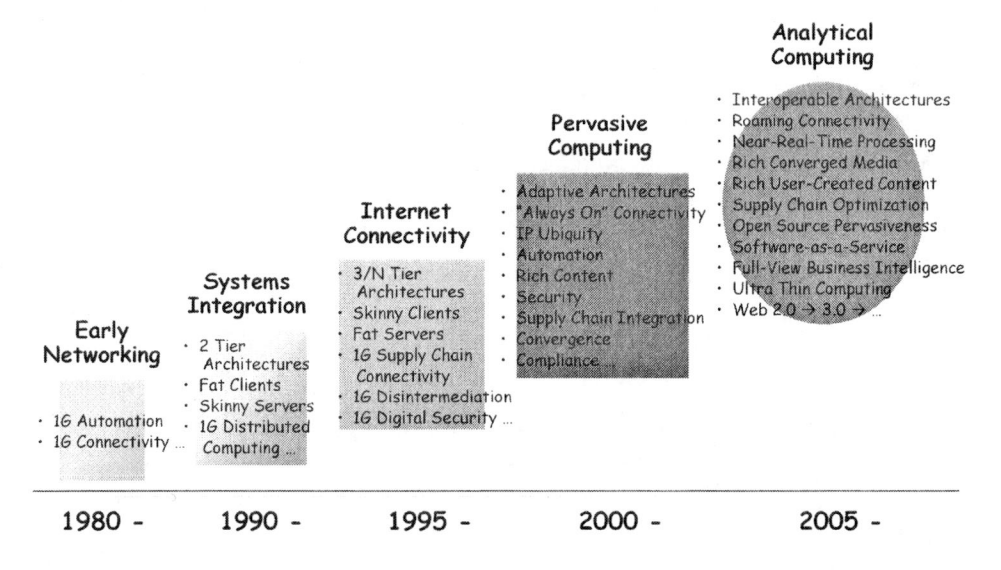

Early Networking (1980-)
- 1G Automation
- 1G Connectivity

Systems Integration (1990-)
- 2 Tier Architectures
- Fat Clients
- Skinny Servers
- 1G Distributed Computing

Internet Connectivity (1995-)
- 3/N Tier Architectures
- Skinny Clients
- Fat Servers
- 1G Supply Chain Connectivity
- 1G Disintermediation
- 1G Digital Security

Pervasive Computing (2000-)
- Adaptive Architectures
- "Always On" Connectivity
- IP Ubiquity
- Automation
- Rich Content
- Security
- Supply Chain Integration
- Convergence
- Compliance

Analytical Computing (2005-)
- Interoperable Architectures
- Roaming Connectivity
- Near-Real-Time Processing
- Rich Converged Media
- Rich User-Created Content
- Supply Chain Optimization
- Open Source Pervasiveness
- Software-as-a-Service
- Full-View Business Intelligence
- Ultra Thin Computing
- Web 2.0 → 3.0 →

on Intranets (internal corporate networks based on Internet technology) and the Internet, 24/7. Computing is now "always on." We were also sending and receiving video and other forms of rich content. Devices were converging—cells phones into PDAs into MP3 digital music file storage and playing devices. Business is now continuous; instant messaging is everywhere.

By 2005 everything had changed again. But this time the changes were measured in different ways. For example, corporate firewalls were not only under attack by viruses and other malware, they were also under attack from business models that assumed no distinction between what happens inside and outside of the enterprise. The whole Web 2.0 phenomenon suggests that not only is the Web a viable transaction platform, but it is also a continuous business model that enables companies (and individuals) to find products and services they like in some previously unanticipated ways—like through the "wisdom of the crowd." Social networking is not just for kids: customers, suppliers, partners, and employees are forming their own social networks for professional purposes. Smart companies are not only mining this interaction but they're fostering it as well.

Integration and interoperability have also hit all time highs since 2005. We have evolved from enterprise application integration (EAI) to Web services to service oriented architecture (SOA) in a relatively short period of time. The era of interoperable architectures has begun—and has forever changed the way software is designed, deployed, and supported.

Similarly, we are always connected now regardless of time of day, location, or activity. Even transaction processing is almost real-time through the supply chain. Media is now richer than it ever was and is delivered to lots of devices that didn't exist five years ago. And now there is user-created content—much of which is also "rich."

Supply chain planning (SCP) and management (SCM) have also matured since 2005. The other-than-financial ERP supply chain modules that used to be weak have gotten progressively better and transaction processing transparency has risen at least 50% in the past 5 years. We are well on our way to real-time forecasting, replenishment, and inventory management.

The whole software design and delivery model has also changed since 2005. Open source software was too often perceived as odd or somehow threatening to corporate enterprises—even as these same enterprises schizophrenically deployed Linux and Apache servers. Vendors like Red Hat, Sun Microsystems, and Google have sharpened open source opportunities and the march toward software-as-a-service blurs the distinction of open source vs. proprietary software. It is quite possible that even the largest enterprises will not care what software satisfies their requirements or even where it sits so long as it is reasonably priced, reliable, and secure.

Business intelligence (BI) is the result of all of the investments we have made over the years in data and information management—or at least that is the thinking that lots of nontechnology executives have about "benefits realization." Time and time again we hear business managers and executives complaining about the lack of friendly/ accessible/timely data given all of the money they have invested in data and information architectures. Many people believe that BI is the answer. The big change here is that BI is becoming pervasive or "full view," the dashboard through which internal and external analyses are conducted. Finally, the analytical computing era will embrace thin client computing.

The nature and trajectory of technology trends is relevant to all technology investors. Technology buyers need to make sure that the applications, communications, data bases, infrastructure, and support technology they are evaluating is consistent with the general directions of the field, what their competitors are doing, and with cost management best practices. Vendors need to make sure that their new offerings are not perceived as radical and can be integrated into their existing products as well as their clients' architectures and infrastructures.

Technology investors play schizophrenic roles. On the one hand, they want to invest in technologies that extend the power of existing technologies and the business models they support, but on the other hand, they also want to create new demand for whole new technologies. For example, groupware—software that enables "threaded" discussions and collaborative computing—was a technology before it was a solution. Vendors like Lotus Development Corporation had to first explain what groupware was and why it was useful before they could get companies to adopt the new technology. Investors love to invest in technologies that are "disruptive" but there is risk in the practice: if companies have to change too much of how they do what they do to accommodate the new technology, they will not deploy it—or will fail in the process.

But there is another dimension to consider. Technology does not develop in a vacuum. Those who create, buy, and invest in technology need to understand the relationship that specific technologies have with related technologies. For example, what is the semantic Web? Is it a technology concept, a prototype technology, or a whole technology cluster? What about voice recognition technology, semantic understanding, and the Segway mini-transportation vehicle? Are they concepts, emerging technologies, or part of larger technology clusters?

Technologies can be segmented into concepts—ideas like "the semantic Web," emerging prototype technologies, like Web 2.0, and technology clusters that include real technologies plus infrastructure, applications, data, standards, a developer community, and management support. Technology impact is related to concepts, prototype technologies and clusters; concepts are wannabes, prototype technologies have potential, and mature technology clusters are likely to have huge sustained impact on business.

Technologies can be mapped on to an impact chart which reveals that many of the technologies about which we are so optimistic have not yet crossed the technology/technology cluster chasm—indicated by the thick blue line that separates the two in Figure 2. Technologies in the red zone are without current impact; those in the yellow zone have great potential, while those in the green zone are bona fide. The chasm is what separates the yellow and green zones—and keeps some technologies from clustering.

The essence of all this is that technologies will have limited impact until full clusters develop around them consisting of all of the things necessary for technologies to grow, all of the applications, data, support, standards, and developers that keep technologies alive and well over long periods of time.

Given all this, CIOs and CTOs have learned to:

- Mostly buy clusters
- Pilot prototypes
- Enjoy (but not invest in) concepts

Vendors have learned to:

- Frequently invest in prototypes
- Float concepts to their biggest clients
- Extract as much revenue as they can from clusters

Venture capitalists (VCs) have learned to:

- Hype concepts
- Over-sell prototypes
- Largely ignore clusters

The nature and trajectory of technology is relevant to all technology investors. Technology buyers need to make sure that the applications, communications, data bases, infrastructure, and support technology they are evaluating is consistent with the general directions of the field, what their competitors are doing, and with cost management best practices. Vendors need to make sure that their new offerings are not perceived as radical and can be integrated into their existing products as well as their clients' architectures and infrastructures.

CRITERION #2: FEW OR NO INFRASTRUCTURE REQUIREMENTS

Technology solutions that require large investments in existing communications and computing infrastructures—like more powerful laptops or more bandwidth—are more difficult to sell and deploy than those that ride on existing infrastructures. If technology managers have to spend lots of money to apply a company's product or service, they are less likely to do so—if the choice is another similar product or service that requires little or no additional investments. VCs and vendors also shy away from technologies that require major infrastructure tweaking. (A possible exception to this vendor/VC rule is during excessive bull markets when technology buyers will spend just about anything, on just about nothing, for strategic advantage; see a later section for more on this phenomenon.)

A quick example. In order to implement Oracle's enterprise financial management system, users have to first install the Oracle data base engine. If the user's current data base management platform is IBM's DB2 or Microsoft's SQL Server, then the move to Oracle is likely to be complicated and expensive. It is harder to sell a technology that requires additional technology investments than one that requires relatively little infrastructure modification. Other examples include the need for greater bandwidth to accommodate additional video processing, or the need to buy all new mobile computing equipment to support a new—but unproven—remote customer relationship management (CRM) strategy.

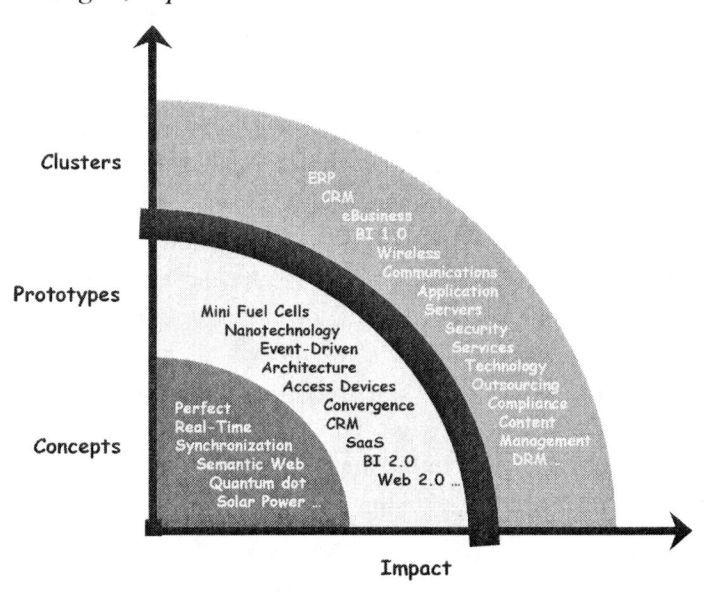

Figure 2. Technologies, impact and the chasm

CIOs are incredibly sensitive to the law of unintended consequences: if an investment chain reaction is suspected as a result of a new technology investment the investment will not be made. Vendors are also sensitive to the need to make sure that their products and services dovetail with existing products and services. At the same time, most vendors are good about telling their clients about major technology shifts they plan to make over time. They are also usually good about helping their vendors migrate from the old to the new—without major disruptions to their technology infrastructures or business processes. VCs understand the need to stay as close to existing infrastructures as possible, but also—because of the nature of what they do—push the envelope more aggressively than either CIOs or major vendors, which is why many of the truly disruptive technologies come from smaller, entrepreneurial companies rather than large established vendors with deep client lists.

CRITERION #3: BUDGET CYCLE ALIGNMENT

It is easier to sell into a new or growing budget cycle than into an older or shrinking one. As sales professionals have known for years, it is tough to sell at the end of the fiscal year. In order to make sales in November or December you have to get creative, often offering to, in effect, defer billing until budgets get renewed. Another aspect of the budget cycle worth noting is the identification of "protected" budget lines, the lines for products and services that just about everyone agrees they need. Today, Sarbanes-Oxley compliance projects are often considered "protected." In the late 1990s the protected projects were Y2K remediation and e-business projects. If we could accurately predict the protected budget lines over time we could make an enormous amount of money.

Budget cycle alignment is about internal timing and externally-driven requirements, such as regulations that must be respected by companies, VCs, and vendors. For example, it is interesting how many new compliance software companies appeared after Sarbanes-Oxley became law, or how many Y2K compliance tools appeared in the mid- to late-1990s. VCs fueled much of this activity with private equity funding, just as new and established vendors have jumped on the regulatory compliance bandwagon over the years, especially in the services area.

Internal budget cycle alignment is often vertical industry sector-driven and driven by the state of the overall technology spending climate. So let us talk a little more about business technology timing. Let us also talk about vitamin pills and pain killers, and let us talk about buying and selling technology and the really big drivers of technology decision-making.

Capital markets drive spending which in turn determines the market for vitamin pills and pain killers. As Figure 3 suggests, both drivers are on a continuum.

Bear markets kill technology (and other) spending. Bull markets make companies lose their heads and buy just about everything they see. Vendors of course hate bear markets—but buyers should love them. Pain killers include those investments that reduce costs and increase efficiency. They are usually made at gunpoint: someone decides that an investment has to be made before some huge technology problem arises. It is usually the CIO that holds the gun to the Chief Financial Officer's (CFO's) head—seldom the other way around.

Vitamin pills—those to-die-for applications that completely transform business—are the elixirs of bull markets. Vendors and VCs love bull markets because the normal business-case-before-we-buy discipline flies out the window propelled by unbridled optimism about business technology success. Bull markets breed "killer apps," "silver bullets," and "disruptive technologies"; bear markets take silver bullets for the team.

All of this is another way of saying that timing is important. There are "good" times to buy, invest, and create technology and there are "bad" times. Capital markets, the financial condition of companies, and even the general condition of the industry all contribute to good or bad timing. In the late 1990s everyone was trying to get Y2K compliant—and spending lots of money to do so. But by 2001 the entire Y2K compliance market disappeared. Similarly in 1997 everyone had to have a Web site. Web site development companies made a fortune in the late 1990s

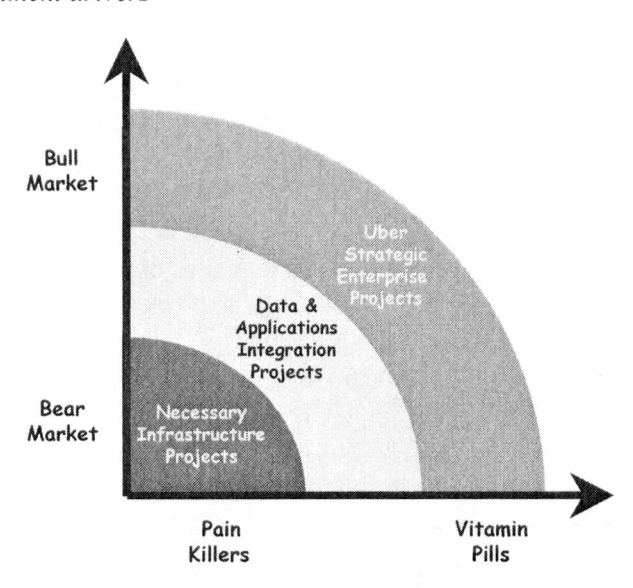

Figure 3. Investment drivers

overcharging their customers for what today can be purchased for $1,000. It is still true: timing is everything.

Pretend for a moment that we are buyers inside a company that spends serious money on business technology every year. Let us assume that we are in a bear market where capital spending is generally—and specifically in technology—way down. First, do not even think about proposing huge "strategic" enterprise technology projects (that might be necessary and even prudent). When the competition for funds is fierce it only makes sense to fight battles you can win. Secondly, work your vendors to a price unheard of during bull markets. They will "cooperate." They have to make their revenue numbers—which supersede earnings in bear markets. Third, take a hard look at your infrastructure with an eye to what it must do when the market turns. In bear markets, we expect relatively little from our computing and communications environments, but as the market transitions from bear to bull, expectations rise. Can your infrastructure handle the transition? When times are tough, it is time to tune-up the infrastructure, assess its vulnerabilities, and get it into good enough shape to scale in anticipation of the transactions it will need to support. Any outsourcing deals you do in bear markets should be shared-risk deals, where the vendors only get paid if they perform well, not if they just "perform."

Now let us pretend we are a vendor. First, we need to embrace total-cost-of-ownership (TCO) and return-on-investment (ROI) calculations. We need to champion tough business cases and pilot applications. We need to offer incentives to our buyers to open their checkbooks; we need to stay close to them during all phases of the work. In short, we need to hustle. But unlike the hustling we do during bull markets, bear market hustling must be quantitative. In bear markets vendors sell pain killers (in bull markets they sell vitamin pills). The trick of course is to morph the same products or services from pain killers to vitamin pills as the capital markets swing. But this is tougher than it sounds. Some products just do not morph well. Customer relationship management (CRM) applications are optimistic, enthusiastic applications that assume more and more customers that have to be handled just right. They are often expensive with long payback periods—tough to sell in bear markets. But other investments—like data and applications integration—can be sold offensively *and* defensively, as investments that can help companies protect and optimize what they have as well as tools that can help them grow.

VCs run and hide in bear markets. Look what happened after the dot.com crash in 2000. The amount of venture capture deployed from 2000 – 2004 was less than 25% of what was deployed in 1999 alone. VCs and the entrepreneurs that they fund are extremely sensitive to capital markets—so sensitive, in fact, that entrepreneurs also hide in bear markets primarily because the valuations for their companies and technologies they receive from private equity venture capitalists are especially anemic when times are tough.

CIOs see things differently. In spite of how far we have come and how elegant and powerful much of our information technology really is, and in spite of technology bear market prices, there is still a widespread perception that we spend too much money on business technology—way too much.

Regardless of where one sits, good timing is essential. It is important to understand market context, what is expected, and what is realistic. Capital markets fundamentally change the buying and selling climate; capital markets determine the popularity of pain killers vs. vitamin pills.

CIOs, vendors, and VCs also all need to appreciate the capital context for their technology investments. What vendors think is a bargain, CIOs often think is highway robbery. Seed and early stage VCs are forever selling promises, so price is often discussed only in vague terms. Context is extremely important for all technology investing.

CRITERION #4: QUANTITATIVE IMPACT

If a product's or service's impact cannot be quantified then one has to rely upon anecdotes to persuade prospective customers that the product or service is worth buying. But if impact can be quantified then it can be compared against some baseline or current performance level. Clearly, if quantitative impact is huge, for example, reducing distribution costs by 40% or increasing customer satisfaction by 30%, then it is relatively easy to persuade customers to at least pilot a product or service. CIOs, VCs, and vendors all like measurable quantitative impact.

Ideally, impact reduces some form of "pain," though at times (during bull markets) the impact of "vitamin pills" can be appealing. Quantitative impact also helps differentiate products and services (see below for more thoughts about differentiation).

VCs are really good at spinning impact stories. They search for—and often subsidize—"lighthouse" customers of the products or services of the companies in which they invest. Some of these lighthouse customers are bona fide, but some are friends-of-friends who are accustomed—like expert witnesses—to saying the same things all the time. CIOs, on the other hand, are born skeptics. They often reject even the most reliable impact data. Vendors are somewhere in the middle: they must speak the impact language of CIOs but they must also spin a good story about just how great their new product or service is.

Impact arguments must fit the audience. CIOs, vendors, and VCs all need to hear, and argue, slightly different things about impact, but the clear requirement, regardless of the audience, is quantitative "proof" that the technology product or service is meaningfully better that the incumbent product or service. The word "meaningful" is critical here. A 10% improvement or reduction in cost may mean little to a CIO, but a 10% improvement in an existing process for little or no additional investment may be a major selling point for a technology vendor. VCs are always looking for monster impact, often in the outrageous 50% - 75% range.

CRITERION #5: CHANGES TO PROCESSES AND CULTURE

If a product or service requires organizations to dramatically change the way they solve problems or the corporate cultures in which they work, then the product or service will be relatively difficult to sell or deploy. Conversely, if a product or service can flourish within existing processes and cultures, it will be that much easier for organizations to adopt. The best example here is the relatively slow adoption of knowledge management software products. Those who sell these products and services assume that organizations will want to share information and collaborate—even if the organization is by nature non-collaborative. Knowledge management has been suffering this conundrum for years: gurus are selling group problem-solving and solutions repositories when organizations may or may not have large appetites for sharing.

Perhaps a better example is customer relationship management (CRM). CRM is not technology, software, or "architecture": CRM is a state of mind, a philosophy, a business strategy. It is amazing just how many companies believe that a CRM (hosted or in-house) software application is the answer to their customer relationship problems. Successful CRM applications that we buy (from companies like Oracle/Siebel) or rent (from companies like Salesforce.com) assume a variety of things to be true before implementation (though the vendors tend to hide many of them in fine print). Newsflash: If your company is not customer friendly, technology will not change a thing (except the technology budget).

CRM, *the philosophy,* vs. CRM, *the technology,* regards customers as life-long clients whose personal and professional lives can be monetized through the proactive management of the client's needs, values, and ability and desire to pay. CRM *the technology* is about applications that leverage customer data, supplier data, company data, and even vertical industry data into actionable information. The disconnection that often occurs is between the corporate and technology views of customers, not different perspectives on how software applications should be acquired and deployed. There are also disconnections among what companies sell, what they charge, and what customers are willing to pay.

Customer-centered companies have wide and deep protocols around customer care—processes designed around customer satisfaction and retention. They also have specific protocols around the acquisition of new customers. Nordstrom department stores get it; Ritz-Carlton hotels get it; Lexus car dealers get it. While far from "perfect" these and other vendors understand that the extra profit they embed in their products and services better be offset by the quality of the service they provide. Many customers are quite willing to pay more than they should in return for over the-top service (at least compared to what lower cost vendors provide). High-end vendors have always understood this and manage their customers accordingly. Some middle-end vendors also treat their customers elegantly; while some others offer alternative value propositions to their customers—like low prices—as a trade-off to mediocre or downright poor service.

The CRM danger zone is reached when companies misjudge the product/value/customer service relationship ratio of customer care and investment. Some companies, for example, are in the middle of the price/value hierarchy but provide horrible service. Other companies are at the very top and provide marginal service. (Any company at the top of the price/value hierarchy that provides horrible service is unlikely to stay there.) The CRM success zone is reached when a company synchronizes its price/value/service ratios with investments in CRM processes and technology.

CRM software applications will not change a company's CRM processes. If a company's processes are hopelessly broken or customer negligent there is a process gap that must be filled through management action designed to specifically to close the gap. In other words, if the customer-centric processes are nonexistent or broken, then investments in CRM applications will require huge new investments in CRM processes and, indeed, the very culture of the company—which must transform itself from a customer-neutral or customer-hostile culture to a customer-friendly one. CIOs, vendors, and VCs understand full well the implications of "culture change." Technology investments that depend even a little upon "culture change" for their success are much less likely to succeed than those already aligned with existing processes and culture.

CRITERION #6: SOLUTIONS

Increasingly, everyone is looking for integrated solutions to broad complex problems. While it is great to sell personal computers, it is better to sell personal computers, asset management systems, break-and-fix support, and desktop/laptop/PDA/cell phone management strategies. Why? Because clients need all of these services and must often work with multiple vendors to assemble services that are difficult or impossible to integrate. It is just plain easier—and often more cost-effective—to work with fewer vendors; sometimes one "strategic partner" represents the best integrated solution. The newest breed of consultant is the "solutions integrator" which purports end-to-end support for whatever (broadly defined) technology problem clients might have.

CIOs are increasingly sensitive of the inter-relationships among all facets of their technology environments, including especially the inter-relationships among applications, communications and data. They are always on the lookout for technologies that cross-cut their infrastructures, for "solutions" that solve as many problems as possible. Vendors are aware of these requirements and try to provide just the right mix of services for the right price. VCs are interested in service business models because the margins can be high in well run solutions companies.

CRITERION #7: MULTIPLE EXITS

Since not all technology investments work perfectly, it is nice when there are multiple paths to success. If a company is selling a vertical solution that can relatively easily move to another vertical industry, then there is contingency built into the plan. If a horizontal technology product can go vertical pretty easily, that is a good "default." The point is simple: if a company's product or service is horizontally and vertically flexible, extensible, and adaptive, it stands a better chance of surviving unpredictable shifts in the competitive marketplace.

CIOs see the default option differently than vendors or VCs. CIOs bundle their default outcomes within larger risk management frameworks. If a major application fails, they think about how to mitigate the impact; for example, smart CIOs will never cut over to a new application until the new application has been thoroughly tested. This means that organizations frequently run two applications as they make sure that the new application does everything it is supposed to do.

VCs also see multiple ways to win, though they focus less on risk management than they do on reward maximization. Ideally, a technology company in which they invest might be ready for a public stock offering as soon as possible. In the go-go Internet days, companies were going public in a matter of months. But if companies do not have public company potential, they might still be sold to competitors—or become profitable quickly enough to be able to stand on their own two feet. Even if a company fails to gain significant traction in a relatively short period of time, VCs can redistribute risk by offering subsequent investment rounds where new investors are invited in to participate—to invest in a round of financing that may or may not be based upon a huge increase in the company's valuation. In other words, there are several ways VCs can monetize their investments; VCs continuously rank-order investment exit opportunities.

Vendors actually have the greatest challenge here. After investing heavily in a new application, a new service, or both, vendors have a great deal to lose if the new product or service fails to impress their clients and the larger market place. There are in fact only a few default options open to vendors. One of course results in the success of a new product or service expressed primarily in terms of customer acceptance. The other is the "in progress" default where problems detected in, for example, Version 1.0 of a new application can be continuously improved over some respectable period of time. This is essentially what Netscape did with their initial browser in the 1990s and what, arguably, Microsoft does every time it releases new products that require multiple "service packs" to get the initial application to work the way it was designed to work.

CRITERION #8: HORIZONTAL AND VERTICAL STRENGTH

Microsoft is the quintessential horizontal technology company: it sells software to anyone and everyone, regardless of their industry. But there are companies that only sell to specific industries, like insurance companies, banks, and pharmaceutical companies, and there are companies—like IBM and many of the larger consulting and systems integration companies—that sell horizontally and vertically, with "practices" that specialize in multiple industries. The best products and services are those that have compelling horizontal and vertical stories, since customers want to hear about industry-specific solutions or solutions that worked under similar circumstances (like for a competitor). Without a good vertical story, it will become more and more difficult to make horizontal sales.

Some vendors have begun to develop simple, cost-effective solutions to vertical industry problems with horizontal technology that is customized for specific industries. For decades vendors offered one-size-fits-all infrastructure technology that gets "tuned" by in-house or outside vertical consultants (also known as "subject matter experts" or "domain experts") after it is deployed. Several applications vendors have offered specific solutions to several industries for years. Manufacturing, financial services and health care all have their technology leaders but they tend to be on the applications—not the infrastructure—side.

There are at least three dimensions to "verticalization":

- The documentation of vertical industry requirements
- Vertically-customized infrastructure technology
- Plug and play simplicity

Vertical industries have mature relationships with information technology. It is not like we are still trying to figure out how banks should use technology to solve their problems. Vanguard, the giant financial services company outside of Philadelphia, knows exactly how to leverage technology on to old and new business models and processes. But they do almost all of it in-house. Like many companies, Vanguard has served itself with wide and deep vertical expertise. It has customized off-the-shelf software to satisfy its financial services requirements. But why could not the vendors (not the consultants) customize it for them? Of course one can always hire an army of consultants to do whatever is necessary, but why bloat the process with more overhead than necessary? Aren't the financial services, pharmaceutical and the chemical industries, among others, big enough to justify their own infrastructure requirements—and solutions?

Just about all of the major vertical industries have sets of requirements that are generalizable across the companies within that industry. Are Fidelity's requirements all that different from Vanguard's? Are DuPont's all that different from Rohm and Haas'? What happens when RFID really takes off? How many flavors will there be? Why would not there by one primary flavor for each vertical industry?

CIOs would love to see the primary vendors of hardware, software, and communications infrastructure develop full vertical suites complete with all of the bells, whistles, and hooks that make it possible to transact business across any number of vertical industries. Issues like privacy, compliance, reporting, business-to-business (B2B) transaction processing, data representation, and security, among others, are approached differently by different vertical industries. Why cannot CIOs get some help here to at least create some intra-industry infrastructure standards?

The chemical industry is a case in point. Just about all of the major companies have standardized on SAP's R3 as their enterprise resource planning (ERP) application. This means that infrastructure providers (without the consulting middleman) can optimize their solutions to back office and front office SAP application modules. How easy is that? Vendors who provide communications, workflow, groupware, knowledge management, messaging, content management, security and data base management, among other capabilities can tailor their offerings to SAP and the chemical industry.

One of the most interesting opportunities for open source providers is the possibility of simplifying hardware and (especially) software architectures. It is unlikely that the major proprietary vendors will travel down this path, but the Linux desktop crowd could move in this direction pretty easily (though not without resistance from the entrenched proprietary vendors). But the real leverage still lies with the optimization of proprietary legacy infrastructure technology. While the open source community wants to simplify its offerings, real progress could be made if the proprietary vendors would lighten and verticalize their applications: we really do not need all the bells and whistles they shrink wrap; instead, we need applications stripped down, and juiced up, to optimally support specific vertical industries.

CIOs expect their vendors to understand their business. Vendors organize themselves horizontally and vertically to appeal to their clients. VCs look to companies with products and services that speak to the requirements of the biggest and richest vertical industries.

CRITERION #9: INDUSTRY AWARENESS

If no one has ever heard of the product or service someone represents, then there is an uphill investment climb. While there are sometimes huge opportunities to create brand new awareness—and in the process become a market trend setter—it is often easier to sell into an area that already has high industry recognition. Perhaps the most obvious validation is from the conventional industry analysts, like Gartner, IDC, or Forrester. If a technology product or service is unknown to this community, then companies have to spend their own money to create awareness before they make a sale. Companies should also understand how to play the awareness game and how to work industry analysts.

CIOs have a tough time buying products or services with little or no name recognition. Most companies avoid early technology adoption simply because there is too much risk in the practice. Vendors work hard to make sure their marketing at least keeps up with (if not surpasses) the capabilities of their products and services. VCs spend a lot of money with industry analysts boosting the awareness of the technologies and technology companies in their investment portfolios.

CRITERION #10: PARTNERS AND ALLIES

It is getting harder and harder for companies to go it alone. Given trends in "solutions integration," outsourcing, and the pace of technology change, it is incumbent for (especially) new companies to form the right channel partnerships and alliances. While direct sales and marketing can often work extremely well, it helps to have the right friends in the right places saying the right things about products and services. Relationships with the management and technology consulting companies, the systems integrators and the support vendors can extend a technology company's reach by orders of magnitude. Companies unaware of this reach are likely to miss important channel opportunities.

CIOs expect a broad network of support. Put another way, they prefer to invest in technology clusters. Vendors need enough partners to optimize their success. Some sell directly and some only through channel partners. Some have relationships with consultancies and some do not. VCs push their companies to strike as many deals as prudent and necessary to get their products and services to as many people in the shortest possible time.

CRITERION #11: "POLITICALLY CORRECT" PRODUCTS AND SERVICES

It is difficult to convince conservative enterprise buyers of technology products and services to adopt something new. No one wants to live on the "bleeding edge." Many technology managers will not risk their careers on what they perceive as risky ventures—even if the "risky" product or service might really solve some tough problems. Buyers also want products and services that will ease real pain. While "vitamin pills" are nice to have, "pain killers" are essential. Reducing costs and staff, measurably improving processes, and improving poor service levels are pain killers that make buyers look smart. This is a good place to be—and invest.

"Politics" has a profound effect on business technology decision-making. Everyone relates to "politics" and the impact it has on corporate behavior.

But politics is one aspect of the overall context that influences decisions. The others include the culture of the company, the quality and character of the leadership, the financial condition of the company, and the overall financial state of the industry and the national and global economies. Figure 4 says it all.

The three most obvious pieces of the puzzle include the pursuit of collaborative business models, technology integration and interoperability, and of course the management best practices around business technology acquisition, deployment and support. Three of the other five—politics, leadership, and culture—are "softer"; two of them are "hard" and round out the context in which all decisions are made.

Let us run through the variables. It is important to assess the political quotient of companies. Some companies are almost completely "political": a few people make decisions based only on what they think, who they like (and dislike), and based on what is good for them personally (which may or may not be good for the company). Other companies are obsessive-compulsive about data, evidence, and analysis. In the middle are most of the companies out there, with some balance between analysis and politics.

Corporate culture is another key decision-making driver. Is the culture adventurous? Conservative? Does a company take calculated risks? Crazy risks? It is an early, or late, technology adopter? Does the culture reward or punish risk takers? When it tells employees to "think outside the box," is that code for "I dare you to challenge the status quo"? It is important to assess corporate culture accurately. Technology investments must sync with the culture (as well as the rest of the variables that comprise the total decision-making context).

What about the leadership? Is it smart? Is it old—nearing retirement? Is everyone already rich? Is everyone still struggling to get back to where they were financially in 1999? Is it embattled, struggling to retain control? Is the senior management team mature or adolescent? Is it committed to everyone's success or just its own? Is it compassionate or unforgiving? The key here is the overall leadership ability of the senior management team. There are some really smart, skilled and honorable management teams out there and there are some really awful ones as well. Trying to sell a long-term technology-based solution to a self-centered team with only their personal wealth in mind simply will not work; trying to sell the same solution to a team that embraces long-term approaches to the creation of broad shareholder value usually works very well.

How well is the company really doing? Is it making money? More money than last year? Is it tightening its belt? Has the CIO received yet another memorandum about reducing technology costs? Is the company growing profitably? Is there optimism or pessimism about the future?

Is the industry sector doing well? Is the company the only defense contractor losing money? Is it the only pharmaceutical company without a new drug pipeline? Or is everyone in the same boat? Is the general economy looking good or are there regional, national, or global red flags? What is the confidence level for the sector and the economy? Where is the smart money going? It is essential to position companies within the larger economic forces that define national and global bear and bull markets.

Figure 4. The whole context

CIOs, vendors, and VCs should all pay very close attention to politics, culture, leadership, the company's financials and the overall national and global economies. If the lights are all red, maybe it is a bad time to propose any changes or any large technology investments. But if there are some red—but mostly yellow and green lights—then perhaps it is time to work the context to an advantage. One thing is for sure: ignoring any of the pieces will jeopardize the chances of success.

CRITERION #12: RECRUITMENT AND RETENTION

Finding truly talented professionals to staff product and service companies is emerging as perhaps the most important challenge facing companies in all stages of development. Companies that have identified employee recruitment and retention as core competencies are more likely to survive and grow than those that still recruit and retain the old fashioned way. Creative solutions to this problem are no longer nice to have, but a necessity or, stated somewhat differently, creative recruitment and retention strategies are no longer vitamin pills. They are pain killers.

CIOs expect their vendors to have lots of really smart, dedicated professionals. If there is evidence to the contrary they are likely to not make the technology investment. Vendors understand this and try to surround their sales, marketing, and delivery efforts will the best and the brightest. VCs expect their portfolio companies to understand how to recruit and retain only the best—or at least people that can be described as "only the best."

CRITERION #13: DIFFERENTIATION

If a technology company cannot clearly and articulately define its differentiation in the marketplace, then a large red flag should be raised about the company's ability to penetrate, let alone prosper in, a competitive market. Differentiation is critical to success and while not every differentiation argument is fully formed when a company is first organizing itself, the proverbial "elevator story" better be at least coherent from day one. The best differentiation stories of course directly address the uniqueness, cost-effectiveness, and power of the new (or old) product or service.

CIOs need a lot of help here. In order to sell a technology investment, especially a large one like an ERP implementation, they need a business case that unambiguously describes how informed their choice is from all of the alternatives. Sometimes the marketplace itself helps with differentiation—or the lack thereof. Industry consolidation usually means that there is not enough differentiation among the players in an area to sustain competitive advantage among the players. This is why there are only a handful of PC manufacturers left, or just a few major data base management vendors out there. The nature of differentiation tends to change as markets mature. Later stage differentiation is more about execution than the measurable technological differences among products. Sometimes differentiation is only about execution and service, where CIOs knowingly settle for a relatively inferior technology product from a vendor that consistently offers extraordinary pre- and post-sale support.

Vendors will kill for credible differentiation data. Sales and marketing teams work extremely hard to craft just the right, and impenetrable, differentiation story. VCs live and die by the differentiation sword. It is all about "better," "faster," and "cheaper" than everyone else.

CRITERION #14: EXPERIENCED MANAGEMENT

The key here is to see the right mix of technological prowess and management experience available to develop and deliver a successful product or service. Ideally, the management team has "been there and done that," and is mature enough to deal with all varieties of unpredictable events and conditions. There are other ideal prerequisites: experience in the target horizontal and/or vertical industry, the right channel connections, the ability to recruit and retain talented personnel, the ability to work industry analysts, communicate, and sell. To this list we might all add a number of qualities, but the key is to find experienced managers knowing full well that past success is not necessarily a predictor of future success.

Lots of people strongly believe that business technology alignment depends almost completely on the quality and availability of the right people at the right time. Take a look at Figure 5. It is a tool to help profile people according to their capability, energy, ambition—and a few other characteristics. Clearly, the goal is the assembly of a bunch of smart, sane, energetic, and appropriately ambitious professionals.

Figure 5 presents at least three kinds of knowledge which, of course, need to be integrated. *Generic, structured knowledge* includes facts, concepts, principles, and formulae that describe what things are and how they work. Finance is a good generic, structured field. Computer Science is another one. College students major in these fields. *Industry specific knowledge* comes from different sources. A little comes from colleges and universities, but most of it comes from on-the-job experience, training and industry certifications. *Company specific knowledge* comes from time spent in the trenches of particular corporate domains.

When we talk about "smart," we are talking about depth in the three knowledge classes as well as the ability to integrate them into insights, inferences and decision-making. But while "intelligence" is fed by integrated knowledge combined with raw intellectual horsepower, energy, and ambition are measured independently.

Some people are really smart and some are not. Some work at understanding existing and emerging business technology trends, and some do not. Some even work at increasing their natural energy levels, but most do not.

Some want their boss's job; some are clueless. Some are evil; some are sweet. Some got where they are mysteriously; some really earned it. Who are the "keepers"?

What about the "jerk factor?"

If you're new to an organization after a week or so of "observation" you begin to make mental lists. One of them is a list of the people who are so far over the top that you find yourself slipping into a state of buyer's remorse, wondering how you could have been so stupid to accept the new position. People of course fall into all sorts of categories. Some are hopelessly rude and arrogant. What do we do with these people (that have buddies just like them all over the place)? What do we do about people that disrupt and undermine? People that complain all of the time? People that have nothing to offer but bitterness, anger, and jealousy?

People can be smart, ambitious, and energetic, but arrogant and caustic. Who do you want on your staff? To whom do you entrust major business technology initiatives? How do you make the trade-offs?

Companies that have lots of smart people with no energy or ambition, have a problem. But it is actually more dangerous if they have dumb people with tons of energy and ambition. The military has known for centuries that officers can be smart *and* arrogant, but never arrogant *and stupid*.

When companies are making lots of money, people (and companies) suffer fools amazingly well, but when times get tough, tempers and patience grow short. When times are good, they should find what the jerks do best and isolate them accordingly; when times get tough, they should prune them from the organization.

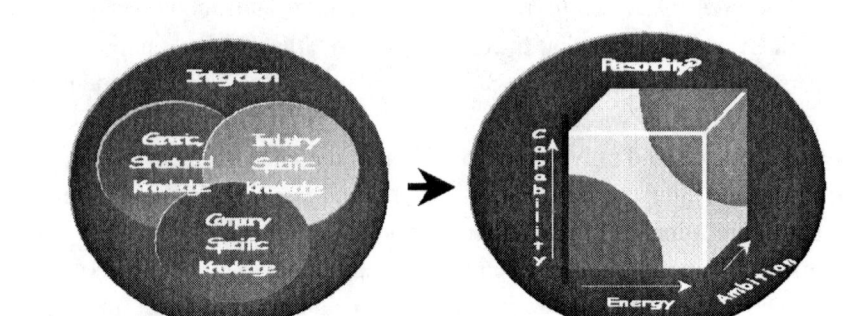

Figure 5. The talent profiler

CIOs expect their technology vendors to have solid management teams. Vendors understand that if the teams look and act stupidly they will not sell very much product or service. VCs understand just how important good management teams are; in fact, many VCs weight the management criterion more heavily than any of the other due diligence criteria.

CRITERION #15: "PACKAGING" AND COMMUNICATIONS

While it may seem a little strange to acknowledge the primacy of "style" over "substance" and "form" over "content," the reality is that "style," "form," and "sizzle" sell. Product and service descriptions and promotional materials should read and look good, and those who present these materials should be professional, articulate and sincere. Companies that fail to appreciate the importance of form, content and sizzle will have harder climbs than those who embrace and exploit this reality.

What are the pieces of a good technology marketing strategy?

First, consider what's being "sold." Hardware, software, services, *image and perception*. When everything goes well everyone thinks that the technology people are really pretty good, that things work reasonably well—and for a fair price. If the hardware and software works well, but the image is poor, technology is perceived to be a failure, just as bad hardware and software—but good perceptions—will buy you some time. Like everything else, we are selling hard and soft stuff, tangible and intangible assets, and processes.

Next consider who is being sold, noting from the outset that we sell different things to very different people. Yes, we are selling hardware, software, and services (along with image and perception) to everyone, but the relative importance of the pieces of the repertoire shifts as we move from office to office. Senior management really does not care about how cool the network is or how we have finally achieved the nirvana five-9s for the (.99999) reliability of the infrastructure. They care about the 20% lopped off the acquisition project just launched, or how company data is finally talking to each other and that the company is now able to cross-sell its products. Yes, the content *and form* of the message is important.

What is the brand of the technology organization? If it were a professional sports team, what would be a good name for it? Would it be the Innovators? The Terminators? Put another way, if you asked the analysts who cover a stock to word-associate technology and the company, what would they say? Disciplined? Strategic? Weak?

Is there a technology "road show"? A consistent message about the role that technology plays in the company, how technology is organized, what matters most, the major projects, and technology's contribution to profitable growth, among other key messages, is indispensable to a company's success.

Clearly technology vendors must have compelling stories to tell their clients with supporting collateral materials. CIOs expect to see and feel the buzz of the vendors' products and services. Vendors spend a lot of money on these materials and VCs force their companies to have the very best promotional materials. The fact is that the right mix of reality and hype is important to all of the parties involved in a technology investment decision, regardless of where they sit.

In summary, there are 15 criteria that apply to all technology investments—regardless of where you sit or the nature of the deal. The criteria frame investment questions. They include:

- Products and services that are on the right technology/market trends trajectory
- Products and services that have the right infrastructure story
- Products and services that sell clearly into budget cycles and budget lines
- Products and services whose impact is quantitative
- Products and services that do not require fundamental changes in how people behave or major changes in organizational or corporate culture
- Products and services that, whenever possible, represent total, end-to-end "solutions"
- Products and services that have multiple exits
- Products, services, and companies that have clear horizontal and vertical strategies
- Products and services that have high industry awareness recognition
- Products, services, and companies that have the right technology development, marketing, and channel alliances and partnerships
- Products and services that are "politically correct"
- Companies that have solid people recruitment and retention strategies in place
- Products, services, and companies that have compelling "differentiation" stories
- Company executives that have wide and deep experience
- Products and services companies that have persuasive products/services "packaging" and communications

Taken together, the criteria represent an investment profile. They represent a way to understand an investment's potential, risks, and uncertainties. They also segment into categories that address the technology itself, the positioning of the technology in the marketplace, the technology's "packaging," its management and its ability to compete in the marketplace through its ability to differentiate itself among its competition.

CRITERIA WEIGHTING

The 15 clusters of questions and issues constitute the framework we use to examine technology investments of all kinds. CIOs buy technologies all the time. Private equity venture capitalists invest in technology companies all the time. Entrepreneurs are constantly looking for opportunities. Technology companies must prioritize their research and product or service development budgets. Small and medium-sized companies constantly invest in computing and communications technology. How should they all proceed?

The criteria clusters are, however, weighted differently depending on who is using them. Due diligence criteria are extensible to different kinds of investment decisions. The key lies in the rules around their applicability to alternative investment opportunities.

There are 15 questions that can be used to analyze technology investment opportunities. As Figure 6 suggests, they include the 15 due diligence criteria framed as "yes," "no," and "why"?

The answers to these questions can be "green" (good), "red" (bad) or "neutral" (yellow), as suggested in Figure 7.

Criteria can also be weighted according to their relative importance. How would you weight the 15 criteria? How would you process relative weights? Figure 8 suggests how the weighting process works. Note the questions, the weights, the scores, and the total (calculated from the weighting factor and the score).

Criteria weighting can be accomplished in a variety of ways. The way we suggest is by far the simplest way possible. Why so simple—and what are the alternative weighting schemes you might consider?

Simplicity is always a friend, at least that is how we have come to see the due diligence process. The objective of due diligence is risk reduction, not perfect outcomes. The simple weighting suggested here provides a relative ranking of alternative investments—nothing more, nothing less. It also provides the ability to tweak the process and play "what-if" games—"what if the weights were changed … what if the scores change … how does that change the overall assessment?"

The analytical hierarchy process (AHP) is the multicriteria decision-making methodology we use to evaluate alternative technology investment opportunities. AHP is a powerful multicriteria decision-making tool that structures decision problems into a hierarchy that permits alternatives to be evaluated according to criteria weights and alternative scores.[1]

Figure 6. 15 key questions

		Yes	No	Why?
1.	"Right" Technology Trend?			
2.	Low Infrastructure Change Requirements?			
3.	Aligned Budget Cycles?			
4.	Quantitative Impact?			
5.	Small Changes to Process/Culture?			
6.	End-to-End "Solution"?			
7.	Multiple Defaults?			
8.	Clear Horizontal & Vertical Stories?			
9.	High Product/Service Industry Awareness?			
10.	Right Sales Partnerships & Alliances?			
11.	"Politically Correct" Products/Services?			
12.	Solid Recruitment/Retention Strategies?			
13.	Clear Differentiation?			
14.	Management that Gets It?			
15.	Good "Packaging" & Communications?			

According to Wikipedia,

The analytic hierarchy process (AHP) is a structured technique for dealing with complex decisions. Based on mathematics and human psychology, it was developed by Thomas L. Saaty in the 1970s and has been extensively studied and refined since then. The AHP provides a comprehensive and rational framework for structuring a problem, for representing and quantifying its elements, for relating those elements

Figure 7. The good, the bad, and the neutral

		Yes	No
1.	"Right" Technology Trend?		
2.	Low Infrastructure Change Requirements?		
3.	Aligned Budget Cycles?		
4.	Quantitative Impact?		
5.	Small Changes to Process/Culture?		
6.	End-to-End "Solution"?		
7.	Multiple Defaults?		
8.	Clear Horizontal & Vertical Stories?		
9.	High Product/Service Industry Awareness?		
10.	Right Sales Partnerships & Alliances?		
11.	"Politically Correct" Products/Services?		
12.	Solid Recruitment/Retention Strategies?		
13.	Clear Differentiation?		
14.	Management that Gets It?		
15.	Good "Packaging" & Communications?		

Figure 8. Criteria weighting

	Weight	Score	Total
Right Technology Trend?			
Low Infrastructure Requirements/Low Change?			
Aligned Budget Cycle?			
Quantitative Impact?			
Small Changes to Process & Culture?			
End-to-End "Solution"?			
Multiple Defaults?			
Horizontal/Vertical Stories?			
High Industry Awareness?			
Right Partnerships & Alliances?			
"Politically Correct"?			
Recruitment/Retention Strategies?			
Differentiation?			
Good Management?			
Packaging & Communications?			

to overall goals, and for evaluating alternative solutions. It is used throughout the world in a wide variety of decision situations, in fields such as government, business, industry, health care, and education. Several firms supply computer software to assist with the steps of the process.

Users of the AHP first decompose their decision problem into a hierarchy of more easily comprehended sub-problems, each of which can be analyzed independently. The elements of the hierarchy can relate to any aspect of the decision problem—tangible or intangible, roughly estimated or carefully measured, well- or poorly-understood—anything at all that applies to the decision at hand.

Once the hierarchy is built, the decision makers systematically evaluate its various elements, comparing them to one another in pairs. In making the comparisons, the *decision makers can use concrete data about the elements, or they can use their judgments about the elements' relative meaning and importance. It is the essence of the AHP that human judgments, and not just the underlying information, can be used in performing the evaluations.*

The AHP converts these evaluations to numerical values that can be processed and compared over the entire range of the problem. A numerical weight or priority is derived for each element of the hierarchy, allowing diverse and often incommensurable elements to be compared to one another in a rational and consistent way. This capability distinguishes the AHP from other decision making techniques.

At the end of the process, numerical priorities are derived for each of the decision alternatives. It is then a simple matter to pick the best alternative, or to rank them in order of relative preference.

While it can be used by individuals working on straightforward decisions, AHP is most useful where teams of people are working on complex problems, especially those with high stakes, involving human perceptions and judgments, whose resolutions have long-term repercussions. It has unique advantages where important elements of the decision are difficult to quantify or compare, or where communication among team members is impeded by their different specializations, terminologies, or perspectives.

The first step in the analytic hierarchy process is to model the problem as a hierarchy. In doing this, participants explore the aspects of the problem at levels from general to detailed, then express it in the multileveled way that the AHP requires. As they work to build the hierarchy, they increase their understanding of the problem, of its context, and of each other's thoughts and feelings about both.

A hierarchy is a system of ranking and organizing people, things, ideas, and so forth, where each element of the system, except for the top one, is subordinate to one or more other elements. Diagrams of hierarchies are often shaped roughly like pyramids, but other than having a single element at the top, there is nothing necessarily pyramid-shaped about a hierarchy.

Human organizations are often structured as hierarchies, where the hierarchical system is used for assigning responsibilities, exercising leadership, and facilitating communication. Familiar hierarchies of "things" include a desktop computer's tower unit at the "top," with its subordinate monitor, keyboard, and mouse.

"In the world of ideas, we use hierarchies to help us acquire detailed knowledge of complex reality: we structure the reality into its constituent parts, and these in turn into their own constituent parts, proceeding down the hierarchy as many levels as we care to. At each step, we focus on understanding a single component of the whole, temporarily disregarding the other components at this and all other levels. As we go through this process, we increase our understanding of whatever reality we are studying.

An AHP hierarchy is a structured means of describing the problem at hand. It consists of an overall goal, a group of options or alternatives for reaching the goal, and a group of factors or criteria that relate the alternatives to the goal. In most cases the criteria are further broken down into subcriteria, subsubcriteria, and so on, in as many levels as the problem requires.

The design of any AHP hierarchy will depend not only on the nature of the problem at hand, but also on the knowledge, judgments, values, opinions, needs, wants, and so forth, of the participants in the process."

AHP is but one of several techniques for criteria-based/weighted analysis. Multicriteria decision-making is a popular methodology for deciding among competing alternatives.[2] Some of these alternatives include:

- Multi-attribute global inference of quality (MAGIQ)
- Goal programming
- ELECTRE (Outranking)
- PROMETHÉE (Outranking)
- Data envelopment analysis
- The evidential reasoning approach
- Dominance-based rough set approach (DRSA)

All of these methods represent variations on the multicriteria theme. Suffice it to say here that this is a discipline unto itself—and not the focus of this analysis. AHP performs more than adequately for the due diligence purposes intended here.

ORGANIZATION AND EXECUTION

Successful technology investing requires discipline. Many CIOs, vendors, VCs, and other technology investors get it right some of the time, but not nearly enough implement repeatable processes that improve the odds of an investment paying good vs. anemic dividends. Like so many business processes, success often depends on how routine they really are. If every time a technology investment is approached it is evaluated differently, it will be impossible to assess how well or poorly the investment decision-making process really is. Consistency and repeatability are key.

There are several steps necessary to improve the likelihood of success. They include:

- Organizing the right team
- Finding and leveraging savvy consultants
- Scheduling the right schedule
- Developing powerful business cases

ORGANIZING THE RIGHT TEAM

It is all about the team. CIOs need to make sure that the team is comprised of professionals who can objectively represent short- and longer-term corporate requirements. They also need to populate the team with rocket technologists–people that really understand the technology in which they plan to invest. They also need people that can translate business technology requirements into technology investment options with special reference to good business outcomes.

CIOs need to understand the impact that the investment might yield. Vendors need to understand how a new product or service will be accepted by their existing and future clients. VCs need to determine if the investment will yield lots of money over a short period of time—or whether there is just too much financial uncertainty around the opportunity.

CIOs need to organize a team that can translate and integrate requirements and technological capabilities. Vendors need incredibly diagnostic insight into what their clients are willing to buy and deploy. VCs need teams that can translate and integrate requirements and technological capabilities as well as estimates about the likely size and timing of acceptable investment payback. VCs are the least patient; CIOs the most; vendors are in the middle.

CIOs, vendors, and VCs all need solid financial modelers. "Spreadsheet jockeys" model expectations about the returns on investments (ROI) as part of complicated assessments of market size, revenue, and costs. CIOs model initial cost, longer-term cost, and payback schedules. Vendors calculate market size, cost-to-develop new products and services, sales and marketing costs, and conversion rates.

The question of perspective is critical. Technology investors that intend to deploy technology inside of a company to achieve some result see the world differently from those seeking a quick financial return on their investment. Technology managers should focus on the pain that business managers feel. The really good ones keep a running list of the most difficult problems—the sharpest pain points.

Business pain comes in many forms. Some comes in the form of cost control, such as headcount and overhead cost reduction. Other pain relief comes in the form of improved business response and control, such as improved management effectiveness, employee productivity, and supplier relations. The search for business pleasure should also occupy a leader's time and energy. Some pleasure includes revenue generation, up-selling, cross-selling, organic growth, acquisitive growth and, of course, increased profit. The whole pleasure/pain exercise focuses on business success. Technology investors understand what makes people heroes, what the organization values.

Venture investors are looking for very different outcomes. While they would like the customers of the companies in which they invest to see the offerings as pain killers or sources of business pleasure, they're fine with the perception of value—so long as they can time their investment and exit profitably. The simultaneous challenge of vendors is to set trends and satisfy pain while dispensing pleasure.

Due diligence teams should be comprised of professionals with the right investment perspective and, perhaps most importantly, professionals capable of seeing the world objectively. Objectivity is so important that some organizations actually assign a designated "devil's advocate" to make sure that the team does not fall in love with the technologies and people that surround the deal. One of the classic mistakes that VCs make, for example, is to overweight management teams over the robustness of the technology being hawked. Many VCs fall in love with management teams because they may have had some success in the past. I remember a conversation about how incredibly persuasive a technology company CEO had been over time—and therefore worthy of an investment; my response was that even Barnum and Bailey could not sell punched cards in the technology industry today. The fact is that most of the due diligence criteria offered here need to be satisfied before making technology investments. Due diligence processes should be data-driven, not driven by relationships with buddies, brothers-in-law, or managers that come "highly recommended." The due diligence team should adopt the right perspective as it evaluates alternative technology investment opportunities.

FINDING AND LEVERAGING SAVVY CONSULTANTS

Everybody needs help. The primary reason to enlist consultants to support the due diligence process is to gather and analyze facts objectively. Consultants have no vested interest in the decision or its outcome (or at least that's how the relationship with the consultant should be structured). Paid by the drink, they only have incentives to tell it like it is, to challenge the on-going assumptions and, when necessary, play an aggressive devil's advocate role.

Consultants can also fill some critical technology gaps. How informed are any of us about service oriented architectures (SOAs), RFID standards, and wireless WiMax technology? Consultants specialize in vertical and horizontal areas and should be leveraged for what they know.

Relationships with specialty consultancies and individual consultants should be nurtured. Larger consultancies have conflicting objectives: while they of course want to help you with your due diligence process, they also are under pressure to extend the engagement for as long as possible. Build a rolodex of objective consultants and use it as frequently as possible. If you buy a lot of technology then you might consider retaining consultants to make sure they are available when you need them. You should also assess their productivity and the quality of their analyses continuously.

As suggested above, it is important not to arrange your relationships with due diligence consultants around the outcome of the deal. Instead, pay them a fixed fee for their services. If you create a financial incentive for consultants based on the outcome of an investment, they will make their assessments based on how much money they stand to make across the spectrum of possible outcomes not according to what might be best for the primary investor.

SCHEDULING

Technology investors should realistically schedule all due diligence analyses. "Realistic" means that they should find ample time to understand their current and longer-term requirements, their competitive positioning, to conduct pilots and to roll out the technology across their enterprises.

Vendors need to understand the timeline of their current offerings, that is, understand full well how their customers see the life cycle of their technology investments. For example, some companies see investments in ERP applications as long-term. Others have adopted a three year refresh cycle for hardware. CIOs try to squeeze as much life as possible from their hardware, software and communications technology investments. If vendors propose a more aggressive migration schedule, their customers will balk. If they attempt to punish their customers for not moving fast enough to new versions of software or new hardware architectures, they run the risk of alienating their customers.

VCs need to move fast—especially in a hot venture market. They need to apply the 15 due diligence criteria as quickly and thoroughly as possible—but should not short-circuit the process. Sometimes experienced VCs believe too much in their own instincts: there is no substitute for objective data. The key to rapid due diligence is the simultaneous pursuit of data around all of the criteria. This usually means that VC due diligence teams segment into specific areas of expertise and sometimes actually cross paths with other teams. "Drive-by" or "fly-by" due diligence is always dangerous.

Due diligence requires lots of structured analysis and discussion. Technology managers need to understand what the technology can really do for them, and what it cannot. They need to talk to the creators of the technology as well as the early adopters. They need to model the costs and benefits of the technology. They need to work with the prospective users of the technology. All of this takes time and should be sequenced to maximize the benefits of the process. Diving in is instinctive but significant background work is necessary before discussing the specifics of a technology implementation project.

The same is true for vendors creating new technology and venture capitalists looking to invest in technologies with broad appeal. Lots of groundwork should occur before inspecting the management teams, facilities or the technologies themselves. The simple rule of thumb is analysis → contact → more analysis → more contact → final analysis → decision.

Due diligence can last weeks or months; it is silly to think that serious due diligence can be accomplished in days.

DEVELOPING POWERFUL BUSINESS CASES

The argument to invest (or not to invest) in a technology lies in the business case that CIOs, vendors, and VCs develop and defend. The 15 criteria should form the essence of the business case for investing, not investing or requesting more information. Are the lights green, yellow, or red? If heavily weighted criteria are yellow or red, it probably makes sense to delay the investment, but if the heavily weighted criteria (and perhaps some others) are all green, then an investment should be made.

How do you sell a technology project to a battered senior executive team that is in no mood to spend a ton of cash on a new project, especially after you've educated them about the 25% - 30% chance of success? How do you sell an investment committee at a private equity firm? How do you tell a technology vendor that they need to spend a ton of money on another new initiative—that all of their customers will die for?

Every single project should be based on a rigorous analysis of a solid business case (based on data collected during the due diligence process). Every project should be conceived and evaluated with three possible outcome decisions: go/no go/we need more information. All three should have an equal chance of winning. Business case development is all about the identification of real and political reasons to create, buy, or invest in technology. This can be a very tricky process, since (because of the perennial competition for funds) there will always be project enemies, those just waiting to say "I told you so" when the project goes south. Technology buyers, especially in large enterprises, have to make sure they have covered their flanks. The "business case" is therefore as much a "real" document as it is a political one. When you read, or write, them, make sure you use these lenses. In other words, pay attention to the total context of the technology investment.

Business cases should be derived from data collected around the 15 criteria clusters listed above. Let me say up front that the answers to these questions should be in a 10-page document (with appendixes if you must) with a one page executive summary: it you generate a long treatise on every investment, you will never get the time or respect of busy senior executives. Would you read a 50 page business case?

The business case should also identify at least two "accountable" people, people whose reputations rests to some extent on the success of the due diligence project and the outcome of the investment. One should be from the technology side of the organization and one from the business. If there are no investment champions, it is time to go home.

THE BUSINESS OF BUSINESS TECHNOLOGY CASES

How do you sell a technology project to a battered senior executive team that is in no mood to spend a ton of cash on a new project, especially after you have educated them about the 25% chance of success.

Projects come from lots of different places: cCocktail parties, directors, in-flight magazines, ball games, dinner parties, Gartner reports, advisors, spouses, loss of market share, falling margins, arguments and security breaches, among other predictable and unpredictable sources. Every single project should be based on a rigorous analysis of a solid business case. Every single project should be conceived and evaluated with three possible outcome decisions: go/no-go/we-need-more-information. All three should have an equal chance of winning.

Business case development is all about the identification of real and political reasons to buy something or engage a consultant. This can be a very tricky process, since (because of the perennial competition for funds) there will always be project enemies, those just waiting to say "I told you so" when the project goes south. Technology buyers—especially in large enterprises—have to make sure they've covered their flanks. The "business case" is therefore as much a "real" document as it is a political one. When you read—or write—them, make sure you use at least these two lenses.

Business cases are typically organized around questions designed to determine if the investment makes sense. Let's look at the key questions that you should answer before buying hardware, software, communications, or consulting services. Let me say up front that the answers to these questions should be in a 10-page document (with appendixes if you must) with a one page executive summary: it you generate a long treatise on every investment, you will never get the time or respect of busy senior executives. Would you read a 50 page business case?

The first step is to identify, and answer, collaborative business value questions or, put another way, answer one simple question: how will this project help the business collaborate profitably?

Here are the questions. Think of these as offering a summary of the due diligence criteria assessments converted into a more or less standard business case—the kind that many companies already use.

Key Collaborative Business Questions

What collaborative business processes are impacted by the investment/application(s)?

The correct answer here identifies a process that is broken, a process for which you have empirical benchmarking data. For example, if you were serving customers with call centers, you would have to know what you are spending now and that the costs were rising dramatically, or customer satisfaction levels were falling. You would then need to know exactly what your performance target should be (e.g., reduce costs by 25%; increase satisfaction by 25%).

How pervasive is the product or service among your traditional and unconventional competitors?

Many decisions about business technology adoption are driven by what the competition is doing: it is always easier to sell something in-house when your competitors have already adopted the product or service (of course, this assumes that your competitors are smart). Be prepared for the contrarian argument here: "if those people are doing we sure as hell shouldn't be!"

What competitive advantages does/will the product or service yield?

In other words, why is this product or service so great and how will it help you profitably grow the business?

How does the new product or service interface with existing collaborative business models and processes?

This is a big question, since the "wrong" answer (like: "it doesn't") will kill a project. If deployment means ripping out or enhancing infrastructure, then you have just significantly increased the price (and given your internal enemies ammunition). You have also raised a legitimate yellow flag.

Key Technology Questions

How mature is the product or service?

The point of these questions is to determine if the technology or service actually work—and work well. You will need quantitative data here: it will not help to tell everyone about how great your brother-in-law's company is. Additional questions here concern scalability, security, modifiability, usability, and so forth.

What share of the market does the product or service own?

If the answer to the question is: "well, 1%," then a major explanation will be necessary. Remember to stay mainstream, anyway. Why do you want to be an early adopter of someone's half-baked product?

How does the new product or service integrate with the existing or planned communications and computing infrastructure?

This question is, of course, the second most important question you need to ask, and answer, because it identifies any problems downstream that might occur because of decisions already made (that cannot be undone, or would be prohibitively expensive to undo).

Key Cost/Benefit Questions

What are the acquisition, implementation, and support costs and benefits?

Here you need to look at obvious costs, like software licenses, and less obvious ones, like training, indirect user and help desk support, as well as the expected operational and strategic benefits like expense reduction, increased market share, improved customer service, increased cross- and up-selling, improved customer retention, and so forth. Here's where total-cost-of-ownership (TCO) data gets inserted.

What are the support implications? How complex, how costly, how extensive, how timely?

Support is extremely important. You need to know, empirically, what the support requirements and costs will be defined in terms of money, people, and time.

What are the migration issues? How complex, how costly, how extensive, how timely?

This may or not be relevant, but if another tool is in place you have to answer questions about how to get from one to the other.

Key Risk Questions

What are the technical, personnel, organizational, schedule, and other risks to be considered? What is the risk mitigation plan?

The risk factors that everyone will worry about include scope creep, cost and time overruns, incompetent or irritable people, implementation problems, support inadequacies, training problems, and the like.

If risks are assessed as medium or high then a mitigation plan must be developed, a plan that either moves the risk down to the "low" category or eliminates it altogether. If the risks remain high, the project sale is dead.

THE RECOMMENDATION

The recommendation is go/no-go/we-need-more-information. The business case should also identify at least two "accountable" people, people whose reputations rests to some extent on the success of the project. One should be from the technology side of the organization and one from the business side (if you can find the right hybrid). If there are no project champions, it is time to go home. Who owns all this? If there's no Process Officer in your company, no anointed process, and no business case requirement, you are in some trouble. It is amazing how crazy all this is. Business cases are not that difficult to create: what is difficult is swallowing them. Cultures that are especially political (see Figure 4) have the worst digestion problems—and make the most business technology investment mistakes.

Figure 9. Business case drivers

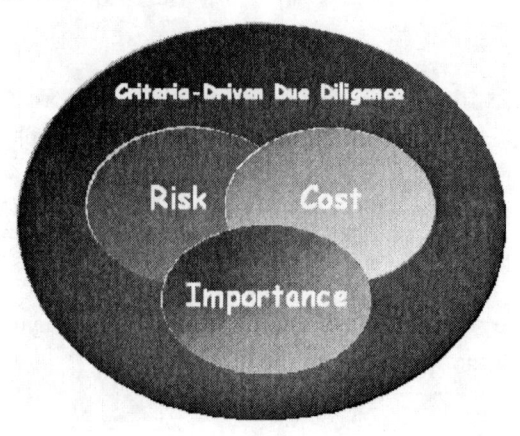

EXECUTION

There are things to buy, outcomes to expect and criteria to be analyzed as part of the technology investment due diligence process. The open question, as always, is about the amount of discipline we're prepared to accept, and repeatedly practice. I have also argued the extensibility of the 15 criteria that can serve those who create, deploy, and invest in technology. The criteria represent the center of the due diligence process and while criteria weighting will differ depending on who is making the investment, the criteria themselves work nicely across all technology investors.

Is this all foolproof? No. Many of the assumptions that we make—and even empirically validate—can be wrong. Assumptions about the quality of the technology itself may be unwarranted. Assumptions about the people involved in the process may be generous. And there is always the intangible, unpredictable, and unfathomable. In other words, it is impossible to engineer perfect outcomes. The discipline described here can reduce, not eliminate, risk. It is a prudent methodology, not a magic potion. CIOs, vendors, and VCs need to reduce risk to increase the returns on their technology investments. The application of the due diligence criteria described here can help a lot, so long as we all know what we need to buy, what we can buy and how to manage our investment expectations.

OUTCOMES

Expectations are always a challenge to manage. CIOs have different expectations of their technology investments than vendors and VCs. VCs are the easiest to decode:

money, money, money. They are not interested in long-term business relationships (except with "serial entrepreneurs"). They want quick strikes. They want massive returns on their investments. Vendors see the world a little differently. They want money, of course, but they also want a vendor/client relationship that might pay long-term dividends: they really want to monetize their customers for as long as possible. CIOs want impact on business processes, models, cost structures, and people. They want reliability, security and scalability. They want to contribute to operational efficiency *and* strategic success.

What CIOs Want From Their Technology Investments

CIOs and CTOs ideally see the world through the eyes of the businesses they support. They should speak the language of business. But most importantly, they should focus on the pain that business managers feel. The really good ones keep a running list of the most difficult problems—the sharpest pain, the things that keep business managers up at night.

Business pain comes in many forms. Some comes in the form of cost control, such as headcount and overhead cost reduction. Other pain relief comes in the form of improved business response and control, such as improved management effectiveness, employee productivity and supplier relations.

The search for business pleasure should also occupy the time and energy of CIO's and CTO's. Some pleasure includes revenue generation, up-selling, cross-selling, organic growth, acquisitive growth and, of course, increased profit.

The whole pleasure/pain exercise focuses on business success. It also focuses on what individual business professionals will personally find exciting—and rewarding. Effective CIOs and CTOs understand what makes people heroes, what the organization values.

Figure 10 identifies three paths in the business technology alignment-to-partnership journey. CIOs have to appreciate business pain and pleasure, have to become more than just credible, and have to define business value around operations *and* strategy. If CIOs understand these paths, they can redefine the business technology relationship.

Remember that the business expects technology to reduce its pain—almost always defined around cost reduction. But it is more than that. Business managers worry about their supply chains, their competitors, their manufacturing, distribution and, of course, their profit margins. The technology agenda needs to speak directly to their pain points—which, when relieved—can become the sources of wide and deep pleasure. If CIOs become dispensers of pleasure as they reduce pain, their credibility will rise, which will enables the second path to business technology partnership.

Hopefully when technologists walk into a room the business managers do not run for cover or, worse, attack them mercilessly for their sins (network crashes, Web site debacles). Nirvana here is influence—defined in terms of how the business thinks about how and where technology can help. Does the business respect its technology managers enough to confide in them, to commiserate with them, to invite them to brainstorm about business strategy?

If a CIO is influential, the CIO can shape both operations and strategy. If CIOs get operations straightened out, they can spend most of their time—with their new partners—thinking about competitive advantages, revenues, and profitability. There is no better place to work, no better way to spend your time. Effective CIOs seek this influence.

The following summarizes the business plan/please objectives.

- **Business Pain**
    - Cost Reduction
        - Headcount Reduction
        - Cost Management
        - Overhead Cost Reduction
        - Other Cost Reduction ...
    - Improved Business Response & Control
        - Improved Management Effectiveness
        - Employee Productivity/Effectiveness
        - Improved Supplier Relations
        - Organizational Awareness ...
- **Business Pleasure**
    - Revenue Generation
        - Customer Satisfaction & Retention
        - Up-Selling & Cross-Selling
        - Organic Growth
        - Competitive Advantage
        - Profit ...

Figure 10. Paths to business technology partnership

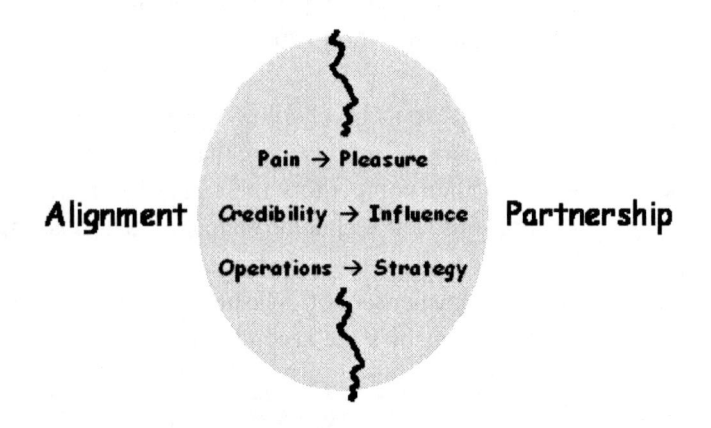

Alignment — Pain → Pleasure / Credibility → Influence / Operations → Strategy — Partnership

The distinction between operational and strategic influence is important to CIOs. The best outcome of technology investments is both operational and strategic quantitative impact. As Figure 11 suggests, however, we tend to associate business pain with the operational environment and pleasure with the strategic environment.

Discussions about the commoditization of technology are accurate to a point. There's no question that PCs, laptops and routers are commodities; even some services—like legacy systems maintenance, data center management and certainly programming—have become commoditized. But the real story is not the commoditization per se but the bifurcation of business technology into operational and strategic layers. Operational technology is what is becoming commoditized; strategic technology is alive and well—and still a competitive differentiator, in spite of what some recent articles have been arguing.

What is the difference? Operational technology supports current and emerging business models and processes in well-defined ways with equipment and services whose costs have become stable and predictable and have generally declined significantly over the past decade. Hardware price/performance ratios are perhaps the most obvious example of this trend but there are others including what we are willing to pay for programming. Strategic technology on the other hand is the result of creative business technology convergence where, for example, a Wal-Mart streamlines its supply chain, a Starbucks offers wireless access to the Web, and a Vanguard leverages its Web site to dramatically reduce its costs. There is no limit to how creative business technology convergence can become; there is no limit to how strategic the business technology relationship can be.

Figure 11 draws the important line. Above the line is where management lives. It is also where big front office applications—like CRM applications—live. Below the line are the back office applications and the infrastructure that enables digital contact with customers, suppliers, partners, and employees.

CIOs need to optimize above the line and below the line. Their technology investments must speak directly to one or the other, or, ideally, both.

What Vendors Want From Their Technology Investments

Vendors spend billions a year on research and development (R&D). Microsoft and IBM alone spend well over $10B annually. Vendors look for the next new thing, but they also simultaneously look back to make sure that their new technologies integrate and interoperate with their existing technology.

In general, the industry is getting much better at integration. A significant trend is the industry's increasing ability to make disparate pieces—data bases, hardware systems, applications—work with each other. Good progress to an even better end game: ubiquitous real-time transaction processing through seamless technology

Figure 11. Operational and strategic technology impact

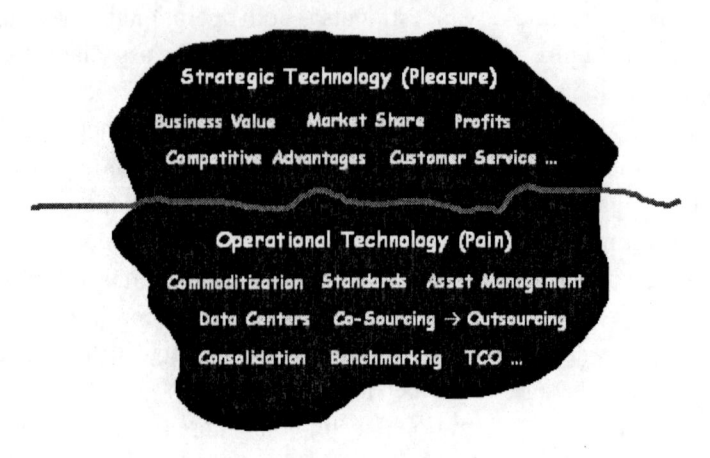

> **Strategic Technology (Pleasure)**
> Business Value   Market Share   Profits
> Competitive Advantages   Customer Service ...
>
> **Operational Technology (Pain)**
> Commoditization   Standards   Asset Management
> Data Centers   Co-Sourcing → Outsourcing
> Consolidation   Benchmarking   TCO ...

integration. Integration is now a core competency in many companies and the industry has responded with a barrage of new tools to make things cooperate. This trend will continue and threatens to dramatically alter software architectures and technology infrastructures.

We are now able to wrap older legacy systems in newer standards-based technologies. We are able to integrate supply chains with Web services, and we are able to think holistically about application integration and the transactions it supports—like up-selling and cross-selling.

Vendors also want competitive positioning in their markets. They invest in technology to differentiate themselves. They nibble at new markets with creative (yet interoperable) offerings. They also want revenue and profit. R&D investments have their own life cycle. Vendors seek as much return from these investments as possible.

What VCs Want From Their Technology Investments

The obvious answer is money. But the path toward the money depends on the stage of the investments that the VCs prefer as well as the particular mission of the venture fund in question. For example, some VC funds prefer "exits" that do not involve public offerings. Others live for public offerings. Seed and early stage investors—investors that work with very young companies and immature technologies—have different expectations than VCs who invest in later stage companies.

There are a number of specific outcomes that VCs pursue. The first is simply the rapid ramping of a company where revenues increase dramatically over a short period of time. When this occurs additional options appear, such as selling their investment in the company, borrowing against future revenues, taking the company public, or engineering the acquisition of the whole company (which enables all of the investors to exit the investment).

Revenues can come from sales of products or services or from licensing revenue from the intellectual property (IP) that their investments make possible.

Along the way, VCs expect the companies in which they invest to develop partnerships, alliances, and other revenue-generating relationships. They expect the companies to invest in their own products and services and recruit and retain the best people. Ultimately, of course, VCs seek financial exits from their investments. They are quite predictable here. They are not long-term partners. The typical private equity venture fund has a 10-year life with limited partners and general partners who want to see their investments grow.

REFERENCES

Andriole, S. J. (2005). *The 2nd digital revolution*. IGI Publishing.

Andriole, S. J. (2007, March). The 7 habits of highly effective technology leaders. *Communications of the ACM, 50*(3), 67-72.

Bing, G. (1996). *Due diligence techniques & analysis*. Quorum/Greenwood.

Breitzman, A., & Thomas, P. (2002, September 1). Using patent citation analysis to target/value M&A candidates. *Research-Technology Management, 45*(5), 28-36(9).

Camp, J. J. (2002). *Venture capital due diligence*. John Wiley.

Cullinan, G., Le Roux, J. M., & Weddigen, R. M. (2004, April). When to walk away from a deal. *Harvard Business Review, 82*(4), 96-104, 141.

Gerson, S. P. (2000, May). Patent law: Strategic due diligence. *The National Law Journal*.

Harvey, M. G., & Lusch, R. F. (1995). Expanding the nature and scope of due diligence. *Journal of Business Venturing, 10*(1), 5-21.

Howson, P. (2003). *Due diligence: The critical stage in acquisitions and mergers*. Gower Publishing.

Lajoux, A. R., & Elson, C. M. (2000). *The art of M&A due diligence: Navigating critical steps & uncovering crucial data*. McGraw-Hill.

Lawrence, G. M. (1994). Due diligence in business transactions. *Law Journal Press*.

Marlin, S. (1998, December). Technology due diligence. *Bank Systems + Technology*.

McGrath, R., Gunther, Keil, Thomas, & Tukiainen, T. (2006). Extracting value from corporate venturing. *MIT Sloan Management Review, 48*(1), 50-56.

Perry, J. S., & Herd, T. J. (2004). Reducing M&A risk through improved due diligence. *Strategy & Leadership, 32*(2), 12-19.

Prahalad, C. K., & Krishnan, M. S. (2002). The dynamic synchronization of strategy and information technology. *MIT Sloan Management Review, 43*(4), 24-33.

Weill, P., & Aral, S. (2006). Generating premium returns on your IT investments. *MIT Sloan Management Review, 47*(2), 39-48.

Zacharakis, A. L., & Meyer, G. D. (1998). A lack of insight: Do venture capitalists really understand their own decision process? - Heuristics and biases. *Journal of Business Venturing, 13*(1), 57-76.

ENDNOTES

[1] McCaffrey, J. (2005, June). The analytic hierarchy process. *MSDN Magazine, 20*(6), 139-144.

[2] See Gal, T., Stewart, T. J., & Hanne, T. (Eds.). (1999). Multicriteria decision making: Advances in MCDM models, algorithms, theory and applications. Kluwer Academic Publishers.

Chapter II
Acquisition Targets

This chapter looks at the range of investment targets ranging all the way from software applications to communications networks to new advanced technologies like Web 2.0 and Web 3.0. The chapter talks about the opportunities in succession describing the range of opportunities available to CIOs, vendors, and VCs. The chapter also highlights the individual opportunities for all three investors through investment guidelines that appear at the end of each major section.

APPLICATIONS TARGETS

Everyone uses software applications almost every day. Those who shop on Amazon.com or bid on eBay use applications. Everyone who uses e-mail or searches for optimal driving routes uses software applications, and of course everyone who sits in front of a screen anywhere in companies around the world use applications to calculate sales, check inventory, or just close the books use software applications. Some of these are very well-known, such as Microsoft Office which contains software applications for word processing, presentation graphics, and spreadsheet calculations. Some are less known and more specialized, like Manugistics' supply chain planning and management applications.

Those who invest in technology invest in software applications; they are everywhere. Applications "face" those who use computing and communications technology; they are what we interact with every time we log on to a network or the

Internet or boot up our PCs. Applications permit us to communicate with data and through preprogrammed instructions for communicating with people, information, companies, customers, and suppliers.

There are a variety of applications we use all the time. There are mainstream applications—like the aforementioned Microsoft Office—and less popular ones, like tools we use to manage our time, ideas, and contacts. I have developed a matrix for describing the range of applications that populate small, medium and large companies, and excite, depress, and confuse, all kinds of investors in applications that can revolutionize the way companies transact business. The applications that excite companies are of course what the major, and entrepreneurial, technology providers seek to design, develop and sell, and what private equity venture capitalists task their various "deal flow" mechanisms to find.

From the perspective of a buyer of technology (a CIO, for example), the matrix is intended to guide an assessment and then a selection process. Companies need to profile their current applications portfolio and then plan for the next one. Nothing radical is implied here: it is impossible to throw away all of the old to make room for the new. At the same time, companies need to migrate constantly and efficiently. From the perspective of investors (a venture capitalist, for example), the matrix is intended to help identify investment opportunities, and from the perspective of technology vendors (for example, an enterprise software developer like SAP AG), the framework will help spot enhancements and whole new applications that their customers might buy.

The matrix appears in Figure 1 and indicates the range of available applications. The boxes represent opportunities and challenges for CIOs, venture capitalists and vendors—for anyone who creates, deploys, and supports software applications.

If you are looking at the applications matrix in Figure 1, what do you see? It depends on the lens you use to assess the opportunities and challenges within its cells. CIOs see the basic applications that enable business processes, the applications that just about everyone in their companies touch every day. Venture capitalists see the cells as containers of marketable technology, and vendors see the list as opportunities to sell more software to existing and new customers.

The first stop is personal digital assistant (PDA) and related "thin client" applications and, by extension, desktop and laptop applications.

There are a series of clear trends that define this area. We know that most desktop and desktop computers run Microsoft Office and that Office will dominate well into the 21st century but will do so by creating, reacting, and adapting to increasingly open source software standards. There will be a significant challenge to Office's dominance from the Open Source vendors but Microsoft will likely respond by modifying its pricing policies to create enough cost-benefit ambiguity that most users will ultimately stay with Office. On the other hand, if Microsoft fails to respond to either pricing or open standards pressure, it could lose market share.

The desktop browser—the application we use to get to the Internet and the World Wide Web—will become the ubiquitous front-end to desktop applications and applications integrated with "local" (desk-top, laptop, PDA/thin client) computing and the computing that occurs externally when users go on to the Web to search for information or assemble transaction components.

As applications really do get "thinner" (with more and more computing power moving to servers and away from the access "clients"), then the desktop/laptop operating system (OS) and the primary desktop applications that run on the desktop/laptop will become relatively less important, but unless all users are thin-client users, there will still be a huge need for a major desktop OS that is "standard."

From a vendor/company/VC investors' perspective, several of these trends deserve additional attention. The thin vs. fat client debate is one that was actually launched by Larry Ellison, the founder and CEO of oracle Corporation, when he appeared on the Oprah Winfrey show a decade ago to discuss "network computers." This was when a lot of ahead-of-their-time ideas were out there, like Apple's Newton and IBM's voice recognition applications. Larry was wrong then because

Figure 1. The range of application targets

	Application Targets
PDAs/ Thin Clients	• Web Access • Email • Calendaring • Contact DB ...
Desktop/ Laptop	• Web Access • Word Processing, DB, Graphics, Spreadsheet • Portals ...
Enterprise	• Legacy • Packaged ERP • SFA, CRM • DB Management ...
Management	• Network & Systems Management • Applications Management • Synchronization ...
Services	• End-to-End Services • "Hertical" • "Vorizontal" • Maintenance ...

our networks were not reliable, secure, or ubiquitous enough to support "thin client" architectures. Never mind that the devices themselves were also a little weird and way too proprietary. But if Larry reprised his appearance on Oprah tomorrow, he would be dead right. The concept was always right.

Let us look at several trends that point to why "thinfrastructure" makes sense. First, we have segmented our users into classes based on their frequency and depth of use: some people use very expensive and hard-to-support laptops pretty much for e-mail. We have begun to deploy a bunch of devices—like PDAs, cell phones, and pagers—that are converging. Many of these devices are pretty dumb, but that is OK because we do not ask much of them. Multiple device synchronization is, however, difficult and expensive. Desktops and laptops are way overpowered for the vast majority of users (as is the software that powers them). Enterprise software licensing is getting nasty (if you do not renew your Microsoft software every three years, you get punished). Desktop/laptop support is still complicated and expensive; large enterprises—even if they have network and systems management applications—still struggle with updates, software distribution , and version controls, and the larger the organization the greater the struggle.

Network access is ubiquitous today: we use desktops, laptops, personal digital assistants (PDAs), thin clients, and a host of multifunctional converged devices, such as integrated pagers, cell phones, and PDAs to access local area networks, wide area networks, virtual private networks, the Internet, hosted applications on these networks as well as applications that run locally on these devices. These networks work. You can be on if you want to be on.

Many companies provide multiple access devices to their employees, and employees often ask their companies to make personal devices (like PDAs) compatible with their networks. All of this is made even more complicated with the introduction of wireless networks which make employees more independent and mobile (though less secure). The cost to acquire, install and support all of these devices is out of control. What if I told you that the annual support cost for each wireless PDA in your company was more than $4,000? You would either laugh or run and hide. Run and hide.[1]

How should companies approach the network access device problem? What are the viable alternatives—today—and 3 to 5 years from now? Should companies prepare to support multiple devices and access strategies for each employee? Should you standardize on one device? Should you skinny it all down?

Small, cheap, reliable devices that rely on always-on networks make sense. Shifting computing power from desktops and laptops to professionally managed server farms makes sense. Moving storage from local drives to remote storage area networks makes sense. Fat clients should lose some weight—as we bulk up our already able servers. The total-cost-of-ownership (TCO), not to mention the return-on-investment

(ROI), of skinny client/fat server architectures is compelling (to put it mildly). Is this about control? Is it about reliability? How about security? Yes. Since we all agree that timing is everything, given all of the new devices appearing and how many of them are converging into wholes greater than the sum of their parts, isn't this a great time to think thin? If you do not, then the only people you will make very happy are the vendors you have to hire to keep all the toys humming. If you get control of the servers, you control everything that plays on every access device; and if you skinny down the access devices, you get control, flexibility, standardization, reliability, and scalability for your users. No more software conflicts, instant software upgrades, quick and easy new application deployment. Sound too good to be true?

If networks and remote access procedures were foolproof, thin clients would begin to appear at a much faster rate than they are appearing today. Individual users control the applications that reside on their personal computers (PC) and are—when they choose to be—independent of network or other connections to be productive. But if network and other connections were reliable and pervasive, and personal productivity was unaffected by the location of the application itself, then users (and their managers) would accept thinner clients connected to reliable networks.

Why? Because costs can be lowered and productivity can be increased. The major obstacles to thinner clients has been the inability of networks to behave like "utilities." In a very short time, they will, and we will have more application deployment options available to us.

Desktop, laptop, PDA, and thin client trends are significant drivers of investment prioritization. For example, investments in a Microsoft Office-killer application—one designed to displace Microsoft's dominance in the personal productivity space—would not appeal to venture capitalists who, more often than not, search for easier prey. A CIO might look at alternatives to MS Office especially given the cost of proprietary enterprise software (open source alternatives, like OpenOffice and Sun Microsystems StarOffice are far cheaper than MS Office), but there are always compatibility and migration issues. What about the vendors who design and develop personal productivity software? Most of them look for ways to plug into the dominant platforms which, in this case, means plugging into MS Office.

Investments in thin clients and the software that powers them represent sound investments in software applications for all sorts of reasons. Thin clients are simpler and cheaper to operate than fat clients (the PCs and laptops that we use today). There are new opportunities to design and develop software applications that will support the increased use of thin clients. Technology managers in medium-sized and large corporations should be piloting thin clients; venture investors are interested in new applications that power thin clients and vendors are already investing in applications that will run on the servers in thin client networks as well as the "skinny" software that will reside on all varieties of thin clients.

The Next Stop (in Figure 1) is Enterprise Applications

Many organizations look at enterprise applications as solutions to their heterogeneous, expensive applications problems. Enterprise resource planning (ERP) applications, like SAP's R/3 and Oracle Financials, were designed to relieve proprietary applications pressure by "standardizing" organizations on single applications that served the entire "enterprise." Instead of 20 financial reporting systems, ERP financial applications offered the promise of a single integrated application. Even better, these applications were described and sold as "backbone" applications that could be enhanced with "modules" that would add functionality to the ERP application whenever one needed it. More recently, so-called "back-office" applications—like financial accounting and reporting—are being integrated into "front-office" applications, like customer relationship management (CRM), to increase the functionality of the enterprise application backbone.

The reference to back-office and front-office applications is important. ERP vendors are moving very quickly to bolt on front office applications—like sales force automation (SFA), supply chain planning (SCP), customer relationship management (CRM), and business process management (BPM) applications—to the existing back-office applications like enterprise financial management tools. Companies planning to move in the ERP direction should step back and think more holistically about the application as a platform. In fact, they should regard the ERP application as a "backbone" application to which they attach more and more back-office and front-office application modules.

Corporate technology managers should look at enterprise applications as pain killers. Venture capitalists should regard them as opportunities so long as they're consistent and compatible with mainstream applications and the market share of the ERP vendors. Today, ERP is dominated by just a few players, the largest of which is SAP. Should VCs look to displace SAP and Oracle with a new generation ERP application that will entice users to switch from their installed applications? Should they invest in applications that hook into the major platforms? What should the major vendors do? What new capabilities should they create?

Enterprise software applications represent important investment targets for everyone who deploys, creates, or invests in technology. One of the steps that more and more companies are taking is the conduct of "optimization audits." Several companies I work with have requested them because they have made major enterprise applications investments and would like to know if they are getting the bang for the buck they expected (and were told to expect by the vendors who provided the hardware and software and the consultants who assisted in the implementation of the monster applications or rejuvenated infrastructure).

Optimization audits look at existing applications and assess their potential value to sales, marketing, growth, and profitability. These audits are different from the more conventional total-cost-of-ownership (TCO) or return-on-investment (ROI) assessments that companies often make before they approve business cases for new technology projects. Optimization audits focus on unexploited business value from investments already made. Put another way, they are designed to answer the question: "what the hell did we get for the $100M we just spent?" and "you better not tell me that all we got for all that cash was an efficient way to close the books."

The greatest need for Optimization Audits is in companies that have made major investments in enterprise resource planning (ERP), customer relationship management (CRM), and network and systems management (NSM) applications. The price tag for these investments can easily exceed $100M. But there is a life cycle problem with these mega applications: implementations tend to consume so much time, money and effort, that payback tends to be tactical and operational for way too long—and to the relative neglect of strategic payback. For example, let us assume that a company implements an ERP system to integrate disparate financial reporting systems. Most of the effort is devoted to the consolidation of financial data and the standardization of financial reporting. While operational efficiency is obviously valuable, the existence of a common transaction processing platform enables much more than standardized reporting of financial data. An ERP application, for example, can integrate back-office, front-office, and virtual office (Internet) applications. If the data bases are also standardized across these applications then cross-selling and up-selling may be possible, along with supply chain integration and even dynamic pricing. CRM applications can help define "whole customer" models with long-term life cycles that can be monetized year after year. These are the kinds of dividends that enterprise applications can yield if they're pushed to their full capacity—capacities that even the vendors themselves often fail to stress. In their desire to sell operational solutions they sometimes fail to sell the longer-term strategic impact of their technology.

Optimization Audits are designed to answer the following kinds of questions:

- Now that we have a standardized (ERP, CRM, and NSM) platform, what are the quantitative tactical and operational returns on the investment?
- What are the potential tactical and operational benefits we are *not* seeing?
- What strategic benefits are we seeing?
- What strategic benefits are we *not* seeing?
- How can the business be transformed by the efficiency of the platform and its possible extensions?

Optimization Audits take a top-down approach to holistically model your company's information, product, and service processes and their relationship to the standardized platform they have implemented. The top-down profile is informed by the existence of the enterprise platform (which, more often than not, is implemented from a set of bottom-up priorities). The last step is the connection to business value metrics, like sales, growth and profitability.

Optimization Audits should be conducted by companies who have implemented large enterprise applications or massive infrastructure platforms for primarily tactical or operational reasons. While these reasons are solid, they are incomplete. Additional strategic payoff should be defined, and pursued, as vigorously as they pursued tactical and operational payoffs. But remember that strategic payoff is only meaningful when it is defined around business, not technology, metrics. The days are long gone when a CIO can authorize a $100M project to make some applications talk to one another or manage technology assets more cost-effectively. It is no longer about just cost management: it is now—and forever—about growth and profitability. Optimization audits are all about finding the optimal path from technology to profitable growth.

All of this reflection and analysis of enterprise applications defines the enterprise applications target space. What investments should be made? Should you spend money to optimize existing investments? Should vendors develop software tools to optimize software investments? Should venture capitalists fund companies that create optimization technologies? Should whole new applications be developed? Perhaps simpler, less expensive ones that are easy to implement and support?

The Next Stop in Figure 1 is Applications Management

In order to keep applications humming—especially in large enterprises—companies need some kind of central administration of technology management processes. Companies can go with a major framework—like IBM's Tivoli, Computer Associates Unicenter, or HP's OpenView. Or companies can go with smaller enterprise frameworks from any number of vendors (such as Mercury Interactive). Or they can go with "point solutions," individual applications that address aspects of their network and applications management problems, like security intrusion and software distribution.

"Frameworks," large application suites that do everything from manage help desks to distribute new releases of software applications, are coming along, but we are still a few years away from a "seamlessly integrated" suite of products that holistically support the full range of network and applications management.

Regardless of the solutions, applications management requirements will not go away. Companies have to manage their requirements and collect performance data on their effectiveness and support costs. They need products to do this or a relationship with an outsourcer who knows how to do it (who will also use products).

What are the investment priorities here? Network and systems management frameworks represent major investments by corporate CIOs—which are very careful about when to pull that particular trigger. The vendors that create and sell these frameworks try to include features that CIOs will find attractive. The large framework marketplace is already well defined, consequently VCs are unlikely to be that interested in deploying large amounts of capital to build yet another major framework (though they might well be interested in technologies that can offer enhanced features to existing frameworks). The dominant vendors continuously enhance their frameworks for their existing and prospective customers.

The Next Stop in Figure 1 is Application Services and Support

By "support" we are talking about the entire applications life cycle from requirements to design to development to testing to maintenance and redesign. If companies "home grow" their software applications (only because they cannot find a solution in an off-the-shelf package), they need support to test, integrate, and deploy applications. If they outsource the design and development of applications, they have lots of decisions to make about how to procure the necessary services.

Vendors that offer the right products and services as a package of offerings often make good money. "Solutions vendors" who have relationships with other vendors so they can offer integrated solutions, are stronger than vendors with only slices of service offerings. Ideally, service and support vendors understand the vertical business their clients are in, and do not treat all business the same. The substance of vertical businesses is unique and totally horizontal technology solutions are not nearly as effective as those applied through a deep vertical domain filter. As suggested in Chapter I, we sometimes think of these vendors as "vorizontal" vendors—vendors who optimize horizontal technology for specific vertical industries.

Today's calls centers will evolve into tomorrow's customer relationship management (CRM) control rooms where all customer "touches" are monitored and measured. Vendors who understand the CRM model are likely to provide valuable help while those that don't will continue to sell disembodied solutions to old problems. Similarly, e-business vendors need to be end-to-end in their services offerings. They also need to know a whole lot about Web-based transactions, competitor business models, and supply chain efficiencies.

The key to assessing these service vendor capabilities is to look at the range of their offerings. Vendors that specialize in a few areas are not nearly as useful as those that can claim expertise across many areas (and ideally across the entire solution space).

Vendor assessments that are end-to-end oriented are of special importance vs. those that treat each phase of the applications life cycle as distinct. Carried to the extreme, you might hire a different vendor to perform each of systems design and development life cycle tasks. Of course that makes no sense, so the filter through which you should pass vendor assessments as they pertain to applications design, development, integration, testing, and the like should assume an integrated process.

These capabilities assessments are important to CIOs, VCs, and vendors. CIOs need broad service organizations capable of providing integrated solutions to their applications problems. VCs are always looking for creative service offerings here and vendors are constantly looking for ways to sell more and more products and services to their customers.

The applications space should be defined—and investments influenced by—scenarios of how your business will use technology to make money. We also need to be ruthless about the business value our applications are really generating. Too often applications come with an entire team, some members of which expect their kids to inherit the code. Applications should be profiled in terms of their life expectancies, as Figure 2 suggests.

Applications should be monitored according to their costs/benefits and then decisions should be made in anticipation of where they will be at some finite place in time. (Had applications portfolio managers took good hard looks at their non-Year 2000 compliant applications in 1990, a lot of time, effort and money would have been saved, since a whole lot of those applications would have been killed, or, as they say in the data centers, "decommissioned").

SOFTWARE APPLICATIONS INVESTMENT GUIDELINES

Here are some guidelines that can help investors in software applications understand the opportunities and risks regardless of which investor hat you're wearing:

Make sure, as always, that you understand what the business wants to accomplish with the applications they will buy and support. If you cannot put your finger on "purpose" then your applications investments will, by definition, be suboptimal. If "to be" business models do not get developed, or the ones that do have minimal applications implications, push the modelers toward emerging requirements. If the team has trouble articulating these requirements, jumpstart the process with alternative business scenarios and competitor case studies.

Look at applications objectively. Which ones contribute measurably to business? Which ones require disproportionate support? It is essential that you assess applications with reference to business strategies and the relative contribution they are making to business processes. If the outcome of that assessment is clear then

Figure 2. Applications assessment

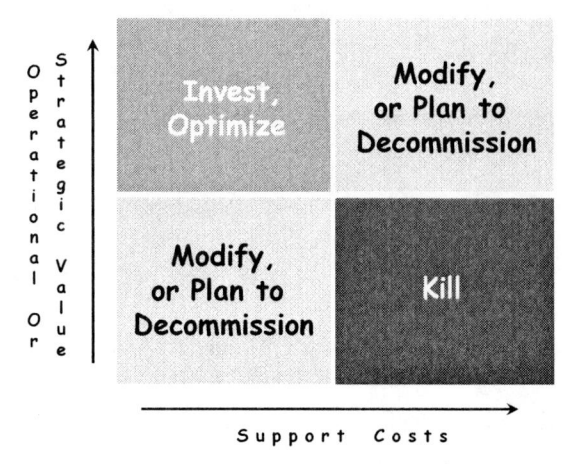

```
Operational  Strategic
    Or        Value

              ┌─────────────┬─────────────┐
              │             │   Modify,   │
              │  Invest,    │   or Plan to│
              │  Optimize   │ Decommission│
              │             │             │
              ├─────────────┼─────────────┤
              │   Modify,   │             │
              │  or Plan to │    Kill     │
              │ Decommission│             │
              │             │             │
              └─────────────┴─────────────┘
                    Support   Costs
```

decisions should be made to decommission applications (in the case of expensive applications that contribute little to the business) or transfer functionality to other, less expensive to maintain systems (in the case of older systems with limited, but still valuable contributions to the business).

You need to assess the variation in your applications portfolio. How many architectures are you supporting? What is the distribution of functionality and architecture type? Do you have your most important applications on the oldest, most expensive to maintain computers? Do you plan to sell applications that require a lot of support?

Stay as standard as possible here. Go with the mainstream desktop vendor—Microsoft—and stay within primary standards with non-Microsoft applications. The primary desktop applications with include the office suite running on a Windows platform. While some of us may find this ubiquity disconcerting, it makes the world compatible. Investors in technology should ride the right technology waves, not swim upstream.

Keep an eye on those enabling technologies most likely to affect your applications performance and costs. Access devices, like biometric fingerprint authentication tools, can help a lot, enabling single-sign-on and reducing security administration costs. Other technologies, like component technologies, can help your applications support and modification processes. They should be watched. Human-computer interface (HCI) technologies should be monitored as well, since speech input/output—as one example—will revolutionize applications performance.

Middleware will continue to be a critical applications technology. As the glue which makes the applications work together, it's important that you understand your overall middleware capabilities.

If you combine these business technology insights with the 15 due diligence criteria, you should make prudent applications investments.

COMMUNICATIONS TARGETS

"Collaboration" has given rise to a whole new set of communications requirements including those driven by changes in business models, processes, and communications devices including:

Changing work models and processes involving:
- Telecommuting
- Mobile computing
- Small office/home office (SOHO) computing
- New customer, employee, supplier and marketplace connections (B2B, B2C, B2E, etc.)
- Letters, phones, faxes, face-to-face; on and off-line, synchronous, and asynchronous communications
- Near Real-Time comparisons of vendor products and services
- Disintermediation and re-intermediation
- "Whole employee management" and "whole customer management"
- Multi-purpose access points: Wired & wireless PCs, laptops personal digital assistants (PDAs), smart cell phones, network computers, Kiosks, local area networks, virtual private networks, wide area networks and the World Wide Web (WWW)
- Anytime, anyplace information sharing

These requirements are driving a new generation of communications technologies and products as well as communications architectures and infrastructures resulting in the need for more bandwidth (and bandwidth management), more access points, better security, improved reliability, scalability, and distributed systems management. The use of intranets and extranets to conduct internal and external business is also increasing dramatically. Five years ago there was serious debate about just how pervasive communications technology might become; today there is no debate about the ultimate outcome. Whole aspects of many businesses will move completely to the Internet, just as those—like Amazon and eBay—will grow comfortably existing only on the World Wide Web. Companies of all sizes and from all vertical industries will continue to rely heavily upon e-mail and messaging, workflow tools, team-oriented groupware and business-to-business (B2B) transaction processing, among other communications-enabled processes and activities.

Those who buy, deploy, and support technology are well aware of the role that communications plays in our personal and professional lives and are forever seeking investments that will pay large dividends. As with the applications space, the value of various communications technologies and systems changes depending upon the specific perspective the investor adopts.

The first communications investment area for consideration is access to networks, applications, and data, and the bandwidth—the size of the communications pipe—that enables access. Right now many companies have a hodgepodge of technology and systems that connect employees, customers and suppliers. Figure 3 presents the range of access technologies that together represent the range of access and connectivity communications investment opportunities.

Remote access (RA) technologies are central to business models. Why? Because as a "minimally acceptable" architectural requirement, remote access is necessary to support employees working increasingly on the road and from home, customers seeking information, service, and transactions, and suppliers who need access to inventories and other resources.

Everyone is investing heavily in wireless connections. Companies have deployed gigabit Ethernet network connections. The world of routers and switches is changing dramatically, with routers getting "smarter" and switches getting cheaper and becoming viable solutions to network efficiency problems. The major vendors—especially Cisco—are working hard to blur the distinctions among network throughput and management tools, offering an assortment of options likely to work together as network requirements change and increase.

Figure 3. The range of communications access and connectivity technology investment targets

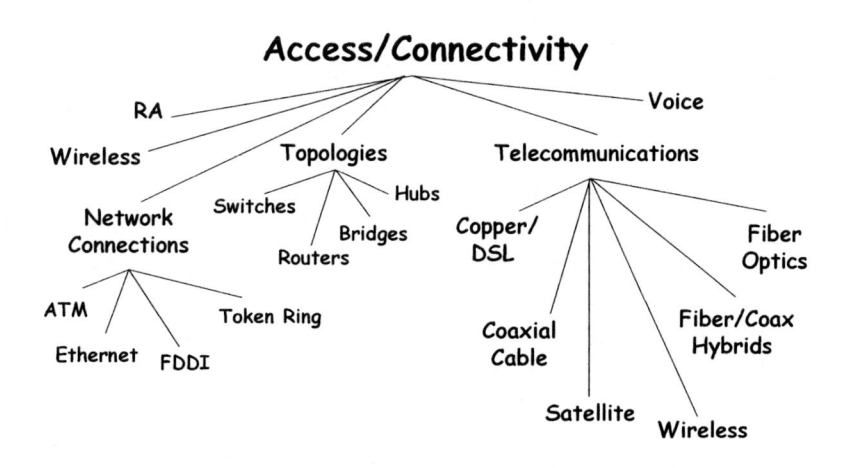

CIOs have known for years that adding more bandwidth to their networks may not be the answer to all processing problems. It may be that spending some money on "bandwidth management" is a more cost-effective solution. Bandwidth management can solve a number of problems, especially if it is architected flexibly. For example, one could design a network capable of adapting to priorities set by the importance of specific applications, like sales force automation (SFA), customer relationship management (CRM) or enterprise resource planning (ERP) applications—so that queries from these applications bump queries from others. Similarly, bandwidth management can give priority to classes of employees so networks won't clog from e-mail or instant messaging (IM) activity.

Long-term we will move to fiber optic connections but in the short term we will continue to exploit copper and cable connections. Many of the technologies have been optimized for copper and some—like gigabit Ethernet—run on copper or fiber (though better on the latter). Migration is the watchword here, but only if requirements justify faster-rather-than-later investments in fiber optic connectivity.

VCs love the communications space because it is so necessary: no one transacts business without communications. Vendors love the space for the same reason. CIOs are wary of the space for the same reasons and because they are responsible for designing, deploying and supporting reliable, secure communications networks that support all kinds of communications and transaction processing.

The next aspect of communication is the collaboration quotient of companies and the need to move toward a collaborative infrastructure capable of supporting anytime/anyplace communications.

There is a distinction between shared communications and collaboration that's important. For example, when we e-mail lots of people—and "carbon copy" even more—we are sharing communications, but when we create a "thread" of communication based on action/reaction, such as what distributed teams do, we are moving toward collaborative computing. In the near future, for example, Internet bidding will become commonplace triggering round after round of action/interaction. As that practice becomes widespread companies will need to support asynchronous collaborative computing – not just shared communications.

The real questions have to do with the kind of collaborative environment the industry needs to create and support and the standards it will need to adopt. While we have de facto groupware standards today, we do not yet have definitive supply chain standards that will make collaborative computing investments standards-proof. Everyone needs to pay close attention to external supply chain standards. Internally, there are lots of options—the key ones being IBM's LotusNotes and Microsoft's Exchange—for supporting threaded discussions and workflow. Lots of additional vendors provide workflow solutions designed to work within the larger collaborative messaging applications.

The range of collaborative communications activities appears in Figure 4. The key for CIOs, vendors, and VCs is decision-making that is holistic. CIOs are careful about not selecting workflow, e-mail, groupware, or messaging technologies and products independently—because they should all be related. CIOs have long accepted arguments for standardization and arguments against best-of-breed products: as communications environments get more and more complicated time and effort is better spent on optimizing communications processes rather than making a bunch of disparate products work together. Vendors understand this as do VCs who love to fund communications "solutions."

The third area is e-business. Collaborative communications infrastructures determine what companies can and cannot do in e-business. But here the need for an explicit e-business strategy (driven as always by a clear overall business strategy) is absolutely critical. The pace of business change is so rapid that e-business communications infrastructures better be ready to satisfy business-to-business (B2B) requirements yesterday. If companies need to string cable, deploy application servers, and plan for bandwidth and bandwidth management, among other important, time-consuming activities, there is no way they will be able to satisfy business requirements.

The use of intranets and extranets to conduct internal and external business will increase much faster than the deployment of brochure-based Web sites in the mid- to late-1990s. Five years ago there might have been some debate about just how pervasive e-business might become: today there is no debate about the size or speed of the locomotive coming down the track. Aspects of many businesses will move completely to the Internet, while others will move partially onto the World Wide Web. No company will escape this trend.

CIOs are well aware of the need for robust intranets and extranets—and the security requirements that they generate. The more customers, suppliers, and partners

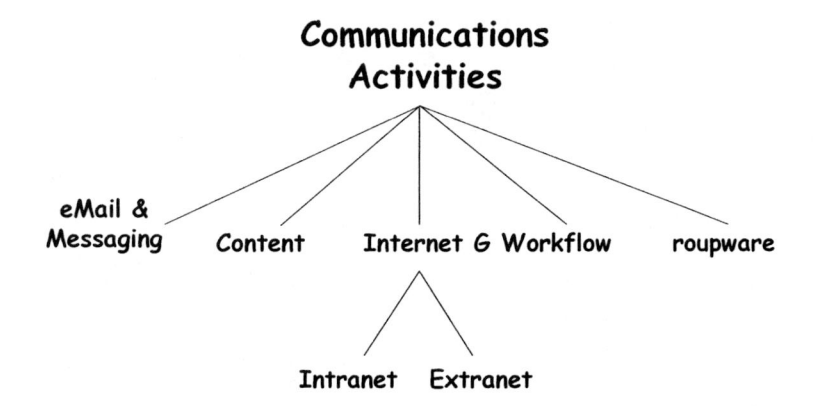

Figure 4. The range of communications activity-based investment targets

Communications Activities
- eMail & Messaging
- Content
- Internet & Workflow
  - Intranet
  - Extranet
- Groupware

companies bring into their physical and digital networks, the more challenging the communications and security requirements become. VCs see gold here. New B2B exchange engines—that support real-time bidding for products and services—killer security products that keep everything safe and whole new online catalog searching are among the many areas in which VCs seek to invest. Vendors need to keep upping the ante in the e-business product and service space as well. ERP vendors are improving their supply chain applications all the time as well as additional tools for tracking e-business processes and efficiencies.

The fourth communications area is not really a technology, but rather a sensitivity to the communications patterns and processes that will drive communications investments. In fact, all investments in communications technology are useless unless there are processes in place to exploit the investments. Why so concerned about processes? Because without them, investments will be wasted. Here are some examples. Consider the company that manages 20,000 desktops that wants to deploy Computer Associate's Unicenter, IBM's Tivoli, or HP's OpenView. The assumption is that everyone will work with the products to enable the key functionality—like software distribution, network event reporting, help desk activities, and the like. But if those processes are ill-defined, then no matter how robust the toolset the project will fail. Process definition, how we plan to do software distribution, for example, who is in charge, and so forth, must be defined and accepted by the organization. If process holes exist in the network and systems management tool rollout, the deployment will fail.

In response to process opportunities and challenges a new set of applications that assist in the modeling of communications and transaction processes have been developed over the past few years. Business process management (BPM) applications have been developed—often with private equity venture capital—to help companies map their transaction and communications processes. CIOs, CEOs, and CFOs are investing in these applications more and more in an effort to better understand the requirements that their computing and communications environments need to satisfy. VCs love the BPM space as well as the related corporate performance management (CPM) space which takes a top-down view of how well (or badly) a company is performing.

The fifth area is communications architecture and infrastructure. Assuming that a company knows what it wants to do and it has defined the processes necessary to make it all happen, it then needs to think top-down about the architecture it has today and the one to which it will be migrating. Figure 5 presents the entire communications architectural "space" in which we live. Based on the overall business model, companies move through this maze—migrating from the old to the new along the way—such that they end up with an integrated architecture. This means that none of the decisions made within the space can be made independently from

Acquisition Targets 63

any and all of the others. The overall business model must guide companies toward an access/connectivity and transaction processing communications strategy that is aligned. It must also guide them toward a strategy that supports migration from where they are today that leads to an infrastructure that works, can be maintained, and can be upgraded over time.

Communications architecture and infrastructure issues are complex because they are operational: everything has to work all the time which means that companies have to define and apply processes that can be implemented and maintained over time. Figure 5 suggests conceptually that transaction processing will be supported by access/connectivity, and that they both need to work together.

The investment opportunities in the communications space are wide and deep. In the late 1990s, more than half of all venture capital dollars were targeted at new communications technologies. As business models have become more distributed and decentralized, and the need for mobile communications has increased, CIOs have focused heavily on remote access, wireless, and other technologies that support their increasingly mobile manufacturing and sales force. Vendors understand how rich their offerings must become. Everyone thinks that communications technology is important. Change is inevitable here. What makes perfect sense today will make no sense tomorrow. Perhaps unlike some of the other investment target areas, communications technology moves at breakneck speed. This requires a nontrivial allocation of trends analysis and monitoring resources: staying abreast of what the

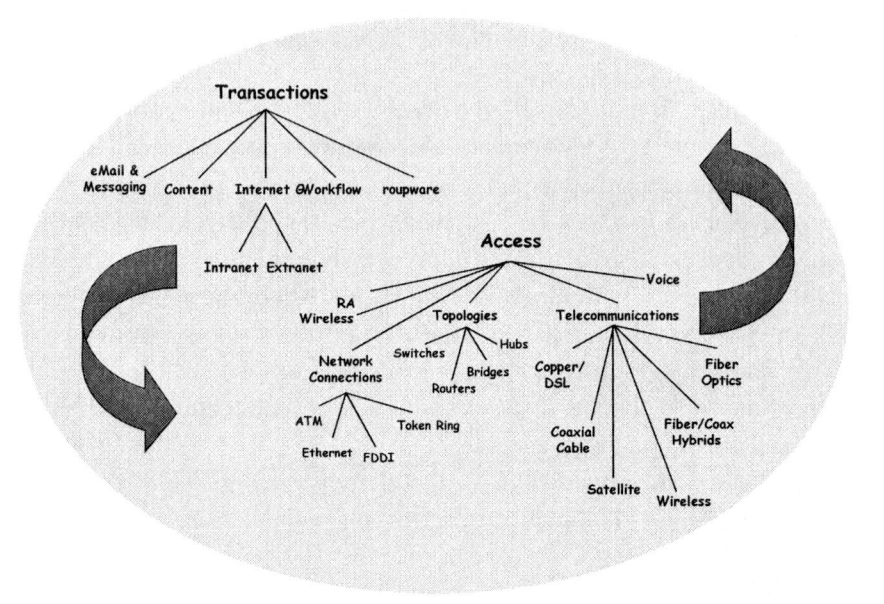

Figure 5. The communications architecture investment space

field is doing—and what the competition is doing with the technology—should become one of your core competencies.

COMMUNICATIONS TECHNOLOGY INVESTMENT GUIDELINES

Regardless of your investment perspective, it is no exaggeration to say that communications technology will make or break your ability to compete. There are all sorts of issues, problems, and challenges that face your organization as it wrestles with its business strategy, its communications response, and its ability to adapt quickly to unpredictable events. The key is to design for flexibility and in the absence of crystal clear requirements and implement minimally acceptable communications architectures and infrastructures.

Having said all this, here are some communications investment guidelines:

Make sure that new business models get developed and that they speak to communications requirements. Some of the communications requirements will be obvious, but others will be subtle, for example, like the ratio of in-house vs. mobile users now and three years from now. If the business models don't get developed, or the ones that do have minimal communications implications, push the modelers toward communications requirements. If the team has trouble articulating these requirements, jump start the process with alternative business scenarios and competitor case studies (and what competitors are doing with communications technology).

Proceed holistically. Make sure that decisions about communications technology are linked. Tilt toward a standardized environment and away from a best-of-breed one: you do not have time to deal with endless integration and interoperability problems.

Identify change drivers. Identify the drivers—like the number of mobile users you'll be expected to support, the amount of disintermediation and re-intermediation occurring in your target industry, and to what extent business will go virtual. Rank-order these drivers in terms of their importance and probability of occurrence.

Summarize your change-driven communications requirements: calibrate current bandwidth and estimate future requirements (for bandwidth and bandwidth management), identify your security needs (as the number of remote users increases), identify the number and nature of the remote access points you will need to provide, and support and plan for the new communications architecture's infrastructure processes and ongoing maintenance.

Measure your current bandwidth against current and anticipated requirements to determine the bandwidth (and bandwidth management) "gap."

From the top-down think about your access and connectivity architecture and infrastructure and—objectively—estimate its ongoing design and support costs: try to develop a total cost of ownership (TCO) model per employee for communications technology.

Against a suite of requirements, like speed, availability, security, adaptability, and configurability, baseline current communications performance and then project effectiveness against anticipated new requirements. Use the gap data to drive products, services and architectures, which will move you to (re-)consider wireless communications, fiber optic connectivity, and fast → gigabit Ethernet network connectivity, and the like.

Devote very special attention to network topology. Lots of action here by the major vendors who are now providing switching and routing options that didn't exist a year ago, and move toward bandwidth management (vs. providing more and more raw bandwidth).

Look into optimizing network and communications architecture for your applications. Cisco—and other vendors—are offering tools that allow companies to optimize traffic for one or another application. If you have deployed an enterprise resource planning application, you might be able to increase its efficiency dramatically via bandwidth allocation and management.

Calibrate collaborative appetites. Do it internally and externally. For example, if you are business model calls for lots of action/reaction/threaded communications, you are moving toward collaboration. Internally, collaboration takes time and effort: you might already be a gregarious culture that likes to "share," but if not you will be faced with some stiff challenges if you try to force-feed sharing.

Consider migrating toward unified messaging where all forms of communications occur via a single application and device. Eventually, it will be commonplace to receive faxes where you receive e-mail where you receive voicemail. Plan now for the infrastructure to support unified messaging.

Your e-business models must be clear yet adaptive (because they change often). You also need to think about the applications you will need, connectivity to legacy systems (since you are not likely to throw out all your legacy data), project and program management, support and metrics to measure your e-business efficiency.

It is also important to rethink customer relationships and the communications that surrounds all of the contact companies have with customers. While that is always good advice, the Web makes it imperative, because your competition will certainly be doing the same thing and the anytime/anyplace possibilities of virtual connectivity must be assessed from whole new strategic vantage points. It is now possible, for example, to sell, service, sell again, service, cross-sell, service, up-sell, service—and then sell the data surrounding all of that activity to affinity marketers!

Make sure to follow the products in the customer relationship management (CRM) and technology-enabled relationship management (TERM) areas.

Without the right processes all is for naught. Well defined, understood, communicated, and approved processes will sustain investments in communications technology. Without process buy-in, technology investments will not pay dividends, and whoever is responsible for those investments will suffer politically. A process guru—not an entire process organization—is necessary to make sure that the processes stay synchronized with the technology investments. Processes are essential to implementation and support. If you build a great infrastructure via an elegant architecture but fail to define and implement the operational processes you will fail.

Communications architectures, the access/technology and transaction processing technologies that you implement, must be integrated. Bottom-up design—or continuously patching an existing architecture—will not yield the results necessary to support virtual enterprises. Top-down design of architecture should yield better, more integrated and supportable, results. Err on the side of standardized architectures vs. those that require massive amounts of customization and integration. This in not an argument against the proverbial "open" architecture but rather a practical recommendation targeted at reliability and maintainability.

Institutionalize a process that reviews, at least a couple of times a year, current and emerging communications technology (driven—ideally—by new business models). Track key communications technologies to determine what your migration plans might look like sooner rather than later. If you combine these insights with the 15 due diligence criteria, you should make prudent communications investments.

DATA TARGETS

Data, information and knowledge requirements come from several directions. On the one hand, companies have existing applications that use data to process transactions every day. That same data is re-used as they extend their business models on to the Internet and as they engage in "whole customer management," or the desire to sell existing customers more goods and services for as long as they can. Data becomes information when it becomes purposeful; knowledge provides the ability to make inferences. There are two aspects of all this: storage and analysis. Storage refers to where we keep the data, information and knowledge; analysis refers to what we do with it all.

Figure 6 maps the storage/analysis space. It offers a matrix that can be used to prioritize data, information, and knowledge requirements. It also offers an opportunity to size data, information, and knowledge investments. Where is the money

Figure 6. A data, information & knowledge requirements matrix

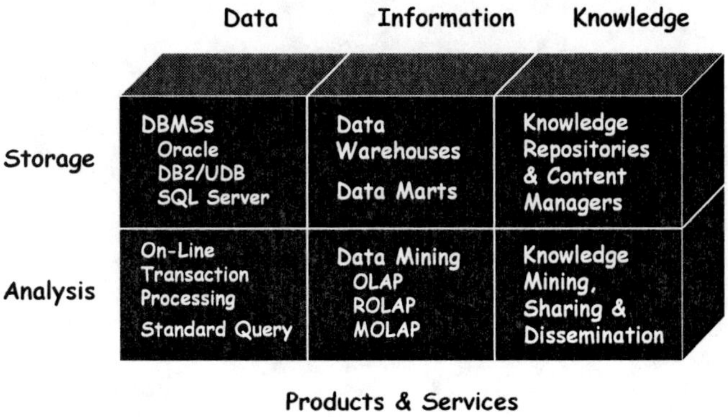

	Data	Information	Knowledge
Storage	DBMSs Oracle DB2/UDB SQL Server	Data Warehouses Data Marts	Knowledge Repositories & Content Managers
Analysis	On-Line Transaction Processing Standard Query	Data Mining OLAP ROLAP MOLAP	Knowledge Mining, Sharing & Dissemination

Products & Services

going? Where are the best opportunities? CIOs, vendors, and VCs monitor the opportunities here all the time. Where should the money go?

The data storage area is interesting because it is changing so fast. Not long ago the list of viable data base vendors included five or six major players, but the data storage market is shaking out pretty clearly: today there are, arguably, only three major data base vendors—IBM, Oracle, and Microsoft. Another major trend is the movement from hierarchical to relational data base management systems and the migration from relational to object-oriented data base management.

Information storage options—data warehouses, data marts and special purpose hybrids—require some serious thinking about where companies think they will ultimately end up, how much money is available for the construction of these artifacts, what users will require, and what the data mining tools will look like.

Knowledge storage is akin to dressing up for a party to which you have no clear directions. Or, if you prefer, investing in a solution in search of a problem. The "knowledge management" (KM) business is just that, but the serious pain it's intended to relieve is perhaps better described by consultant doctors than by CIOs, VCs, or product vendors. Rather than be flip about the field, let us look at some of the assumptions. First, KM assumes there is knowledge to manage, that companies have somehow codified the collective wisdom of their industry's and company's experiences. Second, it assumes that the culture and processes are sharing-centric, that is, capable of exploiting codified knowledge. Next, it assumes that companies have, or are willing to invest in, the tools to make all this happen. Some industries will be in a better position than others to exploit knowledge management. But others will have little or no need for what the consultants are assuring us is the next great revolution in data → knowledge base management technology.

Some potential investment areas are clearly worth serious consideration by technology investors. But others are not so sweet. Knowledge management is a

Figure 7. Data → warehouses/marts → analysis → mining

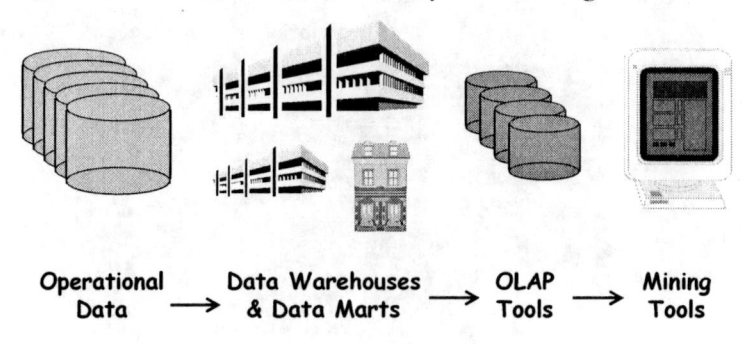

Operational Data → Data Warehouses & Data Marts → OLAP Tools → Mining Tools

great concept but assumes a lot about the companies that may adopt KM tools and techniques. Some VCs like the space but many do not: since the over-hype days of the late 1990s (when it was possible to sell just about anything), many VCs have become much more pragmatic and KM applications often don't make the investment cut. Perhaps the most promising KM tools are coming from the business intelligence space, a space that CIOs, vendors, and VCs all love unconditionally.

Storage is essential to data, information and knowledge analysis. Online transaction processing (OLAP) is what everyone has been doing for a long, long time. Online analytical processing—especially when coupled with data warehousing technology—is how data, information, and knowledge get exploited. It is easy to see why OLAP has fans: it provides flexible querying of data that is relatively untapped by OLTP. In fact, OLAP provides a gateway to information and knowledge analysis. The broad class of "data mining" tools out there exists to service this need. CIOs and vendors must worry about all six cells in the matrix in Figure 6; VCs focus primarily on the lower three—the analysis opportunities—especially the mining opportunities. Why? Because they understand that the real payoff for CIOs is the exploitation of their investments in data base management storage platforms, investments that tend to short-change analysis. VCs also understand that vendors who sell and service mining software that leads to increased revenue and margins are more likely to be well received than those who sell data plumbing. Figure 7 brings it all together.

The role of data, information and knowledge will grow dramatically as the technology itself changes. For example:

- **Data Bases are Moving from Hierarchical to Relational to Object-Oriented**
  - To Deal More Flexibly with Data
  - To Deal with Unstructured Data & "Content," Especially Multimedia Data
- **Data is Becoming "Web-Enabled" to Support**
  - Intranet Access
  - eBusiness ...
- **Data Bases Need to Be Managed More Systematically to Support:**
  - Increased Access
  - More Unstructured Data
  - More Multifaceted Data
  - Security
  - Synchronization
  - Replication ...
- **Data, Information & Knowledge is No Longer Produced by a Single Source**
  - "Stuff" is Generated by/for In-House Applications
  - Is Generated by Access to External Data Bases, Principally Via the WWW
  - Is Generated by Customers
  - Partners
  - Suppliers
  - Employees ...

- **Data, Information & Knowledge Will Need to Be Accessible to Everyone Who Created It, Who Needs to Transact Based on It & Who Needs to Manage It – Including:**
  - Employees
  - Customers
  - Partners
  - Suppliers ...

We can expect to see a rapid increase in data warehousing spending over the next three to 5 years. Data warehousing is becoming increasingly necessary to deal with heterogeneous data and applications environments and to extract information from structured data bases that is otherwise inaccessible. The data warehousing services market will rise even faster and become much larger than the data warehousing tools market. There will also be a major "pull" to data warehousing from the enterprise applications vendors and their integrator partners.

As suggested above, the knowledge storage area will require some clearer definitions of "knowledge management" before vendors and buyers can determine what they will sell and buy, respectively. Vendors are hyping the space but there are relatively few bona fide products in the area and, more significantly, few well-defined problems to which "knowledge management" would be a cost-effective solution. The key here is for vendors to answer the question (posed by CIOs and VCs, among other investors): "knowledge management for what?"

The data analysis area will see continued strong application of online transaction processing (OLTP) embedded within applications. The information analysis area will see the increased use of online analytical processing (OLAP) tools that support the analysis of unstructured data analyses. Data mining—the end user's tools to discover hidden patterns in data—will continue to grow, operating on data warehouses and OLAP servers. The knowledge analysis area will have to wait for clearer definitions and problems well-defined enough to pull technology-based solutions. Currently, the knowledge management space suffers from the "solution is search of a problem" syndrome.

The data space is changing dramatically as developers, users, and vendors all discover how integral data, information and knowledge is to the heterogeneous applications we've created and will continue to deploy.

With the increase in distributed applications, the role of data is growing, as vendors and users seek to, for example, connect Web applications to legacy data bases. The movement in enterprise resource planning, supply chain management, network and systems management, and the maintenance of older (legacy) applications are all contributing the need to make additional investments in traditional data base management, data warehousing, data mining and, eventually, knowledge management.

The big changes are occurring because "data" is no longer just records stored in hierarchical or relational data base management systems. Instead, we are seeing the need to integrate different kinds of data, information and knowledge. We are also seeing the need for "metadata" concepts, standards, and tools. Metadata is data about data, or information about the location, state, quality, and content of data, information and knowledge. Metadata therefore facilitates data searching, responding to queries about data, information and knowledge regardless of where they might be stored.

Overall, we can expect increased problem pull on the data/information/knowledge space. We can also expect to see increasing investments in universal data access through digital asset management, which is the ability to reach any kind of data, information or knowledge from any place at any time to solve structured and unstructured problems.

Everyone has got data in one place or another. Some of it is in an Oracle data base, some in an IBM/DB2 data base and some is still in Sybase data bases. This "operational data"—especially if it is in different forms—often needs to get translated into a form where it can be used by any number of people in your company to perform all sorts of analyses. "Translation" results in the development of data warehouses and smaller data marts which support all varieties of online analysis and ultimately "data mining," the ability to ask all kinds of questions about your employees, customers, suppliers and partners.

Everyone is working on universal data access (UDA) from all tethered and untethered devices. Eventually, structured, unstructured, hierarchical, relational, object-oriented data, information, and knowledge will be ubiquitously accessible. While we are a few years away from all this, it is helpful to understand the Holy Grail and to adapt your business models in the general direction of this capability. Microsoft, IBM, and Oracle all have plans to provide UDA. It is important to stay abreast of their progress—and the implications to your business models and processes. Collaboration will require UDA and integration is the short-term path to that goal. Longer-term, if acquisition decisions are made properly, there should be less need to integrate lots of disparate data bases.

Collaborative business models will drive data. Companies cannot become collaborative unless their data (information, knowledge and content) are integrated. Over the years, you have probably deployed lots of different data base management systems and lots of applications that had specific data requirements (for example, if you have deployed Oracle financial applications you have to run them on Oracle data bases, which may or may not have been your preferred data base platform. Consequently, depending on the amount of data variation in your company, you may be exquisitely ill-positioned for collaboration. Or, if you've had some discipline along the way and only have one or two data base platforms, you are in a pretty good position to collaborate.

Your data integration efforts complement your application integration work. Some of the enterprise application integration (EAI) tools include extraction, translation and loading (ETL), and vice versa. Investments in integration technologies should be driven by the results of your scenario planning exercises that position your company within the collaboration space. These scenarios will determine what applications you need and the extent to which the applications and data must be integrated. But regardless of where you find yourself in the collaboration space, you'll need to invest in data (and application) integration technologies.

Let us summarize. Your data must integrate if you are going to collaborate. If you have lots of different kinds of data in different places then you need to develop a data integration strategy which will probably involve building some kind of data warehouse. Once you build a warehouse you can conduct all kinds of analyses—

Figure 8. The data products and services investment target map

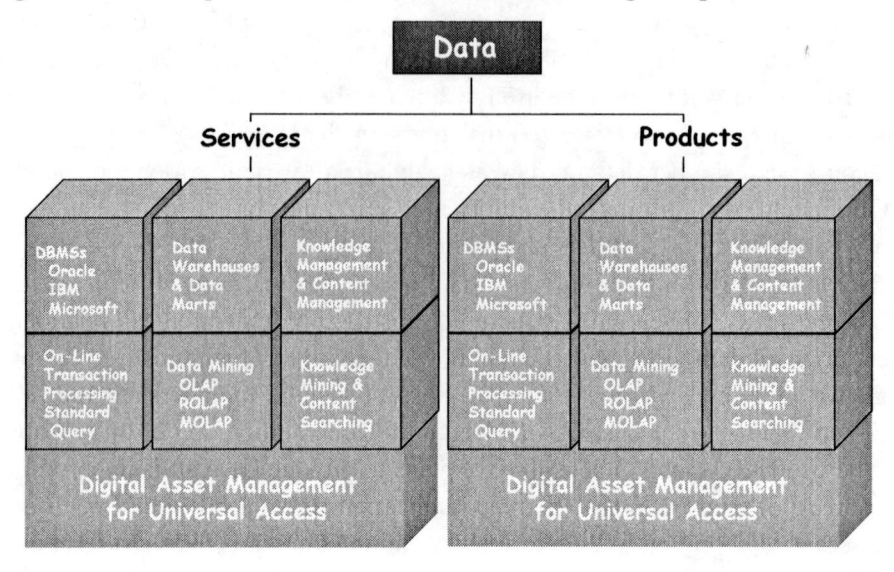

analyses that facilitate collaboration. Over time, you need to reduce the need for all this integration by moving to fewer data platforms and standardizing the analysis tools—the tools you use to mine the data for collaborative insights and models.

Figure 8 summarizes the investment target opportunities in the data area. Note the distinction between products and services: all of the product areas are complemented with services and vice versa.

DATA INVESTMENT GUIDELINES

Regardless of your investment perspective, there are some investment guidelines that can improve your chances of making good investments:

Track the consolidation in the data, information and content areas. The industry is consolidating and it is likely that in a few years there will only be several major data base management vendors. While there will continue to be vendors that support data analysis—especially in the business intelligence area—the smaller vendors will increasingly conform to the standards and architectures of the larger ones

Recognize the interrelationship between data/information/ knowledge products and services. Increasingly, products will be bundled with services and vice versa. Investments will thus be increasingly organized around bundled products and services.

"Hard" data is merging with "soft" unstructured data from, for example, e-mail streams, customer service conversations, and blogs. Investments in data, information, and knowledge should focus on the hard/soft capabilities of product and service vendors.

Data integrity and security are essential to the increase use of data, information, and knowledge that business collaboration assumes. Products and services must demonstrate clear and verifiable data integrity and security capabilities. Avoid investments that cannot speak aggressively to these requirements. Investments in data/information/knowledge "glue" are prudent, since they focus on the integration of (hopefully) decreasingly disparate data, information and knowledge. Vendors that cannot integrate disparate data, information and knowledge are at a competitive disadvantage to those who can.

If you combine these insights with the fifteen due diligence criteria, you should make prudent investments in data.

INFRASTRUCTURE TARGETS

Let's begin with a definition of infrastructure:

- "Infrastructure" Refers to the hardware, software & support services necessary to enable the legacy, e-business and enterprise applications that support business objectives
- "Infrastructure" is the data, computing and communications "Plumbing" that every company must acquire, upgrade & support …

The list below identifies some of the more important sources of infrastructure requirements:

- Legacy applications—Especially after they have been modernized with web services—will continue to support business objectives and therefore need to be efficiently supported.
- Applications that are becoming more distributed and therefore require distributed infrastructure support.
- Users are becoming more mobile; the need to support remote access and re-engineered applications to optimize their use as remotely accessed applications.
- Procurement practices that are under extraordinary pressure to show returns on technology investments and manage the total cost of infrastructure ownership.

- The rise of e-business which has fundamentally and forever changed the way companies do business.
- The need for infrastructures to adapt to these changes by providing security, scalability, interoperability and connectivity.
- The dramatic rise in the deployment of enterprise applications which has expanded the role of infrastructure in many organizations which historically only supported custom applications.
- An increase in major infrastructure outsourcing deals led by growing concerns about desktop, laptop and server management.

Figure 9. Infrastructure layers

Access Levels

Fat Client   Web Browser   PDA   Network Computer   Cell

Coordination Levels

Query Services   Messaging Services   Directory Services
Application Servers   Network & Systems Management
Security Services   Web Servers   Transaction Services

Resource Levels

Business Rules   Distributed Applications Management
eCommerce Applications Management   Application Services
Metadata   Connectivity/QoS Management
Data, Information & "Knowledge"

Figure 10. The range of infrastructure support services

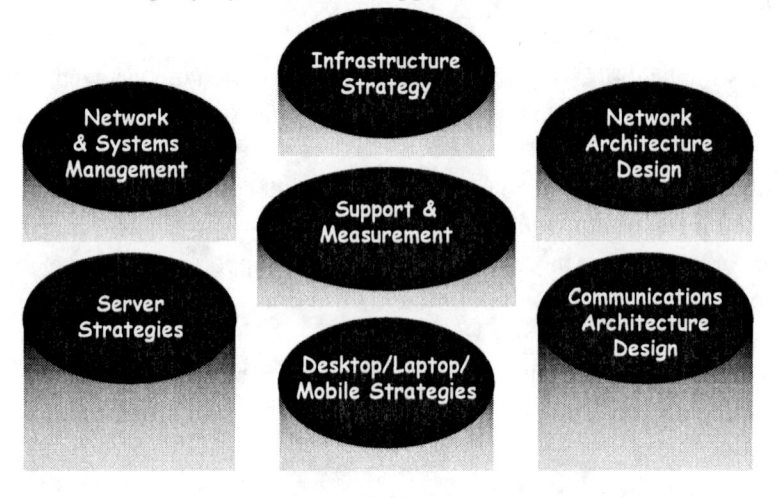

- Infrastructure Strategy
- Network & Systems Management
- Network Architecture Design
- Support & Measurement
- Server Strategies
- Communications Architecture Design
- Desktop/Laptop/Mobile Strategies

Figure 9 can help with an assessment of infrastructure requirements and suggests that there are infrastructure "layers." These layers represent an easy way to think about infrastructure and how infrastructure components interrelate. The access layer refers to the desktops, laptops, browsers and other devices that permit access to data, applications, communications, messaging, workflow, and groupware capabilities. The coordination layer refers to the query services, messaging services, directory services, and security services that comprise infrastructures. It also includes the transactions servers, application servers, and Web servers that support applications. The resource layer refers to the applications themselves as well as the applications management services necessary to keep transactions running. It also includes the data/information/knowledge/ content/metadata resources necessary to support the transactions and applications. Data centers reside in the resource layer of infrastructures.

One of the services everyone needs is infrastructure support services (Figure 10). The perfect outsourcer would be able to develop an infrastructure strategy aligned to a company's business objectives. These services are typically offered by technology consultants, such IBM Global Services, EDS, Perot Systems, and Accenture, among lots of other local, regional, and global vendors.

Network and communications architecture design calls for expertise that can translate infrastructure strategy into network and communications architecture design. Desktop/laptop/mobile strategies include all of those personal devices that make employees more productive. The perfect services organization would identify all of the options, the issues, and the processes, tools and technologies necessary to support, integrate and interoperate these disparate devices. Server strategies include the number, power, location and support of the servers that support distributed environments. The ideal support service provider would also specify a network and systems management strategy. But requirements suggest even more infrastructure support services—like infrastructure integration and optimization—as Figure 11 suggests.

Some VCs like the infrastructure space. Those that do subscribe to the "sell-shovels-in-a-gold-rush" theory of investing, that there is always money to be made in infrastructure products and services. Others feel that the space is becoming way too commoditized with profit margins under continuous attack. Vendors understand the dependency that their customers have on infrastructure, just as CIOs know that they must "run the engine" cost-effectively.

Overall, the infrastructure area is a solid investment area regardless of where investors sit. Companies will always be dependent upon their computing and communications infrastructures; they continue to spend over 70% of their technology dollars in the infrastructure space. The pricing pressure in the area is "good" for CIOs, but not so good for vendors and VCs.

Figure 11. Infrastructure integration

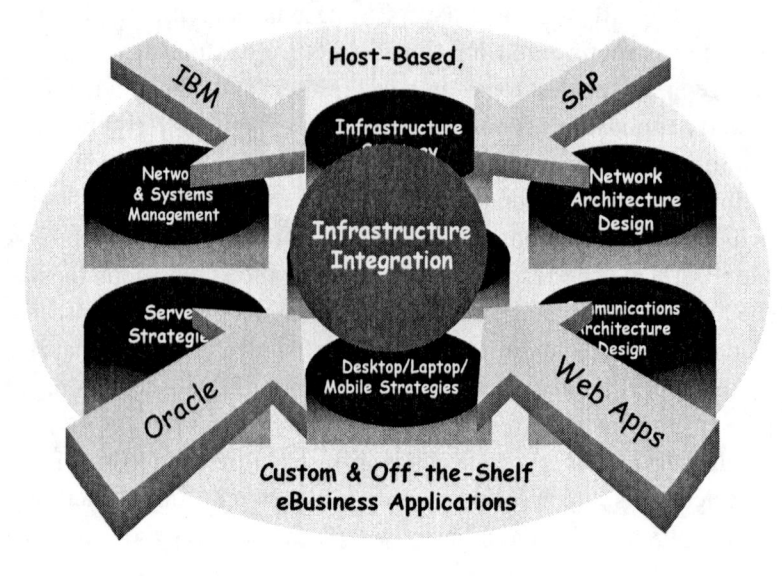

INFRASTRUCTURE INVESTMENT GUIDELINES

Here are some investment guidelines for the infrastructure area.

All decisions must be passed through the business model(s) filters. If they do not exist, then you have to assist in their development and validation. If none exists and you cannot get the organization to make the investments in business modeling necessary to optimize infrastructure investments, then make minimally acceptable investments until the models clarify. Minimal investments include a layered "access," "coordination," and "resource" infrastructure strategy optimized for e-business and enterprise applications.

The business strategy will no doubt require more distributed (e-business and enterprise) applications which, in turn, trigger the need for infrastructure capabilities that may or may not exist today in your organization. There may be a significant "infrastructure gap" between your applications and business strategy. Investors need to identify and define this gap: business strategy cannot be developed without reference to your applications and applications cannot be conceived without reference to infrastructure support.

Figure 12 identifies the range of infrastructure products and services investors need to assess. The framework should help prioritize investments and can be used as a filter through which to pass products and services options. The key, however, is to fill in every cell in the matrix since infrastructure must be top-to-bottom and be supported by efficient products and services.

Figure 12. Infrastructure investment framework

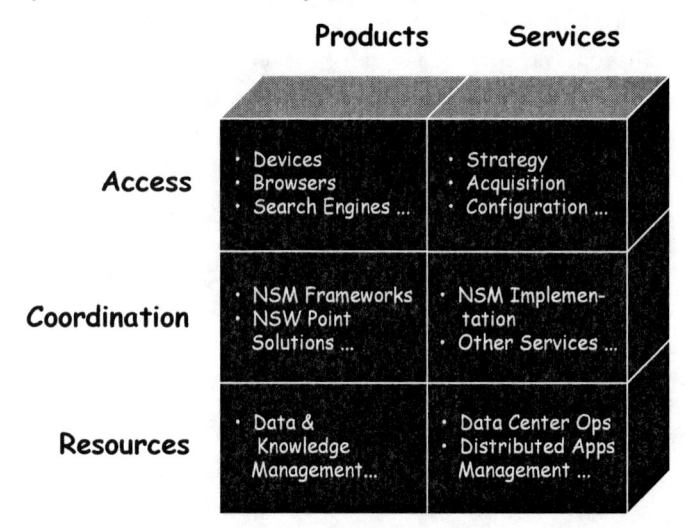

	Products	Services
Access	• Devices • Browsers • Search Engines ...	• Strategy • Acquisition • Configuration ...
Coordination	• NSM Frameworks • NSW Point Solutions ...	• NSM Implementation • Other Services ...
Resources	• Data & Knowledge Management...	• Data Center Ops • Distributed Apps Management ...

You need to profile your current access "assets," including your desktops, laptops, PDA, and other devices used to access your applications and data bases. You need to determine how skinny or fat your access devices need to be: skinny clients—where applications are primarily network-based—may make sense if your users are applications-centric; fatter clients, the traditional desktops/laptops with lots of applications on them, make sense when your investment in fatter clients (and support) is large and when computing autonomy is a high priority for your users. You need to plan for an environment that will support an increasing number of skinnier clients and one that uses all computing devices as remote access devices.

You need to standardize on browsers and on an applications architecture that uses the browser as the common applications interface, that is, the primary way users (employees, suppliers, and customers) access applications and data bases.

While perfect standardization seldom works, the goal should be to standardize on as few access devices, directory services, messaging systems, applications servers, and the like that make your applications work. Standardization can be vendor-specific or best of breed. Increasingly, large enterprises are moving away from best-of-breed and toward a more vendor-specific standardization strategy. Obviously Microsoft owns the desktop application world as well as the desktop operating system world. Windows XP and Vista may well dominate the server operating system world for the foreseeable future.

A closely related issue is the standard-vendor vs. best-of-breed approach to infrastructure configuration. It's important to assess the strengths and weaknesses and costs and benefits of the alternatives. Before you decide, look at the infrastructure trends in your vertical industry. If all of your suppliers and customers are likely to use a single vendor for most of their computing and communications infrastructure needs you have little or no freedom deciding what to do.

You will need a network and systems management strategy which can be based on individual point solutions or on an integrated framework. The implementation of network and systems management frameworks is complex and expensive. While you might be tempted to implement IBM's Tivoli or Computer Associates' Unicenter yourself, you might need some outside help to do it cost-effectively. Also make sure you develop some effectiveness metrics for your network and systems management processes and tools. Without metrics, it's impossible to determine if your investments are sound or ill-advised.

The applications and data bases need to be supported in data centers. But data centers will evolve to distributed data centers that (virtually) house distributed applications and data/information/knowledge/content as well as legacy applications and data bases that all must co-exist in the same infrastructure. It's difficult to find professionals who can support such heterogeneous environments. Make sure that your in-house personnel are up to the task. If they are not, then consider outsourcing to a company that has the right mix of skills and experience. While there may be some reluctance to outsource your current data center, remember that legacy data centers have been outsourced for years and that as the number of deployed e-business and enterprise applications rises, it is likely that current data center management processes will have to be substantially modified to integrate and support the newer applications. All of this may make outsourcing the smart move.

The range of infrastructure support includes:

- Infrastructure strategy
- Network architecture design
- Communications architecture design
- Desktop/Laptop/Mobile (access device) strategy
- Server atrategy
- Network and systems management strategy
- An effective performance measurement strategy

Together these services constitute infrastructure solutions integration—the objective of everyone's search for the perfect range of services. The integration of access, coordination and resource layer support services represents an end-to-end infrastructure support strategy that should be driven by assumptions about:

- The rise of e-business applications
- The rise of enterprise applications

- Consolidating platforms
- Opportunities for the creative acquisition outsourcing of infrastructure services, especially shared risk outsourcing

Infrastructure needs to be sold as an enabler and accelerator. It should be sold as e-business and enterprise applications "multipliers" likely to make existing investments in these applications pay larger dividends.

These guidelines will help organize the due diligence criteria necessary to evaluate alternative infrastructure investments.

SECURITY TARGETS

Security requirements and investment opportunities come from lots of directions. The list below identifies some of the more important ones:

- The rise in number and complexity of distributed applications
- Increases in denial-of-service attacks, worms, and viruses
- The need to connect more and more people to data bases and applications
- The perception that wireless computing is unsafe
- Regulatory pressure regarding privacy, confidentiality, and financial reporting

This list suggests that the issues and challenges surrounding security are as much social and political as they are technical; in fact, one might argue that it is impossible to address security problems without a context that extends far beyond the obvious technical challenges.

Also interesting is the source of security problems. Perhaps contrary to perceptions, lots of problems can be traced to viruses, not sabotage or other widely perceived likely causes. But is this trend likely to continue? We can expect that security breaches will occur much more frequently as e-business continues to increase and become strategic. This means that some form—and some that might be very nasty—of information warfare will break out roughly at the time when we're all feeling pretty good about how far we've come with virtual business processes.

While we have yet to hear too many stories about how saboteurs destroyed viable businesses, catastrophic digital events will occur. Some may be more publicized than others but there will be major disruptions in business caused by people who use technology as their weapon of choice. The U.S. federal government has recognized the threat for some time now; private business will soon have to pay increasingly attention to information warfare.

Figure 13. Integrated security requirements and investment opportunities

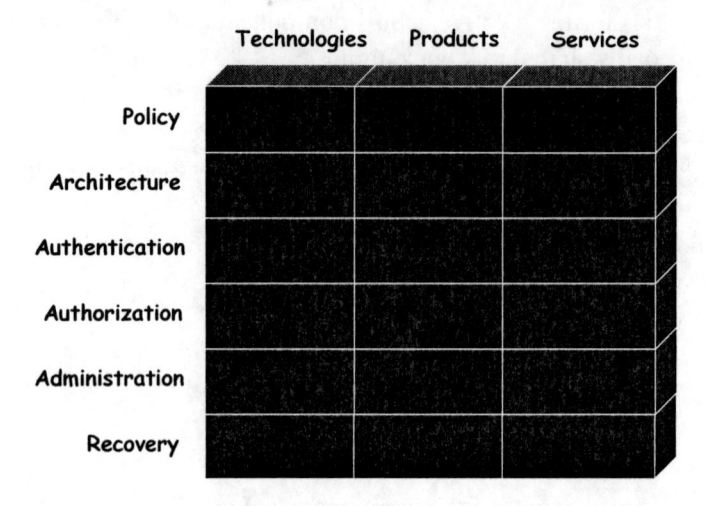

There are a set of technologies, processes, and services that together constitute your security investment strategy.

Figure 13 maps the security landscape. There are lots of cells in this matrix. Every one of them demands attention and they all need to work together.

First, there is *policy*. Here are the guidelines:

- You have to write it down. All of your security and privacy policies and procedures must be codified, communicated and updated on a regular basis.
- Your security policy should not be a bible, but it needs to be specific enough to reduce ambiguity. Rather than spend 2 years writing the perfect security or privacy policy—one that covers every aspect of your security environment—spend a lot less time and get one out that works, and make sure that you offer training around it.
- If you cannot write a credible policy in-house, outsource it.

The document should include, at a minimum, policies that address:

- Data access
- Applications access
- Network access
- Software
- Privacy
- Business resumption planning

- Systems design and development
- Risk assessments

Authentication is absolutely key:

- Passwords work, but can be expensive to administer and complicated for users to use on a regular basis, especially if they must remember multiple passwords.
- If your password policies are well developed (for example, requiring users to change them every 30 days or so, or the systems automatically deny them access) then they might work for you for the next few years. But if they are complicated and cumbersome to manage, you might consider alternative authentication methods, tools and products like smart cards and biometric devices.
- Single-sign-on to networks and applications remains a worthwhile objective. Take a look at the available tools and the processes that support them. As we deploy more and more networks and users traverse more and more networks and applications, you'll need single-sign-on capabilities.
- Make sure you investigate the range of available firewall technologies since they're changing all the time. Note the trend to embed more and more functionality into firewalls, functionality that includes lots of flexible authentication (and authorization) techniques. Over time, you should be able to off-load lots of functionality onto your firewall (and other hardware devices and software applications).

Who gets to do what? *Authorization* follows authentication:

- Once users are authenticated, they need to be monitored according to some predefined authorization schema. Access to networks, applications and databases needs to be defined and individual and classes of users need to know where they can go and what they can do once they get there.

All security (and privacy) policy, authentication, authorization and recovery require *administration*:

- Make sure that you ask questions about administration each time you consider a method, tool, technique or process.
- Develop some metrics against which you can track the effectiveness of your administrative procedures. Track the data over time to determine the cost-effectiveness of whatever administrative processes you put in place.

Some basic administrative reports include:

- User sign-on error reports
- User policy violation reports
- Resource activity reports
- User access reports

Recovery is as essential to security as authentication. Make sure you don not short-change your security strategy by under-cutting investments in systems recovery and business resumption planning. Business disruption and resumption planning simulations should be conducted on a regular basis—at least twice a year—to determine if your business resumption planning policies and procedures will actually work when a major business disruption occurs. The basic elements of a business resumption plan should include:

- Plan activation policies and procedures
- Individual, group, and team recovery policies and procedures
- On-Site/Off-Site resumption policies and procedures
- Administrative policies, procedures and responsibilities
- Contingency planning

Supporting your security and privacy policies and technologies is tricky. Unless you have a lot of in-house security talent, you might have to look outside for end-to-end security solutions integration. This decision must be made carefully, since there is a tendency to think that the in-house staff—who may have managed security in a host-based, data center enclosed environment pretty well—can manage a growing number of distributed applications that link employees, partners, and suppliers.

The range of necessary skills is broad and deep. You will need to make sure you cover all of the bases, and well. The key is the integration of the services into an adaptive solution. If your security strategy consists of lots of elegant pieces that do not fit well with one another you don't have a viable security strategy.

The threat of information warfare must be taken seriously. If you misarchitect your security architecture and infrastructure your competitors will find ways into your networks, your applications, and your data bases. Is it easier to spend years working to increase market share or spend weeks destroying the competition's data bases? Information warfare is a real threat. You should invest as much as necessary to protect your business, and in business recovery processes and technology should your defense break down.

You need to let go of the notion that security is a "step" you take (when designing, developing, and deploying networks and applications), or that security can be serviced by lots of tools and techniques surrounded by distinct processes. While these notions are theoretically correct, they miss the long-term point: security is not a part of a network or an application, it is embedded in networks and applications.

In other words, security is as much a part of a network as a router or switch, or as much a part of an application as a user interface or database. When you stop looking at security as a disembodied part of your network and applications infrastructure—but rather as an integral ingredient in an otherwise pretty complex soup—then you have achieved the next level of distributed applications development and management.

Figure 14 will help you implement a security investment strategy. It offers cells in a matrix that you can use to prioritize your security requirements and investments. But note the distinction between products and services. Some of the answers to your security questions are product-based while some require on-going services. Figure 14 identifies the range of support you will need. A key decision is how to procure these services.

As always, you have the option of in-sourcing or outsourcing the procurement. Depending on your industry, the politics of your organization, your employment culture, and a variety of perspectives on the subject, it is really unlikely that your internal staff has all of the answers. If you're predisposed not to outsource, then get some smart consultants to look over the shoulder of your internal consultants. There are also new—and rapidly changing—classes of technologies that we paid relatively little attention to just a decade ago, technologies that we really need to understand in order to optimize our security investments.

Figure 14. Integrated security requirements and investment opportunities

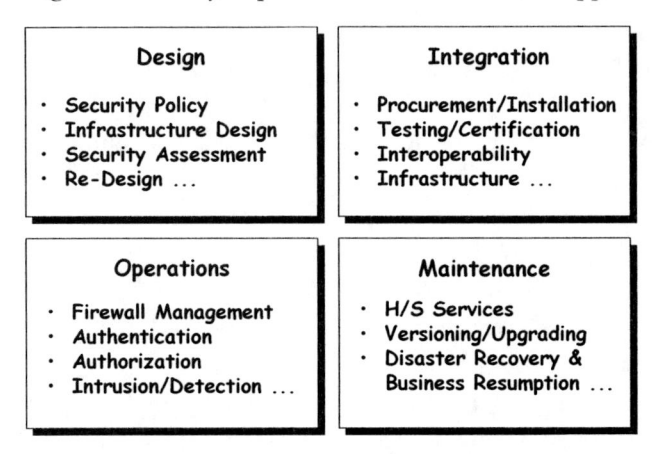

Design	Integration
• Security Policy • Infrastructure Design • Security Assessment • Re-Design ...	• Procurement/Installation • Testing/Certification • Interoperability • Infrastructure ... |//
Operations	**Maintenance**
• Firewall Management • Authentication • Authorization • Intrusion/Detection ...	• H/S Services • Versioning/Upgrading • Disaster Recovery & Business Resumption ...

The list consists of at least the following:

- Encryption
- Firewalls
- Antivirus
- Certificates
- Privacy

Encryption technology is complex, evolving and moving toward standardization. In its simplest form it represents "scrambling" technology. The more sophisticated technology is based on some standards. Encryption technology exists in lots of pretty common applications and is invoked often when users aren't even aware of its use.

Firewall technology is another class of security technology that bears watching. Firewalls help with authentication and authorization; they also support security administration. Firewall technology should become a core competency of yours since firewalls embed lots of security technology that you can activate with specific commands. In other words, firewalls come loaded with security capabilities that you can apply to your environment. But the firewall marketplace is large and complex; it's also likely to segment into vertical industry specific solutions, solutions that may divide again according to the size of the organization they're intended to protect.

It was noted above that viruses still account for a huge number of security problems. As more and more distributed applications are deployed and as the Internet becomes a "platform" for virtual business, we can expect the number of viruses to increase.

The virus problem is real, growing, complicated, expensive to fix, and continuous. You will need to allocate significant resources to the problem by deploying antivirus technology directly or via the tools and frameworks you deploy. Vendors like IBM have developed some automated solutions for you to consider and Java is largely believed to have some natural immunities to viruses.

Another security technology that bears watching is certificate technology. As we search for more sophisticated ways to authenticate and authorize, a third party, honest broker-based technology solution—certificates—is gaining popularity. An increasing amount of e-business will be certificate-based, all as described below:

So how does it work? Figure 15 illustrates how certificate "authority" processes requests for information:

Privacy will grow in importance as the flip side of security. As we bend over backwards to "guarantee" secure business transaction processing, we will need to start looking closely as personal privacy guarantees.

Government regulations are already forcing the technology industry to adopt privacy standards. The P3 standard (platform for privacy preferences) is defining

Figure 15. The certificate process

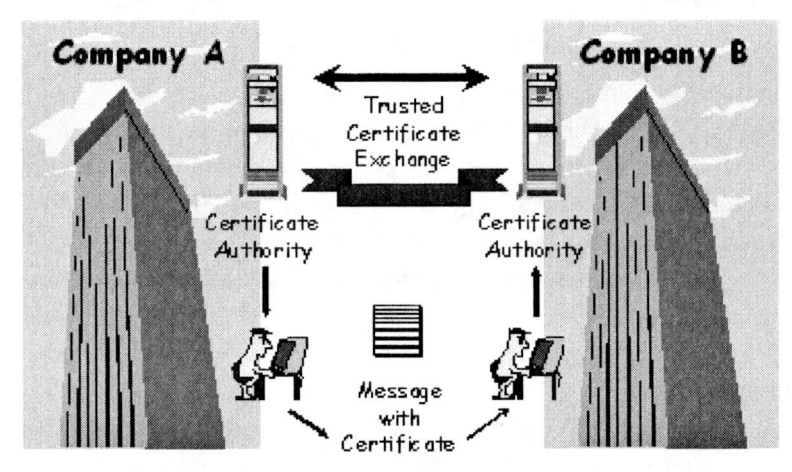

privacy best practices for Internet computing. The Open Profiling Standard (OPS) developed by Microsoft and Netscape is another example of how the industry is responding to emerging privacy requirements. Some countries, like Italy and Germany, have been particularly aggressive passing privacy laws. Pay attention to the regulations and the standards, and make sure that your computing environment is as private as it is secure.

SECURITY INVESTMENT GUIDELINES

Here are some investment guidelines for security:

The framework in Figure 13 should help you prioritize investments and can be used as a filter thought which to pass security products and services options. The key, however, is to address every cell in the matrix since security must be top-to-bottom and must be cost-effective supported by a variety of products and services. Regardless of your investment perspective, all of the products and services must integrate and interoperate: the key is the integration of the services into an adaptive solution and if your security strategy consists of lots of elegant pieces that don't fit well with one another you don't have a viable security strategy.

There are at least five enabling technologies you should track for possible investment:

- Firewall technology
- Antivirus technology
- Certificate authority technology

- Encryption technology
- Privacy

While there might have been some skepticism five years ago about declaring distributed security a core competency, today the opposite argument should raise eyebrows. It is imperative that you understand the technology and how it interacts with your business model, applications portfolio, communications architecture, and other dimensions of your infrastructure. Investments here are unavoidable.

Privacy will become increasingly important, increasingly the focus of government regulations and, therefore, the object of computing and communications standards. Pay close attention to these trends, since the flip side of business security is personal privacy.

Look at the security products and services provided by the major vendors. Security is not an investment target area where untested products and services should be deployed. At the same time, powerful new technologies that address especially persistent problems, like virus protection, should be tracked for their potential.

If you combine these insights with the 15 due diligence criteria, you should make prudent security investments.

ADVANCED TECHNOLOGY TARGETS

Advanced technology excites all technology investors—though for very different reasons. There are specific new technologies, like WiFi and WiMax wireless communications technology, that change the way we communicate with our employees, partners, and suppliers, and there are technology mindsets and "clusters" (as discussed in Chapter I) that change the way we think about solving business problems with computing and communications technology. Context is important here. There are changes occurring that redefine the role that technology plays in business and the very definition of "advanced technology." Let us look at some of the trends that drive our understanding and application of advanced technology.

As already suggested, there is no question that PCs, laptops and routers are commodities; and many services—like legacy systems maintenance, data center management and certainly programming—have already become almost fully commoditized. But the real story here is not the commoditization per se but the bifurcation of business technology into operational and strategic layers. Operational technology is what is becoming commoditized; strategic technology is alive and well, and still a competitive differentiator. What is the difference? Operational technology supports current and emerging business models and processes in well-defined ways with equipment and services whose costs have become stable and predictable and have generally declined significantly over the past decade. Hardware price/performance ratios are perhaps the most obvious example of this trend but there are others as well including what we are willing to pay for programming. Strategic technology on the other hand is the result of creative business technology convergence where, for example, a Wal-Mart streamlines its supply chain, a Starbucks offers wireless access to the Web, and a Vanguard leverages its Web site to dramatically reduce its costs. There is no limit to how creative business technology convergence can become; there is no limit to how strategic the business technology relationship can be.

Advanced technology can be strategic or operational. Technology that represents whole new ways to solve strategic or operational problems is of enormous interest to technology managers, venture investors and, of course, vendors who are always looking for new products and services to sell to their customers.

Another significant trend is our increasing ability to make disparate stuff—data bases, hardware systems, applications—work with one another. We have seen the field move from application programming interfaces (APIs) to data extraction, translation, and loading (ETL) to enterprise application integration (EAI) to Web services—all in about a decade. Good progress to an even better end game: ubiquitous transaction processing through seamless technology integration. Integration is now a core competency in many companies and the industry has responded with a barrage of new tools to make things cooperate. This trend will continue and threatens to dramatically alter software architectures and technology infrastructures—and influence the nature and content of advanced technology.

Another major trend is our willingness to optimize sourcing. As more and more technology gets commoditized, we will see more and more hybrid sourcing models. Some companies outsource lots of processes while others have adopted a cosourcing model where their own people work closely with the outsourcer. The trend, however, is clear: companies are re-evaluating their sourcing strategies and have lengthened the list of potential candidates for full or partial outsourcing. Some of these include help desk support, production programming and application maintenance. If we extend this trend it is likely that we will see a lot more hosting of large applications that companies will increasingly rent instead of wrestling with implementation and support challenges.

Advanced technology may be created, implemented, and supported by people who do not work inside small, medium, and large companies but instead lease it. This trend (among others) will change the dynamics around advanced technology investing. End-user companies may well follow much more than they lead, leaving the creation of advanced technology to technology vendors and the venture investors that provide them. We are faced with an interesting optimization challenge that is ultimately tied to core competency assessments. The decision matrix we all need to master appears in Figure 16. Advanced technology may well be delivered by outsourcers not developed by or implemented by end-users.

Small, cheap, reliable devices that rely on always-on networks make sense. The trend toward shifting computing power from desktops and laptops to professionally managed servers is well under way. Moving storage from local disk drives to remote storage area networks makes sense. Fat clients should lose some weight—as we bulk up our already able servers. The total-cost-of-ownership (TCO), not to mention the return-on-investment (ROI), of skinny client/fat server architectures is compelling (to put it mildly). Anyone updating their infrastructures should think about thinfrastructure. Migrations should be filtered with thinfrastructure opportunities. Pilots should be launched to collect TCO and ROI data, and you should begin the education process in your organization to raise consciousness about the potential for thinfrastructure in your company.

Advanced technology? Thin client architectures are advanced as are the applications designed to run on increasingly thinner devices. The trends around thinfrastructure speak directly to advanced technology investment opportunities.

There is a trend toward renting software from vendors who "host" proprietary applications. In fact, when we step back and look at a variety of trends in the industry, we see even more evidence that hosting will expand—with a vengeance. Web services, utility computing, thin client architectures, service oriented architectures, and even the semantic Web are all conceptually consistent with the hosting trend.

Are we sliding toward to new implementation standard? Are large scale enterprise application projects dinosaurs? Will paying-by-the-drink become the new revenue model for enterprise software vendors? Should you buy or rent?

The technology industry itself will also have to re-configure its software licensing and pricing models—something it has been unwilling to really do. Many companies have already experienced the pain of shelfware; hosted applications open the door to customer expectations about "paying by the drink."

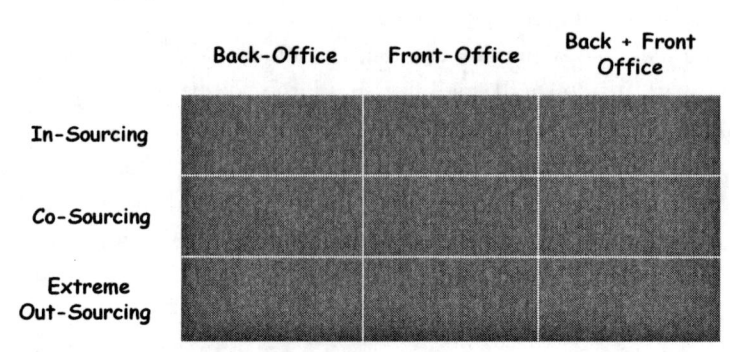

Figure 16. Sourcing optimization matrix

	Back-Office	Front-Office	Back + Front Office
In-Sourcing			
Co-Sourcing			
Extreme Out-Sourcing			

The technology—advanced or otherwise—that supports hosting is evolving quickly. Not surprisingly, network security technology is a large part of the technology solution offered by hosting companies.

All of these trends are defining the advanced technology area. CIOs, vendors, and VCs understand the role that these trends are playing in the advanced technology area and the role that advanced technology itself plays in their respective strategies. CIOs rely on the technology industry to invest in advanced technology to make its products and services better and cheaper. Vendors spend enormous amounts of money to create "killer apps" and "disruptive technology." VCs live for advanced technology-based business models that promise to provide CIOs with just what they're looking for at the right time and for the right price.

It's important to list advanced technology clusters on the investment options list (Chapter III focuses on technology trends analyses and Appendices A, B and C provide example of technology trends analysis reports). All of that said, there some areas that bear watching. Four of them include:

- Adaptive, intelligent agents
- Natural interfaces
- Vertical knowledge integration
- Software components

The first area, adaptive, intelligent agents, represents how an advanced technology area might revolutionize how we do lots of things.

Figure 17 lists but a few of the areas likely to be impacted by adaptive, intelligent agent technology. The key here is the recognition that this technology will automate whole classes of activities that we now conduct manually. For example, it's now possible to launch agents to find the cheapest CDs on the Web; in time, the entire supply chain will become automated through the application of intelligent agents that learn.

Figure 17. Adaptive, intelligent interface technology

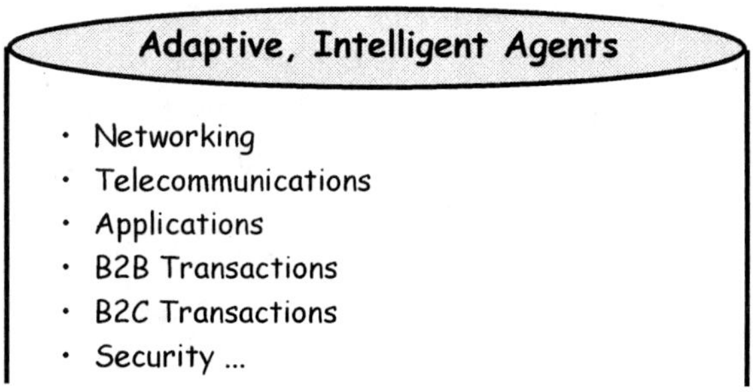

Adaptive, Intelligent Agents
- Networking
- Telecommunications
- Applications
- B2B Transactions
- B2C Transactions
- Security ...

Figure 18. Emerging natural interface technology

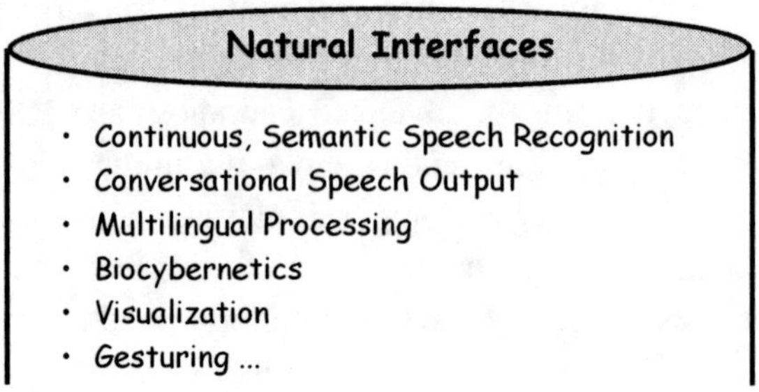

Natural Interfaces
- Continuous, Semantic Speech Recognition
- Conversational Speech Output
- Multilingual Processing
- Biocybernetics
- Visualization
- Gesturing ...

Figure 19. Vertical knowledge integration

Vertical Knowledge Integration
- Automated Data Mining
- Intelligent Document Management
- Knowledge Representation & Organization
- Knowledge Visualization
- "Vinfrastructure" ...

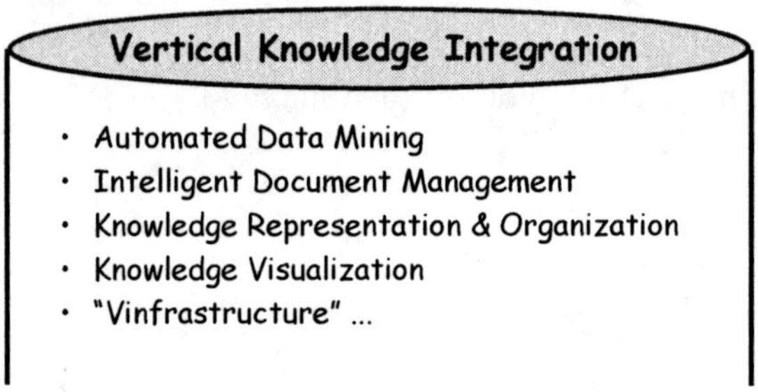

The second area—natural interfaces—refers to how we interact with machines of all kinds. If natural language systems were widely deployed, for example, training budgets would disappear. Similarly, the ubiquitous application of biometric-interface authentication technologies could dramatically improve online security.

When we can talk with machines—and when they talk back intelligently—things will change dramatically and overnight. Whole new classes of users will emerge and when the power of natural human-machine interfaces is combined with intelligent agent technology, we will witness a revolution. This is another technology area that warrants close attention.

Vertical knowledge integration is another area that is emerging as an advanced technology opportunity. As suggested above, however, "knowledge management" has already had a short but troubled history. But we also believe it will mature primarily vertically; companies should track progress here as it applies to their vertical industries.

Component architectures have been discussed for nearly a decade but are finally beginning to emerge as relatively fast ways to assemble applications. The trend here

Figure 20. Vertical knowledge integration

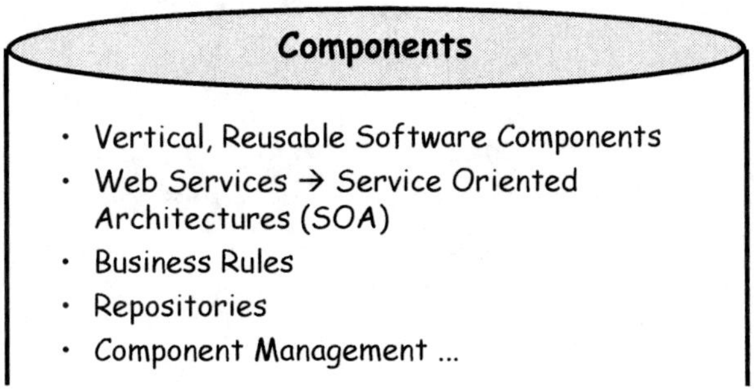

Components
- Vertical, Reusable Software Components
- Web Services → Service Oriented Architectures (SOA)
- Business Rules
- Repositories
- Component Management ...

promises to permit users to deploy a backbone architecture— like an ERP or network and systems management architecture—and then add or subtract functionality (through the addition or deletion of components) as requirements dictate. The use of embedded business rules will also increase and provide application flexibility that we've not had to date. The most important development in recent years is the emergence of Web services and service oriented architectures. Web services is a collection of integration and interoperability standards that the industry is adopting; service oriented architectures (SOAs) represent the exploitation of Web services to create incredibly flexible software architectures that permit the assembly of applications on-the-fly from registries of software components that each represent specific functionality.

These areas are representative of technology trends that should be tracked over time. The purpose of technology monitoring and assessment is to develop lists of technologies likely to have the most impact on particular businesses. "Hit lists" are excellent devices for rank-ordering and screening technologies. They also focus attention on specific technology opportunities.

CIOs should allocate resources to technology trends analyses. Vendors invest billions of dollars every year in new computing and communications technology. VCs track both groups very closely to try to determine what the next wave—like pervasive computing (see Appendix A)—will be.

ADVANCED TECHNOLOGY INVESTMENT GUIDELINES

Here are some guidelines for investments in advanced technology:

Appreciate the big distinction is between tracking advanced technology and investing in advanced technology: tracking trends and key technologies requires relatively little effort and resources, but meaningful investments require lots of resources and attention. The problem is that in order to track you need to set up a semi-formal organization so you can exploit technology potential in formal, measurable ways.

There are a number of key technology areas for you to track. Segment the technology areas (services, data, communications, applications, security, infrastructure, people, and so forth). Develop a list that works for your organization and set up processes for tracking technology trends in the areas.

Pay special attention to the industry's macro trends and the emerging technologies in each vertical area as well as technology generally: often there are nuggets worth understanding. For example, the area of adaptive intelligent agents, as discussed above, is full of promise and deserves close scrutiny.

Specific lists of technology investment opportunities should be developed and communicated throughout the organization. The lists should be reviewed at least twice a year.

Pilot projects are an effective way to determine if a technology is ready for widespread application. But pilots should be managed as real projects with project managers, start dates, milestones, and the like. They should also be compact projects, lasting no more than 6 months. Pilots are an effective way to show the technology flag in many organizations, especially if the right people are behind the pilots. Make sure that there is technology and business support for the pilots you undertake.

Make sure you identify metrics capable of measuring the effectiveness of your technology investments. The metrics should be quantitative and measurable and managed by professionals not involved in the pilots that lead to large scale investments.

It is important to model the impact that advanced technology might have on your business models and processes.

Scenarios that describe good technology investment outcomes and bad technology investment outcomes should be developed and routed throughout the organization.

SERVICES TARGETS

The nature of available strategic and tactical consulting services is changing dramatically. The overwhelming trend is toward the development of processes and technologies that will support business models and processes distributed across employees, customers, and suppliers. These distributed applications require computing, communications and support infrastructures fundamentally different from those that supported legacy mainframe-based and even pure client-server-based applications. Related to this trend is the important services trend toward solutions integration, where customers need providers to offer "end-to-end" services in specific horizontal and vertical areas. Investors in technology services are keenly aware of these trends and requirements.

Another trend is convergence, where the classic vertical industries—like insurance, financial services, pharmaceuticals, retail, and manufacturing—are converging into hybrids. For example, the insurance and financial services industries are converging, as banks sell insurance and insurance companies move into funds management. Similarly, technology companies that offer technology consulting, systems integration, and on-the-ground technology support are moving to integrate their services as the distinctions among management consulting, technology consulting and systems integration blur. The net effect of this convergence is an array of new service industry capabilities, so-called "vorizontals" and "herticals," that cross-cut the pure verticals and horizontals that dominated services in the 1980s and 1990s, including:

"Vorizontal" Service Capabilities

- Data solutions integration
- Communications solutions integration
- Applications solutions integration
- Security solutions integration
- Infrastructure solutions integration
- Enterprise applications solutions integration
- Mergers and acquisitions (M&A) technology solutions integration

"Hertical" Service Capabilities

- E-business solutions integration
- Supply change solutions integration
- Call center solutions integration
- Automated sales and marketing

Here is a brief description of the kinds of services provided by the various vendors out there. All of these providers represent opportunities for CIOs, vendors, and VCs.

Data solutions integrators specialize in either in a converging vertical—like insurance and financial services—or across verticals offering end-to-end solutions. In this case, data solutions integration, the service provider would offer services that included data structure and quality assessment, data warehousing, data mining, document management, knowledge management, information management, and data migration, among other areas that represent the full range of data—information—knowledge services that includes the strategic use of data/information/knowledge in multiple vertical industries.

The same capabilities exist in the communications area, where companies offer communications strategy services, communications technology services, network and communications architecture services, messaging, groupware and workflow services, and services to support the whole communications environment. Increasingly, these services are offered by telecommunications providers but conventional technology consultants are moving to offer these and related services.

Applications solutions integration service providers provide applications portfolio assessment services, application development services, services to manage applications performance and applications support, and services to modernize (design, develop and integrate) applications aligned with future business models and processes. As companies move from host-based applications to distributed applications (with the Internet as the primary application platform) the demand for end-to-end applications services is growing dramatically.

Infrastructure solutions integrators specialize in the planning, design, development and support of computing and communications infrastructures including the desktop, laptop, PDA, and server environments, the mainframe (data center) environment, and the network and systems management area. Companies that specialize in infrastructure solutions integration for converging vertical industries work closely with the applications and communications solutions integrators who provide the infrastructure requirements to the infrastructure solutions integrator.

As the number of distributed applications grows, so does concern over security. Security solutions integrators provide data security/access assessments, firewall technology services, authentication services, disaster recovery services, business resumption planning services, and security administration services. Some converging vertical industries have an especially high demand for these services as well as vertical industries with large numbers of telecommuters and e-business customers and suppliers.

The massive deployment of enterprise applications has resulted in an unprecedented demand for enterprise applications services—which include planning, implementation, integration, and support. In time, major enterprise applications will be managed under larger network/systems/applications management frameworks.

There are major service opportunities available to those who specialize in services defined around the integration of technology infrastructures and applications after mergers and acquisitions occur. As the number of M&As increase, the need for M&A technology integration services increases. There are currently very few service providers with legitimate expertise in this complicated—though huge—area. The vertical slant here is even more profound, since technology integration requirements vary dramatically across vertical industries. Finally, as unlike companies—representing converging verticals—combine, M&A technology integration services will increase even more.

The *e-business services* area of specialization is still timely, though becoming less so as companies blur the distinction between "business" and "e-business." While there are Web site development companies, companies that specialize in Internet transaction engines, Internet security, and Web-to-legacy data base connectivity, there are those that have skills that cross-cut all of these and additional areas. This is where there are still opportunities for existing and new service companies.

All organizations need call centers. But the nature, purpose and organization of call centers are changing as customers get smarter, competition gets digital, and technology enables anytime/anyplace company/customer contact. Call centers are becoming "virtual." Vertical industries have special call center needs: high technology customers, for example, want 7X24X365 access to a slick Web site while customers who purchase complicated financial instruments—like whole life insurance and 401K retirement programs—want access to digital and human agents. The technologies that support all this communication are evolving and complex. Service organizations that support "whole customer management" through call center management will find growing opportunities in the near- and longer-term future.

The need for services *in the supply chain management* area are growing as companies seek to optimize the processes by which they produce, distribute and service their offerings. Supply chain integration will eventually become ubiquitous where retailers will immediately communicate their requirements to suppliers and distributors. Manufacturers will know when product demand increases and will simultaneously notify all of their suppliers. Companies that understand the supply chain, the management of the chain across industries, and the technology that supports supply chain integration, will find growing opportunities across many (converging) vertical industries (such as the convergence occurring in the pharmaceutical, manufacturing and retail industries).

The need for service expertise in sales and marketing is also growing. As industries fundamentally change the way they identify, fulfill, service and retain customers, they will develop "whole customer sales and marketing" expertise. Data will become increasingly important to this process. Consulting in this area will be as essential to the process as the development of Web sites and call centers that track and profile customer behavior. Figure 21 summarizes the service capabilities everyone needs (with a token reference to some service-inspired products). CIOs need integrated services solutions to their problems; vendors try to integrate their offerings; and VCs try to invent next generation service models for their companies to provide.

Figure 21. Technology services map

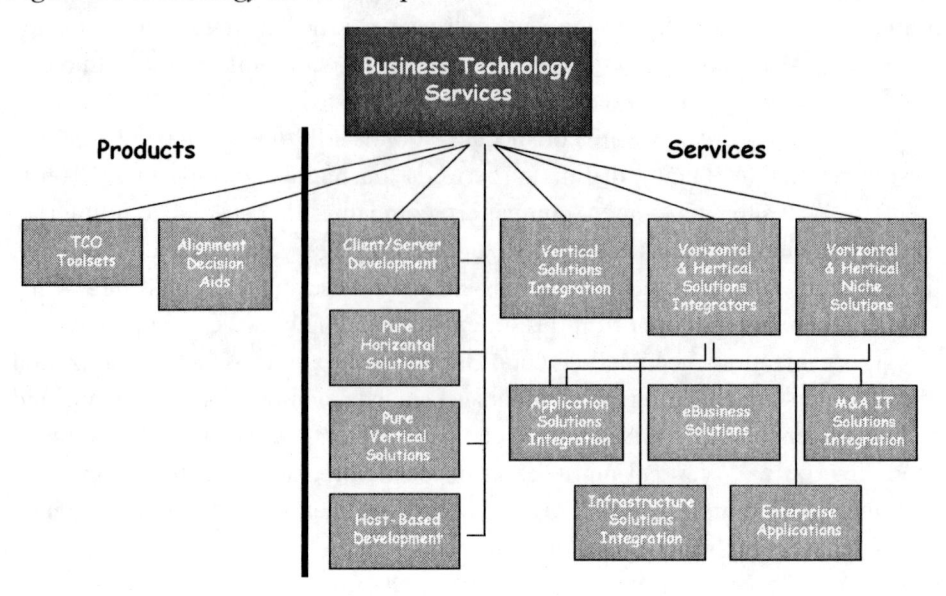

Figure 22. "Food chain" costs and margins

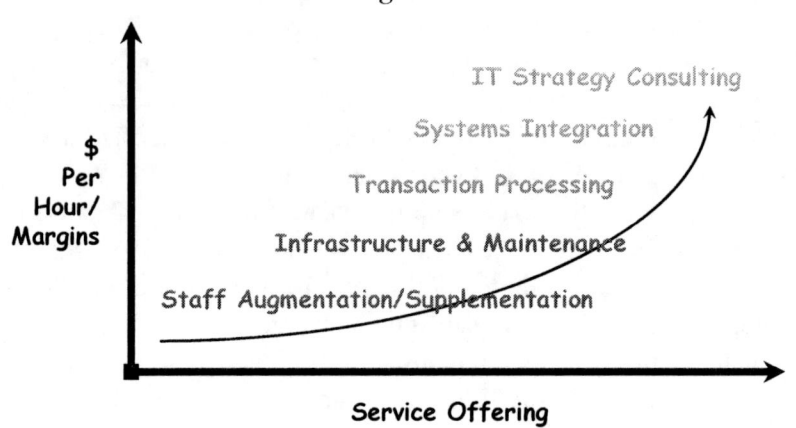

Why is the food chain so important? Figure 22 provides the answer. It is all about costs and benefits. Regardless of whether you do business or technology consulting with in-house or outside professionals you are paying much more for this kind of talent than you pay for data base administrators and—especially—lower level technology support personnel. Figure 22 shows you what you are spending (or charging) whether you realize it or not.

All technology investors are aware of the food chain: CIOs certainly understand what things cost. Service vendors are well aware of what they can charge and VCs are always interested in services at the top of the food chain that generate the highest profit margins.

SERVICES INVESTMENT GUIDELINES

Here are some investment guidelines for the services area.

Pay special attention to the food chain—for obvious reasons.

Services are morphing; many services that were segmented a few years ago are now converging; investors should anticipate the trajectory of convergence

The demand for services is driven to a significant extent by the condition of capital markets generally and the specific financial condition of companies who create or purchase services; investors should assess the need for services in these larger financial contexts.

The rise of technology outsourcing is fueling demand for more and more varied technology services; outsourcing trends are thus necessary to understand in order to optimize services investments.

If you combine these insights with the 15 due diligence criteria, you should make solid services investments.

So let us summarize what we have discussed about the range of investment opportunities available to CIOs, vendors, and VCs. First, they are varied. There are opportunities in the software, communications, infrastructure, data, services, security, advanced technology—and a variety of other areas. They are complex as well, including the basics of each category and fairly exotic technologies. They all speak to a different set of requirements (Andriole, 1996, 2005) as well as opportunities for innovation (Evans, 2003). Some of them speak to process change (Grover & Kettinger, 1995) and some specifically to functional activities like supply chain planning and management (Hanfield & Nichols, 2002). But they all speak to opportunities and risks. The due diligence process is designed to reduce that risk while bolstering leverage opportunities. The investment guidelines at the end of each section are especially important to the opportunity/risk outcome.

REFERENCES

Andriole, S. J. (1996). *Managing systems requirements: Methods, tools & cases.* New York, NY: McGraw-Hill Book Company.

Andriole, S. J. (2005). *The 2nd digital revolution.* Hershey, PA: Idea Group.

Evans, N. D. (2003). *Business innovation & disruptive technology.* Upper Saddle River, NJ: Financial Times Prentice Hall.

Grover, V., & Kettinger, W. J. (1995). *Business process change: Reengineering concepts, methods and technologies.* Harrisburg, PA: Idea Group.

Hanfield, R. B., & Nichols, E. L. (2002). *Supply chain redesign: Transforming supply chains into integrated value systems.* Upper Saddle River, NJ: Financial Times Prentice Hall.

ENDNOTE

[1] The number is actually over $4,300 as reported by the Gartner Group (www.gartner.com).

Chapter III
Business Technology Trends Analysis

Regardless of one's investment perspective, it is essential to understand business technology trends. Some companies assign large teams to track business technology trends to make sure that they do not miss a potentially "disruptive" technology or a suite of technologies that might together revolutionize their industry. Venture capitalists obviously need to understand what is hot and what is not before spending their investors' money. The vendors that create technology also need to understand the trends (that they sometimes actually create). In short, everyone needs a way to track business technology over time. In fact, business technology trends analysis is a necessary core competency for CIOs, vendors, VCs, and all technology investors.

What everyone needs is a technology investment agenda that helps identify the business technologies in which they should invest more and those that should get little or none of their financial attention. The agenda ultimately must be practical: while "blue sky" research projects can be lots of fun (especially for those who conduct them), investors must find the technologies likely to yield the most return, not the coolest write-up in a technical trade publication. But this can be challenging especially when there is so much technology to track—and relatively little certainty about what business models and processes will look like in three to 5 years.

Regardless of where investors sit, business models and processes are the lighthouses to cost-effective technology investments. It is imperative that we develop an understanding of where we expect business models and processes to be in 2 or 3 years so that we can begin a meaningful technology migration. Important here are answers to the big questions about connectivity, supply chains, the percentage

Copyright © 2009, IGI Global, distributing in print or electronic forms without written permission of IGI Global is prohibited.

of e-business we will be conducting, how we expect to configure and manage our infrastructure, and what enterprise applications we expect to deploy and support. We also need to make some fundamental decisions about technology platforms and, especially, how we plan to acquire technology products and services—that is, whether we expect to acquire and deploy with internal technology professionals or whether we expect to outsource the lion's share of our technology efforts. The articulation of "to be" business models and processes provides the necessary insight to these questions. Without that insight, we will likely just extrapolate from where we are today—and quite possibly miss some major business technology shifts that might enable whole new business models and processes.

Vendors need to understand where business models and processes are going as well in order to align their R&D investments with future requirements. VCs need to track all of this but must also try to create (with vendors) new trends that can be "sold" to companies searching for better, faster, and cheaper ways of doing things—without going so far as to be accused of selling "solutions-in-search-of-problems." In short, everyone who makes, buys, sells, and services technology needs to understand business models and processes—today, 3 years out and, if possible, 5 to 7 years out. This chapter introduces the idea of technology trends analysis as an important component of due diligence. It is essential for all investors to remain aware of technology's trajectories—and assess their investment opportunities (and risks) accordingly.

BUSINESS TECHNOLOGY TRENDS ANALYSIS METHODOLOGY

What are the steps? Vertical industry companies and their CIOs, vendors, and VCs all need to understand the following in order to prioritize their technology investments:

- General business trends
- Current → Future business models and processes by vertical industry
- Advanced technology trends

General Business Trends

What are the big changes in the way companies do business? One way to think about the changes is to contrast discrete transactions with continuous ones. Discrete transactions—like selling insurance policies, buying disk drives to be included in PC manufacturing, or buying or selling stock—used to be transactions that one could begin in the morning and compete by afternoon, or the next morning after a good night's sleep. Today these transactions are continuous. One insurance policy is blended with another. Cross- and up-selling are continuous goals. Disk drive acquisition is integrated into PC manufacturing and buying and selling stock is simultaneously linked to tax calculators and estate planners. These transactions are now extended, continuous and (even quasi-) automated.

What if companies could connect all of their suppliers with their manufacturers, marketers, salespersons, and customers? In spite of what supply chain vendors tell us today about how integrated supply chains really are, they are only partially connected. But within a few years they will collapse into webs of efficiency that will exceed anything we have imagined to date. Technology investments today must anticipate a connected tomorrow.

Connectivity among employees, suppliers, customers and partners—though far from compete—is enabling interactive customer relationships, integrated supply chains, and the business analytics that permit real-time tinkering with inventory, distribution, and pricing. The strategies and business models were always there: for years we have imagined seamless connectivity and ubiquitous business. But the convergence of business and technology has made it more than just possible.

The collapse of the dot.coms got way too many of us to believe that the new digital economy died before its time. The irony is that the hype clouded the preview of a classic movie set to be released in just a few short years. But the plot remains what it always was: every year more and more business transactions will be conducted over the World Wide Web.

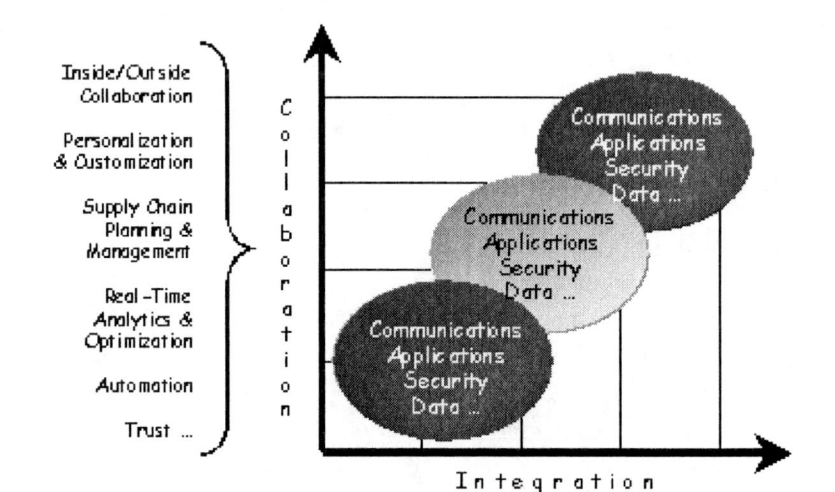

Figure 1. Business collaboration drivers of business technology convergence

Figure 1 identifies collaboration as the business trend everyone should understand (along with technology integration as the enabler of collaborative business models and processes). The best investments serve these two masters: collaboration and integration.

The idea is simple: strategically connect employees, customers, suppliers and partners for competitive advantage. Two things have to be true to make collaboration work. First, collaborative processes have to be defined. For example, Dell assembles computers by exploiting an integrated supply chain that consists of suppliers, employees, customers, and partners. Suppliers provide components that get assembled while employees monitor the quality of the components and the assembly process: if, for example, a batch of disk drives under-performs then a request for supplies is broadcast to alternative suppliers. Customers can customize their computers online where they can buy software and printers (and other devices) from partners. The process defines collaboration; technology enables it. While it is all about collaboration, the process has to be defined, efficient and profitable. In other words, it is possible to define inefficient and unprofitable collaborative processes, just as it is possible to deploy the wrong technology (to enable the wrong collaborative business model).

Take the idea and extend it to customers. Touch them digitally. Obviously this does not mean that companies will *only* touch them digitally, but that digital contact represents another way to stimulate and serve them. Convenience, from a customer's perspective, is the reward here; from the company's perspective, its cost-effectiveness and a way to extend communication with their customers they can use to please them, up-sell them, and cross-sell them.

Now take the collaboration idea and extend it to suppliers. While we will talk more about supply chain planning and management in a moment, supplier collaboration can take various forms. For example, do companies plan with suppliers? Do their suppliers have a window into the production processes of their distributors, wholesalers, and retailers?

Last but not least, extend the idea to partners. Do companies sell their own goods and services as well as goods and services produced by others? Do they subcontract products or services? Imagine a capability that would enable companies to plan with their suppliers and partners to facilitate collaborative forecasting and planning; imagine a capability that would reduce transaction friction, excess inventories, and pricing instability?

Collaboration tightens the value chain of business. If executives and managers are not thinking about the costs and benefits of collaboration engineering, then the company they manage has a problem.

Supply chain planning, management and optimization are a subset of collaboration and enables lots of things including customization, personalization, dynamic pricing, and automated transaction processing. Everyone has supply chains. Even companies whose principal products are ideas. Understanding where the leverage is and how it can be optimized in supply chains is essential to successful collaboration.

As we all know, complex production processes require enormous amounts of coordination. Digital technology can help coordinate complex, distributed manufacturing processes. *Complex* and *distributed* are the key words here. As we outsource more and more component manufacturing, the need for digital glue is rising dramatically.

But the supply chain mantra is much broader, and while we are a few years away from completely integrated supply chains it is time to start thinking holistically now. Think about it as a kind of controlled, vested interest-based partnership among all of the stakeholders in the value chain or, put more crudely, all of the people with which we deal who want to make lots of money. Partnership here is based on a little skepticism and lots of shared communication.

Procedurally, supply chain planning and management requires a reciprocating process. Rules have to be established and followed. Technology vendors have developed software applications that manage the process according to sets of rules specified by the collaborators. The perfect supply chain of course is one where everyone agrees on the processes and everyone then follows the rules. Procedural mapping is critical to success and begins with a mapping of current supply chain processes. Here are some basic questions for early 21st century companies:

- How many supplies are in the chain?
- How many buying processes are supported?
- What does the cost structure look like?
- Which suppliers cost the most, deal the most, and argue the most?
- Where does demand come from and how does it change over time?
- Can demand be predicted?
- Is demand forecasting collaborative?
- Are new products collaboratively designed?
- Is order management and fulfillment integrated across the supply chain?
- Are supply chain events (and transactions) managed and optimized?

Beyond supply chain planning and management is *customization and personalization*. Companies can "personalize" their contact with customers, suppliers, partners, and employees, and can personalize all varieties of messages including sales, marketing, service, and distribution. They can also personalize with paper, telemarketing, advertising, and e-mail. Over time, given how low digital transaction costs are compared to other ways we touch the members of the value and supply chains, and how ubiquitous digital communication is becoming, it makes sense for companies to reassess their customer contact budget allocations. Companies pay for lots of channels to their customers. Which ones pay the best?

It is all about the depth, location, and quality of customer data. It is also about the analyses performed on this data. Some companies have excellent behavioral scientists who analyze the data every which way in search of correlations that explain what customers' value and why they buy what they do, when they buy, and so forth.

Personalization and customization should not stop with customers. The same analytical approaches taken to profile customers can be used to profile employees, suppliers, and partners. In fact, the personalization and customization of employees, suppliers, partners, and customers is a huge part of emerging collaborative business models and processes.

Think about *real-time analytics,* and the optimization of processes and outcomes that it enables, as a corporate dashboard. When you are driving your car there is a ton of active and passive data available to you from a glance or a command. Much of this data, like speed and revolutions per minute (RPM), is displayed continuously while other data is presented when something goes wrong. But what mechanical systems cannot easily do is conduct what-if analyses of seemingly disparate data to discover counter-intuitive correlations among various behaviors. You cannot ask your car questions like: "If I drive you at 100 miles per hour for 11 hours on mountain roads while outside temperatures range from very cold to very hot, which systems are most likely to fail—and in what order?" But companies can ask their customers, through the behavioral data they collect on them, what they like, when they like it, and what combinations of products and services make them happy. Companies can also ask technology questions. They can ask which applications—like ERP, CRM, and other enterprise applications—touch the most customers with the highest (or lowest) margin products or services. They can ask the systems that maintain the technology infrastructure about how well the infrastructure is performing, how much it costs to maintain the infrastructure and which technology within the infrastructure is about to break. Can all of this be done? Yes, if companies invest in the right analytical applications and the right infrastructure technology.

Analytics enable *business intelligence (BI).* Like everything else, there is a process here. The location and quality of data about customers, employers, suppliers, and partners determines just how far a company can analyze business processes. If a company's data is all over the place, ugly, dirty, and in any number of proprietary vendor silos, then management has a problem. Data warehousing vendors are of course happy to sell software, gear and services to fix the problem. But it makes more sense to avoid the nasty integration process altogether—by standardizing on just one or two data base management platforms.

Once the data is accessible, it can be analyzed. We can query data, develop reports, perform "online analytical processing"-based analyses, "mine" the data, visualize the data, export the analyses (to desktops, laptops and PDAs, for example), and ultimately use the data to make decisions—to rethink sales, to tweak the supply chain, to manage the back office (human resources, finance, accounting, manufacturing, and the like).

When we talk about "real-time" analytics we are talking about the ability to convert real-time analyses into immediate action. For example, let us say that the raincoats that a company expected to sell we are still sitting on the shelf. Supply chain visibility permits managers to see that the coats are not moving. The same tools can help predict what sales volume will look like over the next few weeks, given the drought conditions in most of the country. Given how unlikely it is that the company will sell the coats at the current price, managers can roll out price changes and then calibrate sales impact. Predictive analysis can be enormously valuable and it can get pretty complicated involving hypothesis testing, pattern analyses, simulations and what-if sensitivity analyses, among other techniques. But keep in mind that predictive analyses are not just useful for averting disasters; they can be used to proactively optimize prices as well. For example, the real-time generation of demand curves can enable upward price adjustments for goods and services whose prices would otherwise remain static. Optimization software drives dynamic pricing, but lots of things have to be true for it to work as advertised. The supply chain mindset needs to be entrenched, data needs to be clean and accessible, and the technology to access and analyze the data must be reliable. Lots of vendors sell strategies and even more sell technologies embedded in large and small applications that slice and dice (clean, available, organized) data in ways hard to even imagine. More recently, analytics have become the anointed savior of the customer relationship management (CRM), enterprise resource planning (ERP), and e-business applications projects that went berserk. After many millions of dollars and tough questions about return-on-investment (ROI)—questions that were difficult if not impossible to answer—project champions, systems integrators, and strategic consultants searched for ways to justify huge enterprise technology investments. Enter analytics, business intelligence, and optimization. The goal is simple: integrate data from front-office applications—like sales force automation (SFA), CRM applications, e-business applications, and back-office ERP and supply chain management (SCM) applications—and then pump it into a full-blown business intelligence environment that enables the analytics that trigger decision-making.

Does anyone believe that customers, partners, employers or suppliers want to go to the Web to execute the same transactions day after day, week after week? It is silly to think that a procurement officer wants to visit the same Web site every month to order the same number of 55 gallon drums of chlorine, or that undergraduates want to search for Dave Matthews CDs week after week? What about systems? Why cannot their condition and effectiveness be automatically monitored, and fixed when a problem is detected? Back-office systems can also automatically transact all sorts of business. They can be engineered to accept dynamic instructions. What is the difference between a "regular" computer program and one that is "smart?" The regular one computes and recomputers (over and over again) transactions it has been programmed to execute. There is not too much beyond these computations that it can do. But a "smart" application is capable of reacting to outside variables and then—through a preprogrammed set of rules—execute a different calculation each time the outside variables are different. Smart programs can monitor networks, and make corrections, help online users find data, and even adjust prices dynamically if it looks like the new price for raincoats still is not getting customers to the racks. Smart systems also support personalization and customization: your birthday, for example, could trigger all sorts of ads served up to you on your company's intranet or whenever you surf the Web. Smart applications can aggregate searches and then, based on rules you set, execute transactions. If you are looking for the best air fares, for example, a smart application—an intelligent agent—can aggregate data from all of the travel sites and then purchase the "best" one according to your definition of "best." These sites exist today (see, for example, www.sidestep.com).

SFA and CRM front-office applications need automation, which sits at the back-end of the data organization, integration, and analysis process. Once patterns are discovered, once profiles are developed and once rules are specified, and when specific conditions are met, smart applications can, for example, automate a marketing campaign or follow-up a customer service inquiry.

Infrastructure needs automation. Computing and communications infrastructure needs to work together, reliably and securely, but all kinds of things can go wrong. What everyone wants is insight into infrastructure operations with a set of rules designed to react to anticipated and unanticipated conditions. So when the amount of Web traffic is unexpectedly high, a rule fires that says: "when incoming sales traffic is high, and server capacity is low, then reroute or queue nonessential transactions." What does this rule imply? That companies can monitor Web traffic for spikes, that they want sales inquiries to receive priority over other transactions, and that if unusually heavy traffic is detected they want their customers to get through no matter what, even if it means queuing large numbers of nonsales-related transactions. Ideally, the infrastructure knows about itself and how healthy it is. For example, it should know how many computers exist and what versions of which software is running on each one and which ones have too much and too little power. It should know which devices fail most often (and expensively or inexpensively) and take steps—alerts, for example—to anticipate failures. All of these kinds of back-office, front-office and infrastructure tasks can be at least quasi-automated today. The larger argument is that the best management becomes "exceptions management," or the management of unanticipated events and conditions that cannot be easily foreseen.

The evolution of collaboration, supply chain management, personalization, and customization will stimulate automation. As consumer, supplier, employee, and partner profiles deepen it will be possible, and desirable, to automate all sorts of customized transactions. Assuming that our privacy laws and preferences have been worked out, lots of us will empower retailers to execute transactions they know we will like. It is the closest thing to a personal butler that many of us will experience, and the range of transactions—from the complex to the mundane—is theoretically limitless.

Lots of automation is likely to occur through the assistance of hard working intelligent agents. There will be really smart, powerful agents authorized to do all sorts of personal and professional things, and less aggressive agents that will only have basic transaction authority. Instead of debit cards, we will give our kids mini-agents to help them plan their activities and manage their money. As they grow, they will graduate to the smart, powerful agents that can help them manage their lives and plan their futures. We are not talking about deep intelligence here or the inevitability of digital friends or psychiatrists, but rather simple "if-then" applications capable of—and authorized to—transact personal and professional business. Many of us are already part of the automation trend. Our bills are automatically paid each month and we get retail opportunities pushed at us from multiple sources, but we have not authorized agents to transact large chunks of our business. At work we automatically send and receive products every month but here too we are still manually executing the largest transactions.

Collaboration requires automation. The sheer number of transactions that collaboration enables requires degrees of automated processing—unless we want to manually inspect every personal and professional transaction that occurs. While there are still arguments about just how pervasive agents will become, as business becomes more collaborative and the number of transactions grows, we will need to rely on agents to manage these transactions.

What do you need to know about "automation"? First, it is a bona fide trend. Second, it is dependent upon other things like data integration, security, and ubiquitous communications. Third, lots of people are spending lots of money to make devices, infrastructures, and transactions smarter. Fourth, the application of intelligent systems technology is far-reaching from infrastructure, back-office, front-office, and e-business applications; and finally, collaboration requirements will fuel a lot of innovation in this area, innovation that companies should track and pilot.

As the number of digital transactions grows, trust will become even more important primarily because of the physical distance between the transacting parties: if you trust that a meal will be good but it is not, you can immediately complain to the waiter and the chef, and probably get your food replaced or your money refunded. In this instance, trust does not need to extend too far. If it is violated, you have instant recourse. But when you transact business at arms length, trust becomes a necessary and sufficient requirement for a successful transaction. When you buy something over the Web, for example, you need to trust the vendor more than when you buy something in person.

Trust assumes security and security assumes the ability to authenticate users, protect data, control access to networks and applications, and of course the avoidance of annoying viruses. Trust is what companies want their collaborative teams to feel when they interact with each other; if there is any doubt about a company's commitment to privacy or its ability to secure transactions—while protecting the infrastructure from viruses and other problems—its ability to collaborate will fail.

Let us also not forget the role that global economic conditions and government regulation play in business models and processes. Globalization and government regulation are already dramatically affecting the way certain business models and processes play out. For example, governments influence telecommunications which in turn influence continuous transaction processing. Other government policies affect security, privacy and, ultimately, personal freedom. All trends should acknowledge the impact that globalization and regulation will have on all trends.

CURRENT AND FUTURE BUSINESS MODELS AND PROCESSES BY VERTICAL INDUSTRY

The health care industry is an interesting vertical industry to examine. What are today's business requirements and technology solutions? Where will the industry be in a few years? Specific industries should project where their business models and processes will be in three to five years. They should also do so with reference to the technologies that will play larger, and smaller, roles in these future models and processes.

Answers to these and other questions should help define corporate technology investment strategies. Companies need to identify the "as is" and (especially the) "to be" business trends—and broad technology implications that will define their overall business technology environment. CIOs, vendors, and VCs all need to understand the current and emerging business models and processes that will drive computing and communications technology solutions.

Vertical industries all have trajectories. The business models and processes of specific vertical industries—manufacturing, pharmaceuticals, technology, financial services, insurance, and transportation, among others—are all evolving, some more radically than others. It is essential that we track these changes and allow them to drive our understanding of the technologies most likely to enable new business models and processes.

ADVANCED TECHNOLOGY TRENDS

One of the areas that deserves special attention is the advanced technology area where technology areas appear to be on the threshold of major impact. We talked about several of these areas in Chapter I where we also talked about the distinction among technology concepts, prototypes and clusters. One additional way to track the technology trends that matter is to research what the major technology vendors are doing. Microsoft, for example, is focused on the following areas (www.microsoft.com/research):

These areas represent the major focus of Microsoft's multibillion research and develop (R&D) agenda. Microsoft and IBM, among lots of other vendors like Symantec, Sun Microsystems, and Cisco, also have huge R&D programs. IBM's looks like that outlined in Table 2. Accenture's follows.

R&D programs represent billions in annual expenditures—and lots of clues for technology investors. Savvy investors track these investments closely. VCs should track them perhaps closer than anyone since one of their preferred exit strategies is the sale of their companies to large vendors. Everyone, for example, is well aware of how many acquisitions companies like Cisco have made over the years—acquisitions focused directly on advanced technology. Other companies also use acquisitions

Table 1. Microsoft's R&D agenda

Algorithms and Theory

We are working in several emerging fields within theoretical computer science. One is game theory and economics, which includes pricing algorithms and market equilibria. A second is privacy in statistical databases and a third is quantum computing.

Hardware Development

Our research focuses on developing devices that will connect users more intimately, naturally, and efficiently with their computing environment. The devices range from large displays to wearable devices to micro-electro-mechanical systems (MEMS). We collaborate with other groups to build the hardware that will support the next-generation of software. We've developed ideas for new types of microphones, unique data input devices, and we're researching reconfigurable computing hardware.

Human-Computer Interaction

Research on human-computer interaction (HCI) plays a central role across multiple teams at Microsoft Research. Our work is focused on advancing the way users interact with computing devices. This includes search, access, and information management, the display of complex data and information, user modeling and activity recognition, efficient input and interaction, the role of automation and the coupling of intelligent systems with direct manipulation.

continued on following page

Table 1. continued

Machine Learning, Adaptation, and Intelligence

We pursue research on automated reasoning, adaptation, and the theories and applications of decision making and learning. Our research goals include learning from data and data mining. By building software that automatically learns from data, we design applications that have new functions and flexibility. Our research focuses on using statistical methods for the development of more advanced and intelligent computer systems.

Multimedia and Graphics

We focus on new multimedia and graphic experiences that are made possible with the growth in computing power and storage. Our research focus spans the linear and interactive media spectrum across television, broadband, and gaming. We seek to address the challenges involved in the high-computational cost of producing, transmitting, and displaying complex models by researching geometric compression and multi-resolution representations.

Search, Retrieval, and Knowledge Management

Knowledge workers need software that is easy to use and intuitive. They need to find their information long after they've forgotten where they put their documents and what they named their files. Information retrieval and search are a big part of making this happen. We are pursuing research in information retrieval, filtering, and management. Other work has explored the use of classification technologies and the development of systems that will enrich the user experience.

Security and Cryptography

We study various aspects of security related to computer systems. This includes the design of secure systems, the usability, evaluation and certification of security products, the robustness of digital watermarking algorithms, threat analysis for open networks, and database privacy. In addition, we're concerned about security for mobile devices.

Social Computing

We research and develop software that contributes to compelling and effective social interactions, with a focus on user-centered design processes and rapid prototyping. Our projects range from topics in online sharing and mobile applications to trust, reputation, and storytelling. We're interested in how people use computers to enhance their everyday experiences. We are designing interfaces and experiences to make human-to-human communication seamless and exciting.

Table 2. IBM's research program

➢ Multimedia	➢ Algorithms & Theory
➢ Natural Language Processing	➢ Artificial Intelligence
	➢ Communications
➢ Operating Systems	➢ Networking
➢ Performance Modeling & Analysis	➢ Computational Biology
	➢ Medical Informatics
➢ Software Engineering	➢ Computer Architecture
➢ Security & Privacy	➢ Data Management
➢ Storage Systems	➢ Distributed Computing
➢ Supercomputing	➢ Fault Tolerant Computing
➢ User Interfaces	➢ Graphics & Visualization
➢ Mobile Computing	➢ Knowledge Discovery
➢ Web 2.0	➢ Data Mining

Table 3. Accenture's research program

> **Intelligent Device Integration:** Explores the opportunities and potential offered by equipping objects and environments with tiny sensors, communications and tagging technologies. Our research looks specifically at the emerging trends of Sensor Telemetry, Mobile Computing and Intelligent Home Services.
>
> **Analytics and Insight:** Focuses on improving businesses' ability to access, analyze and act upon business intelligence in the emerging real-time economy. Our R&D explores how leading-edge insight techniques, including Predictive Monitoring, could be used as the next key to competitive advantage.
>
> **Human-Computer Interaction:** Examines how emerging technologies can help further enhance the effectiveness of the workforce and includes the areas of collaboration, interaction and visualization. Our R&D looks at new technologies such as the Interactive Wall prototype, a very large, high-resolution interactive screen that puts vast amounts of information at peoples' fingertips, helping them see the "big picture."
>
> **Systems Integration:** Focuses on helping Accenture and our clients achieve advances in software engineering and systems integration. We bring innovation to the design, architecture, building, deployment, analysis and management of large, technology-based business solutions by testing and developing new technologies and tools in several of our experimental software facilities. One prototype, the Repository Navigation Tool, enables large software projects to perform traceability and impact analysis by finding inter-relationships among the large number of documents in a software project repository.

as a way to improve their R&D performance. The purchase of YouTube by Google is another example of the innovation optimization process.

In addition to the major technology vendors, we should track what is happening in the entrepreneurial community as well, always a rich source of advanced technology. Finally, the internal R&D projects within the vertical industries should be tracked for ideas, technologies and pilot projects.

FIVE TECHNOLOGY TRENDS THAT MATTER

As an example of the form and content of technology trends analysis, there are currently five technology trends that matter for the industry. The five broad business technology trends discussed here include:

1. Software development and delivery
2. Web 2.0
3. Master data management for business intelligence
4. Convergence CRM
5. Access devices

The discussion of these five areas illustrates the importance of technology trends as the backdrop to all technology investment decisions.[1]

Software Development and Delivery

Academic textbooks tell us that requirements analyses help us understand business problems. This "understanding" (a.k.a., "requirements definition") is then converted into technology solutions comprised of hardware, software, nd services. We have a pretty good read on where hardware and services will come from over the next 10 years or so (see the section on Access Devices below), but a lot of alternatives when it comes to software.

In the old days we wrote code. Then we installed it. Then we rented it. Eventually we will assemble it on the fly. But code still does not always like other code, in spite of all of the progress we have made with interoperability and integration. While we have made progress with standards, there is still a lot left to do.

But let us assume that we will get it right (and we will—not perfectly right, but essentially right) that software will work together reasonably well. How will we acquire and deploy it?

We have asked a number of CIOs and CTOs if they had a technology do-over would they still install their large enterprise applications. Not one of them said they would. Why not? Because it took them all years to get the software to work and—in some cases— the projects cost hundreds of millions of dollars. Some of these CIOs got fired when they exceeded budgets and schedules; others struggled to realize the benefits everyone promised when they signed the contracts. Some of the major enterprise culprits include ERP, CRM, and network and systems management applications.

Some of the same CIOs and CTOs told us that they were not interested in open source software because it was "too flaky" and they did not want to be associated with the open source crowd. The trends are clearly in the other direction. Just wait until more major vendors endorse open source—or near-open source—software, when open source software is effectively hosted, and when the proprietary software vendors (and software component developers) make it easier and easier to integrate open source with proprietary applications. Some of them do not trust Web services and think that service oriented architectures (SOAs) are still ideas, not reliable software architectures—views that also fly in the face of observable trends. But in all seriousness, the number of technology managers and executives that still believe more in the past than the future is dwindling. The fact is that the new architectures can help companies make money and save money, so their credibility is appropriately high.

Here is how it is likely to evolve:

Relatively few CIOs will launch multiyear, multimillion dollar software implementations. There is too much time, money, and politics involved, and many of the biggest projects have not delivered the goods. Only CIOs in the last year of their employment contracts will attempt multiyear software projects!

If at all possible, CIOs will rent vs. buy-and-install major software applications. They are all secretly hoping that "ASP 2.0" is successful. They really do not want to get back into the enterprise software acquisition, deployment or support business. Most of them are really bad at it and they just do not have the stomach for technology marathons anymore. There is also pressure to reduce capital technology expenditures and move more and more technology costs into operational budgets—a major driver toward renting vs. acquiring-and-supporting.

CIOs will pilot as many SOA implementations as possible to determine where the price/pain/performance ratios lie. They really want all this stuff to work—they need software-as-a-service (SaaS) to become a reality (and they will be happy to sit on hosted applications for as long as they can—as long as someone else is hosting them).

Software-as-full-service will take longer to evolve than anyone thinks, but eventually will mature to the ultimate mix-and-match software architecture. This will take the form of off-the-shelf hosting of packaged applications and the customization of the packages also hosted by the primary software vendor or third party ASPs. In other words, we will stop being so worried about "cracking the box" of packaged applications if we are not responsible for supporting the new customized features or guaranteeing interoperability in an increasingly open world.

The major software vendors will have to decide when—and how—they are willing to cannibalize their own business models. Now they make tons of cash through enterprise licensing and generous maintenance fees but as more and more vendors offer alternative software acquisition models the big proprietary ones will have to completely change their fee structures to accommodate the move away from installed software. Even Microsoft is hosting software these days. Within just a few years, everyone will be hosting their own software and encouraging third-party providers to host and resell the very same applications: while Salesforce.com may have pioneered the ASP 2.0 movement, there will be numerous entrants into the hosting market over the next few years. In fact, it is probably safe to say that nearly everyone will eventually offer a hosted version of their software.

Open source software will penetrate the most inner sanctums of the enterprise because it will meld increasingly easily with proprietary software and the new SOA architectures. In fact, the gap between open and proprietary software will dramatically narrow over time. Does anyone really care that Apache or Linux are behind the curtain? Of course not. Open source software—primarily because of the rise of ASP 2.0 (and 3.0, etc.)—will become increasingly accepted by everyone. Those who host software will exploit the cost and "openness" of open source software and the buyers of hosted services will not care what is running behind the curtain so long as it runs well and securely.

No one will expect software to be "free," but, like hardware, it will commoditize. There will be all sorts of pricing models but the premise will shift from enterprise licensing to multiple flavors of on-demand, pay-by-the-drink models—the opposite of where we have been over the past few decades.

Real innovation will come from small entrepreneurs in small companies, just as it always has. In spite of billions and billions of research and development dollars spent every year to innovate, there continues to be a much great bang for the buck in smaller innovation companies than the largest ones. This conundrum seems to be well understood by the most acquisitive technology vendors (like Cisco) who are increasingly looking to small companies for true innovation.

When someone asks "where will software come from?" you can tell them that software will come from big vendors with creative partners who have finally figured out that their customers would rather pay-by-the drink at someone else's bar and grille.

SaaS, SOA, and ASP 2.0 are all trends with long, strong legs. The trends have legs because the software industry itself is unraveling in multiple directions. Not only is the way we deliver software changing, but the way we build it is also experiencing dramatic changes. "Assembly" is a better word—or at least will be in 5 years—than "development." The rise of packaged applications will be followed by the rise of off-the-shelf components standardized via Web services and service oriented architectures, and because of standardization the packaged applications will live comfortably with open source/near-open source/proprietary components designed to enhance the capabilities of old and new packaged applications.

The economics of the trends will also drive change. As suggested above, renting applications moves software costs from capital to operating expenses. This can dramatically improve cash flow as well as the overall investment posture of some companies. In short, renting changes the technology financial dynamics.

Finally, there is every reason to believe that technology is going the way of a utility—not in the sense that it no longer matters—which is obviously silly, but in the sense that large aspects of the acquisition, deployment, and support process will be routinized, even automated. While huge amounts will elevate to strategic status, a lot of software will be operational, increasingly cheaper, sometimes even free, and susceptible to a variety of charging models, not the least of which is pay-by-the-drink, variations of the so-called "on-demand" offerings that many companies are now providing their customers.

The call-to-action here is simple. If you have not already, start piloting with alternative software acquisition, deployment, assembly, and support models. (See a later section for more specific steps that can be taken to exploit the software development and delivery trends.)

Web 2.0

Publications like Business 2.0, Fast Company, and even Business Week are all writing about Web 2.0—and even Web 3.0—the new net and the next digital gold rush. Is this another bubble? Will Web 2.0 (and then Web 3.0) companies crash and burn like their dot.com parents?

Let us look at the rise of social networks, wikis, blogs, podcasts, RSS filters, folksonomies, mashups, crowdsourcing, and service-oriented architecture and the impact they would have on us all, especially the enterprise. Initially these technologies were destined to support social networking in all its glory. But important changes are occurring, changes that will affect the entire computing and communications spectrum.

Wikipedia defines Web 2.0 this way:

Web 2.0 represents an improved form of the World Wide Web and includes technologies such as weblogs, social bookmarking, wikis, podcasts, RSS feeds (and other forms of many-to-many publishing), social software, Web APIs, Web standards, and online Web services.

As used by its proponents, the phrase "Web 2.0" can also refer to one or more of the following:

The transition of websites from isolated information silos to sources of content and functionality, thus becoming computing platforms serving Web applications to end-users.

A social phenomenon embracing an approach to generating and distributing Web content itself, characterized by open communication, decentralization of authority, freedom to share and re-use, and "the market as a conversation. Enhanced organization and categorization of content, emphasizing deep-linking."

It further describes the innovations connected with Web 2.0.

Web-Based Applications and Desktops

The richer user-experience afforded by Ajax has prompted the development of Web sites that mimic personal computer applications, such as word processing, the spreadsheet, and slide-show presentations. WYSIWYG wiki sites replicate many features of PC authoring applications.

Rich Internet Applications

Rich-Internet application techniques such as Ajax, Adobe Flash, Flex and OpenLaszlo have evolved that can improve the user-experience in browser-based applications. These technologies allow a Web page to request an update for some part of its content, and to alter that part in the browser, without needing to refresh the whole page at the same time.

Server-Side Software

The functionality of Web 2.0 rich Internet applications builds on the existing Web server architecture, but puts much greater emphasis on back-end software.

Client-Side Software

The extra functionality provided by Web 2.0 depends on the ability of users to work with the data stored on servers. This can come about through forms in an HTML page, through a scripting language such as Javascript, or through Flash or Java. These methods all make use of the client computer to reduce the server workloads.

RSS

Protocols which permit syndication include RSS (Really Simple Syndication — also known as "Web syndication"), RDF (as in RSS 1.1), and Atom, all of them flavors of XML.

Web Protocols

Web communication protocols provide a key element of the Web 2.0 infrastructure. Major protocols include REST and SOAP.

Crowdsourcing

Crowdsourcing is a neologism for a business model that depends on work being done outside the traditional company walls: while outsourcing is typically performed by lower paid professionals, crowdsourcing relies on a combination of volunteers and low-paid amateurs who use their spare time to create content, solve problems, or even do corporate R&D.

Wikis

A wiki <WICK-ee> or <WEE-kee> is a Web site that allows the visitors themselves to easily add, remove, and otherwise edit and change available content, and typically without the need for registration. This ease of interaction and operation makes a wiki an effective tool for mass collaborative authoring.

Folksonomies

A folksonomy is an Internet-based information retrieval methodology consisting of collaboratively generated, open-ended labels that categorize content such as Web pages, online photographs, and Web links. A folksonomy is most notably contrasted from a taxonomy in that the authors of the labeling system are often the main users (and sometimes originators) of the content to which the labels are applied. The labels are commonly known as tags and the labeling process is called tagging."

Blogs

A blog is a user-generated Web site where entries are made in journal style and displayed in a reverse chronological order. Blogs often provide commentary or news on a particular subject, such as food, politics, or local news; some function as more personal online diaries."

Podcasts

A podcast is a digital media file, or a series of such files, that is distributed over the Internet using syndication feeds, for playback on portable media players and personal computers.

Here is only some of what it all means:

- Wikis could revolutionize the way companies document policies, processes, and procedures. HR policies, sales manuals, and supply chain management processes can be documented in living wikis that evolve over time from input from in-house and external professionals. Why do we need to hire a consultant to tell us how to sell to our customers when we have countless in-house subject matter experts? There are lots of questions like this that can be at least

partially answered in wikis—and let us not forget how wikis can be used for training.
- Blogs can be used to vet ideas, strategies, projects, and programs. They can—along with wikis—be used for knowledge management. (Do we really need monster applications for knowledge management?). They can also be used as living suggestion boxes and chat rooms designed to allow employees to vent and contribute in attributable and anonymous ways.
- Podcasts can be used for premeetings, in-meetings, and post-meetings documentation. Repositories of podcasts can contribute to institutional memory and together comprise a rich audit trail of corporate initiatives and decision-making.
- RSS filters can be used to fine tune information flows of all kinds to employees, customers, suppliers, and partners. These custom news feeds can leverage information in almost effortless ways. I love the fact that we tell them what to do and they just do it. (Does anyone remember PointCast?).
- Mashup technology makes it easier to develop applications that solve specific problems—if only temporarily. Put some end-users in a room full of APIs and watch what happens. Suddenly it is possible to combine incompatible pieces into coherent wholes with the complements of companies that understand the value of sharing (for profit, of course).
- Crowdsourcing can be used to extend the enterprise via the Web and leverage the expertise of lots of professionals on to corporate problems. If it is good enough for Procter and Gamble and Dupont it should be good enough for everyone. Got some tough R&D problems? Post them on the Web. The crowdsourcing model will change corporate problem-solving, once everyone gets over the fear of taking gifts from strangers.
- Service oriented architecture is the mother of Web 2.0 technologies. The whole notion of mix-and-match with help from strangers tethered to each other on the Web is a fundamental change in the way we think about software design, development and delivery. SOA is actually a decentralizing force that will enable companies to solve computational and display problems much faster than they ever did in the past. What will it be like when we can point to glue and functionality and have them assemble themselves into solutions right before our eyes?
- The good news is that there are legions of Web 2.0 devotees that will accelerate the changes occurring now. The bad news is that for every two steps we take forward we will take one backwards because change always has as many enemies as it has champions. Ignore Web 1.0 Luddites and focus squarely on Web 3.0 while you happily exploit Web 2.0 tools, technologies and perspectives. In less than a decade we will look back on these days as the beginning

of the next new thing, a time when collaboration redefined itself right in front of our screens.

How deep a trend is all this? Note the following Computerworld story about the adoption of Web 2.0 technologies. If the Defense Intelligence Agency is already knee deep in Web 2.0, you can bet that they are already well into Web 3.0. The trend has legs. It also dovetails with the trends we see in software development and delivery. Web 2.0 is part of the overall "decentralization" of enterprise software where the pieces of the software architecture are distributed within and beyond corporate boundaries. Of course this will challenge internal technology staffs and old style technology governance, but for those who embrace the changes there will be huge benefits. Do not underestimate the change in perspective that is necessary for the full exploitation of these technologies. Companies that can break free from obsolete acquisition, deployment, and governance processes and adopt more flexible ones will be in a position to create momentum from these trends.

Placed in the larger context of alternative communication and collaboration tools, Web 2.0 serves the needs of continuous transaction processing better than all of the other methods, tools, and techniques of Web 1. Web 3.0 will extend the current technologies and capabilities via embedded intelligence that will be generic and contextual. It will also deliver plug and play capabilities (remember that phrase?) from published application programming interfaces and full blown components—all built on Web services standards—that unsophisticated business technologists can assemble into functional applications, that will assemble and disassemble on an as-needed basis.

Top Secret: DIA embraces Web 2.0

Analysts are turning to wikis, blogs, RSS feeds and enterprise 'mashups'

February 23, 2007 (Computerworld) - The U.S. Department of Defense's lead intelligence agency is using wikis, blogs, RSS feeds and enterprise "mashups" to help its analysts collaborate better when sifting through data used to support military operations.

The Defense Intelligence Agency (DIA) is seeing "mushrooming" use of these various Web 2.0 technologies that are becoming critical to accomplishing missions that require intelligence sharing among analysts, said Lewis Shepherd, chief of DIA's Requirements and Research Group at the Pentagon.

"The collaboration potential of the social software side is really being thoroughly vetted and is now rapidly being adopted," Shepherd said. "Across agencies, wikis and blogs are becoming as ubiquitous as e-mail in terms of information sharing."

Master Data Management for Business Intelligence

Companies should begin to re-architect their data → information → decision-making strategies with real-time transaction processing in mind. This means that they should begin to consolidate their data, identify mainstream data base management, analysis and mining applications, and extend the collaborative business processes enabled by integrated data and real-time analysis.

Put another way, all of this is about satisfying requests for usable data. If we keep decision-makers happy with timely, accurate, and diagnostic data and information that supports decision-making, we will succeed. Along the way, it appears we will be able to leverage the efforts of several major software vendors to accelerate progress.

Master data management is the analytical infrastructure to which we paid too little attention for years. Business intelligence is the general approach we should take to data and information utilization. Let us look at the processes that will deliver these capabilities.

Wikipedia describes MDM:

Master data management (MDM), also known as reference data management, is a discipline in information technology (IT) that focuses on the management of reference or master data that is shared by several disparate IT systems and groups. MDM is required to warrant consistent computing between diverse system architectures and business functions.

Large companies often have IT systems that are used by diverse business functions (e.g., finance, sales, R&D, etc.) and span across multiple countries. These diverse systems usually need to share key data that is relevant to the parent company (e.g., products, customers, and suppliers). It is critical for the company to consistently use these shared data elements through various IT systems.

MDM also becomes important when two or more companies want to share data across corporate boundaries. In this case, MDM becomes an industry issue such as is the case with the Finance industry and the required STP (straight through processing) or T+1.

MDM is one of three computing types (OLTP transactional computing (typically ERP), DSS (decision support systems) and MDM). These types range from operational reporting to EIS (executive information systems). Master data management is not only required to coordinate different ERP systems, but also necessary to supply meta-data for aggregating and integrating transactional data.

We now find ourselves in the world of increasing applications standardization and data consolidation as more and more legacy applications are retired and replaced by one or two ERP applications. Master data management is emerging as the industry's data/information/knowledge discipline inside and outside of the firewall. *MDM is now strategic, not just operational.* If companies want to expand their supply chains, for example, they need to invest in shared master data.

Consistency is the ultimate objective of MDM. In some organizations, there are multiple sources of data that have to be integrated to enable analyses and decision-making. The data itself speaks to customers, suppliers, employees, and other corporate partners. It may be tied to specific applications or spread across existing data warehouses and data marts.

For years we have focused on data and applications integration. Many of the approaches we have used to achieve integration persist today, like enterprise application integration (EAI), extraction, transformation, and loading (ETL), and, more recently, Web services and service-oriented architecture (SOA) technologies that allow data to be wrapped in quasistandardized interfaces. MDM (and its supporting technologies) involve some form of data consolidation where data from multiple sources ends up in a single place; the propagation of data which facilitates data movement; and the federation of data that enables one view of all of the data and information that an organization needs to integrate and analyze.

Most companies use multiple techniques to achieve MDM. In other words, there are a variety of tools and techniques that companies use to identify, integrate, manage, and "open" their data stores for analysis and decision-making. The key is to settle on a set of methods, tools, and techniques that work for your company and then standardize on the overall MDM process. Some of these tools, techniques and methods—ETL, EAI and even content management (CM)—are sometimes grown in-house but most often are commercial tools, techniques, and methods that have been used for some time in an organization. Some of the more sophisticated approaches require synchronization with source data so that updates flow in both directions. Others are more passive, where data is extracted and then analyzed but there is no change in the source data stores.

As with all important business technology initiatives, MDM requires discipline that occurs within a larger governance framework. The governance around MDM should be defined, discussed, and documented. What are the objectives? What tools will be used? Who owns the data stores? How will integrated data be maintained? What is the approach to federation and propagation? The answers to these kinds of questions will define an MDM governance strategy. But in addition to methods, tools and techniques, MDM governance should embody the overall philosophy around data integration. Put another way, successful MDM requires commitment.

MDM methods, tools, techniques, and governance enable cost-effective business intelligence and customer profiling, among other activities. MDM is simultaneously a philosophy, a discipline, a toolset and a business objective. All four pieces need to exist to make MDM work for companies. Candidly, the four prerequisites are ranked in order of their degrees of difficulty and the first two—philosophy and discipline—are the toughest. Companies often talk a good MDM game but when push comes to shove they have a hard time defining and invoking the discipline necessary to make it all work. The tools and objectives are much easier to identify.

Once the MDM philosophy, discipline, toolsets, and business objectives are in place then all sorts of things are possible. There is definitely a hierarchy of activities, each enabling the other. It is also consistent with the general push to make technology more strategic than operational.

But the real payoff of MDM is business intelligence (BI)—and BI is the Holy Grail of the real-time, adaptive enterprise.

Wikipedia describes business intelligence:

Business intelligence (BI) refers to applications and technologies which are used to gather, provide access to, and analyze data and information about their company operations. Business intelligence systems can help companies have a more comprehensive knowledge of the factors affecting their business, such as metrics on sales, production, internal operations, and they can help companies to make better business decisions.

Business intelligence applications and technologies can help companies analyze changing trends in market share; changes in customer behavior and spending patterns; customers' preferences; company capabilities; and market conditions. Business intelligence can be used to help analysts and managers determine which adjustments are most likely to respond to changing trends.

In the competitive customer-service sector, companies need to have accurate, up-to-date information on customer preferences, so that the company can quickly adapt to their changing demands. Business Intelligence enables companies to gather information on the trends in the marketplace and come up with innovative products or services in anticipation of customers' changing demands. Business Intelligence applications can also help managers to be better informed about actions that a company's competitors are taking. As well, BI can help companies share selected strategic information with business partners. For example, some businesses use BI systems to share information with their suppliers (e.g., inventory levels, performance metrics, and other supply chain data).

BI, like MDM, is part technology, part governance, and part philosophy. It cannot be emphasized more how important philosophy is to the exploitation of MDM and BI. We should no longer have to learn that process investments should precede technology investments, but many companies still ignore this lesson. Investments need to be made in both process and technology for BI to be effective. There needs to be a commitment to making data accessible and therefore to investments that will ready the data and information infrastructure for analysis. Commitments to MDM and BI are often shallow: companies that want to upgrade their BI capabilities must invest in MDM, governance, infrastructure, and architecture. There are no shortcuts here.

Ultimately, BI fuels analysis and monitoring—the objective of business analytics. Analysis includes insight into internal and external operations and transactions. BI tools and techniques enable a variety of tactical and operational analyses including, among others:

- The analysis of large and small quantities of data
- The ability to slice and dice data like sales, customer service, production, and distribution
- The ability to analyze and present data across vertical industries

The internal and external distinction, and the distinction between operational and strategic applications, is important to BI planning and execution—as Figure 4 suggests.

BI should be an all-encompassing strategy. Internal/operational foci include research and development, network efficiency, security, applications management and overall operational performance of the infrastructure. Internal/strategic foci include business process management, business activity monitoring, process re-engineering, strategic performance management and even competitor intelligence. External/operational foci include supply chain planning and management, distribution, sales channel effectiveness and sales and production alliance performance. External/strategic foci include customer profiling, up-selling and cross-selling performance, customization and personalization. There are all sorts of methods, tools and techniques out there to help you achieve these BI objectives, but remember, again, that BI is as much a philosophy-to-process as it is a technology. Without a commitment to analytical processes, BI investments will not return very much (and without a commitment to MDM, the BI goals will not be achieved).

The ultimate impact of BI is dependent upon several maturing streams of innovation. So let us describe what the perfect picture looks like in, say, 2010 or 2015. By that time business will be largely automated with a variety of if-then triggers throughout operations. Information will not be inspected by humans as much as it is assessed by software. Business rules will drive most business processes—rules that are manually changed but much more frequently triggered by the same if-then rules that will together automate key processes and transactions. The more data, information and knowledge brought to bear on operations and transaction processing the closer we all get to real-time optimization, the ultimate objective of dynamic strategy and tactics. When companies know exactly what is happening with their customers, manufacturing processes, and suppliers in real time, when the same companies have elaborate rules engines in place, and when these companies automate the rules, then we will have dynamic real-time optimization, the ultimate expression of business intelligence.

BI matures when there is real-time insight into what is happening, embedded judgments (rules) about what is good and bad about what is happening, and the automated and quasi-automated ability to do something about what is happening. Ultimately, all of this is an optimization equation with descriptive data feeding explanatory data which, in turn, feeds prescriptive data. For BI to be truly effective the loop in Figure 4 must be closed—and continuous. "BI 2.0" is the trajectory of the end game, though the end game may well require BI 3.0 for its actualization. The industry is galvanizing around BI 2.0 as the ultimate BI vision. What this means is that BI has finally gotten some major league buzz, which is not to say that its capabilities have developed dramatically from where they were last year or the year before. MDM → BI enables business analytics at the highest possible level. If the right investments are made it is possible to optimize internal and external business processes.

Figure 2. Internal/external/operational/strategic business intelligence built on master data management

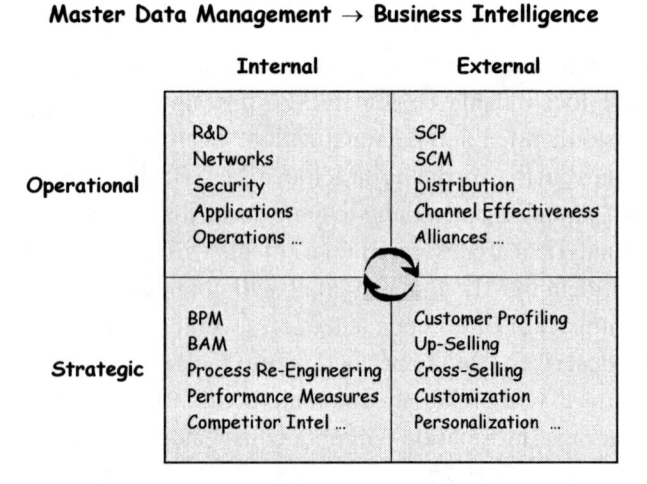

Master Data Management → Business Intelligence

	Internal	External
Operational	R&D Networks Security Applications Operations ...	SCP SCM Distribution Channel Effectiveness Alliances ...
Strategic	BPM BAM Process Re-Engineering Performance Measures Competitor Intel ...	Customer Profiling Up-Selling Cross-Selling Customization Personalization ...

Convergence CRM

I spent some time on hold with—and occasionally actually speaking with—"technical support" representatives from Dell. I listened to the on-hold voice tell me over and over again that I could just go to Dell.com for technical support, since the scripts that the human technical support team used to troubleshoot problems were the same scripts that the digital technical support team used. This advice struck me as peculiar: if I could really get the answers I needed from the Web then why was Dell spending so much money frustrating me with 800-number support? Was the voice implying that I was an idiot to actually want to speak with someone? I bounced from service rep to service rep, ending with a (live) Dell support professional telling me that she did not know how to solve my problem.

What I experienced was the worst of all worlds. After several hours of music, and after being rerouted five times, it occurred to me that maybe we have not come all that far, at least in the technology industry, with customer service. This assessment was punctuated by my being told that if the problem was a software problem, I would be charged a fee for the help, even though I paid for three years of support (learning during the service experience that my warranty only covered hardware), or I could call the software manufacturer myself to discuss my problem.

The experience with Dell was redefined after an experience with the Nordstrom retail chain. I find it hard to imagine a Nordstrom service representative telling me that I had to contact the manufacturer of the shirt whose sleeves were falling off because Nordstrom only supports the boxes in which the shirts are sold, or if I wanted the shirt repaired or replaced, I would have to pay an additional fee or take a trip abroad to solve my problem. If you have shopped at Nordstrom's you know that there is essentially nothing they cannot do for you: the customer is—literally—always right.

Do you pay more for this service? Of course. Truly excellent service is embedded in the purchase price, and for those who want to make the price/service trade-off, the rewards are clear. Dell of course is not the only vendor whose "support" is far from perfect. In fact, given that Dell (and other hardware and software vendors) worries more about being cheap than supported, when one buys a Dell one should expect to receive the same service as one would receive at K-Mart or Wal-Mart, since these chains are often the low-cost retail provider: it would cost Dell, and therefore us, way too much to provide Nordstrom-like support at Wal-Mart, or any retail chain that guarantees low prices every day.

So how should the industry deal with complexity, support ambiguities, ineffective customer service, and service loopholes? As a customer, I think it is simple: so long as desktop and laptop operating systems and applications software is so complex, the product of so many different vendors, and subject to so many failures and conflicts, whoever sells the (hardware + software) *system* should be responsible for supporting what they sell. They should not be able to point a finger somewhere else, charge for fixing problems that are bundled in their own branded boxes, or cut customers loose to solve problems on their own. Is there another retail industry that treats its customers this way? Only the low-cost ones. So what do we want? Cheap prices or great service?

All of this explains why desktop/laptop support is one of the fastest growing outsourcing targets and why technology consultants for personal residences are popping up all over the place. We all need more help with less headaches. If I manufactured the hardware and software, I would try to interrupt these trends; I would try to own my customers. How about a little CRM? Maybe Nordstrom's has the bar too high. How about JC Penny?

CRM can make sense as a business strategy for some companies, but not all companies. Some companies, for example, brand themselves with low prices and not service; others brand themselves around service. Figure 3 tries to suggest how this might work.

But the real trend here, the trend we should define and exploit, extends well beyond what we think of today as "CRM." In fact, CRM as we know it today is already obsolete. The trend is toward converged and continuous CRM where all kinds of data are used to profile, contact, and manage customers throughout their "life cycles."

Brace yourself for an amazing trend. Some think of it as "personalization," some as "customization" and some just as 24/7 selling and servicing. As data bases become more integrated, as shopping becomes more digital, and as "always on" access devices become more pervasive, we can expect to be treated to all sorts of offers. It is another definition of "profiling."

I already get e-mail (and snail mail) from companies that have profiled me. They have analyzed data about where I live, what I earn, and what I buy to determine what I like and what I would pay for what they are selling. This is first generation mass customization, child's play compared to what is coming down the information superhighway. Built on much the same data that mass marketing assumes, mass customization infers beyond the simpler correlations—like age, wealth, time of year—to specific ideas about what you and me would really like to buy, based on inferences about us as part of a larger group **and** as individual consumers. Contact can be "personalized" with customers, suppliers, partners, and employees through all varieties of messages including sales, marketing, service, and distribution. Over time, given how low digital transaction costs are compared to other ways companies touch the members of their value and supply chains, and how ubiquitous digital communication is becoming, companies will reassess their advertising and marketing budgets. They will go increasingly personal.

There is a great scene in the Tom Cruise/Steven Spielberg film, "Minority Report." Cruise is walking in a city in 2054 and as his eyes are (iris) scanned he is immediately pitched a whole slate of personalized products and services. Imagine waiting for a plane or train and receiving countless messages about stuff you could buy that you already like and use all of the time, but is now on sale 12 feet away? What if all employees, customers, suppliers, and partners could be pitched in ways that matched their interests, values and personalities?

How will all this happen? It is all about the depth, location, and quality of customer data. It is also about the analyses performed on this data. Some companies have excellent behavioral scientists who run the data every which way in search of correlations that explain what customers value and why they buy what they do. The power here is amazing. It is possible, for example, to determine:

- When to digitally (via browsers, cell phones, wireless PDAs, pagers) interrupt customers with a deal and when not to interrupt them
- What size discounts need to be, by person, by season of the year, by customer location and time of day
- What combinations of products can be sold; what products do not mix with others
- What short-term and longer-term life events influence which purchases
- What forms and content of customer service each customer prefers

This is just a sampling of the kind of personalized and customized inferences that can be made. The trend is clear: customization and personalization will accelerate as the ability to integrate and correlate preference data, and instantly communicate

Figure 3. CRM scenarios

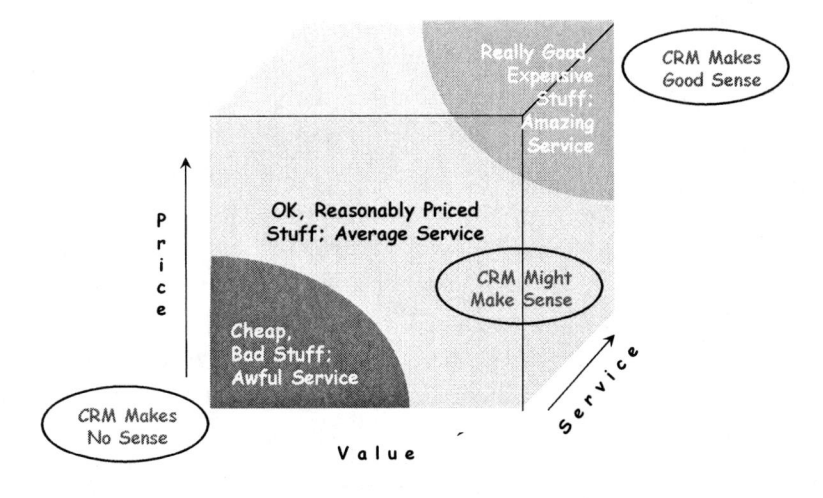

with customers, increases. As inferential data and total access collide with GPS location finders, customers, employees, suppliers, and partners will never be safe again. Full convergence will have occurred.

Is this a good thing or a bad thing? You wake up in the morning in a bad mood. You tell your personal (digital software) agent that you will accept no offers from digital hucksters unless they double the discounts they usually offer—and if that is unacceptable you are going off the air. Or, you are in the middle of an important discussion and your cell phone beeps with an interesting offer—but only if you respond in the next 10 minutes. Or, you are traveling on business in a strange city and it is close to dinner time. Your PDA buzzes with an idea: there is an Italian restaurant—it of course knows you like Italian food (what time it is and precisely where you are)—about a block from where you are standing with a table waiting for you and your colleagues. No reservation necessary; no phone call to confirm; just a quick digital reply on the device—which immediately displays the menu, the specials, and the wine list. As you walk toward the restaurant you and your associates discuss what goes best with what.

Full convergence CRM is upon us and in spite of the "inconveniences" of continuous haranguing (if it is allowed) there is enormous opportunity in the trend for profitable growth. The key is profiling and by this time you have noticed that profiling is dependent upon MDM and BI (and software that is flexible and accessible). The trend has legs because it is the natural result of convergence. The intersection of customer data, buying and selling history, and inventories, among other things, permits statistical analyses of all kinds and, most importantly, monetizable correlations which, in turn, permit sales analyses, Web analytics, customer personalization, product customization, dynamic pricing, inventory management, demand forecasting, and just about anything else you can think of that is based upon empirical data.

Access Devices

Larry Ellison appeared on the Oprah Winfrey show over 10 years ago to discuss "network computers." This was when a lot of ahead-of-their-time ideas were out there, like Apple's Newton and IBM's voice recognition applications. Larry was wrong then because our networks were not reliable, secure, or ubiquitous enough to support "thin client" architectures. Never mind that the devices themselves were also a little weird and way too proprietary. But if Larry reprised his appearance on Oprah tomorrow, he would be dead right.

Let us look at several trends that point to why "thinfrastructure" makes sense. First, we have segmented our users into classes based on their frequency and depth of use: some people use very expensive and hard to support laptops pretty much for e-mail. We have begun to deploy a bunch of devices—like PDAs, cell phones and pagers, and MP3 players—that have finally converged. Many of these devices are pretty dumb, but that is OK because we do not ask much of them. Multiple device synchronization is still, however, difficult and expensive. Desktops and laptops are way, way overpowered for the vast majority of users (as is the software that powers them). Enterprise software licensing is getting nasty (if you do not renew your Microsoft software license every three years, you get punished).

Desktop/laptop support is still complicated and expensive; large enterprises—even if they have network and systems management applications—still struggle with updates, software distribution, and version controls.

Network access is now almost ubiquitous today: we use desktops, laptops, personal digital assistants (PDAs), thin clients, and a host of multifunctional converged devices to access local area networks, wide area networks, virtual private networks (VPNs), the Internet, hosted applications on these networks as well as applications that run locally on these devices. The networks work.

Many (crazy) companies provide multiple access devices to their employees, and employees often ask their companies to make personal devices (like BlackBerrys and iPhones) compatible with their networks. All of this is made even more complicated with the introduction of wireless networks which make employees more independent and mobile. The cost to acquire, install, and support all of these devices is, however, high.

Small, cheap, reliable devices that rely on always-on networks make sense. Shifting computing power from desktops and laptops to professionally managed servers makes sense. Moving storage from local drives to remote storage area networks makes sense. Fat clients should lose some weight—as we bulk up our already able servers. The total-cost-of-ownership (TCO)—not to mention the return-on-investment (ROI)—of skinny client/fat server architectures is compelling (to put it mildly).

Thin is still beautiful. It was beautiful when it was a concept and it is even prettier now that it is a reality. Question: "are you happy with the fat client architecture that runs your life today, with the ever-more-powerful PCs that require more and more support?"

Is this about control? Is it about reliability? How about security? Yes, all of this. Since we all agree that timing is everything, given all of the new devices appearing and how many of them are converging into wholes greater than the sum of their parts, is not this a great time to think thin? If you do not, then the only people you will make real happy are the vendors you have to hire to keep all the toys humming. If you get control of the servers, you control everything that plays on every access device; and if you skinny down the access devices, you get control, flexibility, standardization, reliability, security and scalability for your users. No more software conflicts, instant software upgrades, quick and easy new application deployment.

Wikipedia offers the following regarding the advantages of thin client computing:

Lower IT administration costs. *Thin clients are managed almost entirely at the server. The hardware has fewer points of failure and the local environment is highly restricted, providing protection from malware.*

Easier to secure. *Thin clients can be designed so that no application data ever resides on the client, centralizing malware protection.*

Lower hardware costs. *Thin client hardware is cheaper because it does not contain a disk, application memory, or a powerful processor. They also generally have a longer period before requiring an upgrade or becoming obsolete. The total hardware requirements for a thin client system (including both servers and clients) are usually much lower compared to a system with fat clients. One reason for this is that the hardware is better utilized. A CPU in a fat workstation is idle most of the time. With thin clients, memory can be shared. If several users are running the same application, it only needs to be loaded into RAM once with a central server. With fat clients, each workstation must have its own copy of the program in memory.*

Lower energy consumption. *Dedicated thin client hardware has much lower energy consumption than thick client PCs. This not only reduces energy costs but may mean that in some cases air-conditioning systems are not required or need not be upgraded which can be a significant cost saving and contribute to achieving energy saving targets.*

Worthless to most thieves. *Thin client hardware, whether dedicated or simply older hardware that has been repurposed via cascading, is useless outside a client-server environment. Burglars interested in computer equipment have a much harder time fencing thin client hardware.*

Hostile environments. *Most devices have no moving parts so can be used in dusty environments without the worry of PC fans clogging up and overheating and burning out the PC.*

Less network bandwidth. *Since terminal servers typically reside on the same high-speed network backbone as file servers, most network traffic is confined to the server room. In a thin client environment only mouse movements, keystrokes and screen* updates are transmitted from/to the end user. Over efficient protocols such as ICA or NX this can consume as little as 5Kbps bandwidth.

More efficient use of resources. *A typical fat-client will be specified to cope with the maximum load the user needs, which can be inefficient at times when it is not utilized. In contrast, thin clients only use the exact amount of resources required by the current task.*

Simple hardware upgrade path. *If the peak resource usage is above a pre-defined limit, it is a relatively simple process to add another rack to a blade server (be it power, processing, storage), boosting resources to exactly the amount required. The existing units can be continued in service alongside the new, whereas a fat client model requires an entire desktop unit be replaced, resulting in down-time for the user, and the problem of disposing of the old unit.*

Thin clients are here today and will pervade the market over the next few years. The business case for thin clients is one of the easiest to develop that we have ever seen. The thin client trend will simultaneously save you money while improving productivity. Adoption will be steady for several years and then extremely aggressive: by 2015, at least 50% of all access devices will be thin. Too aggressive? Think again. Companies will exploit the advantages of thin client computing in a variety of flavors ranging all the way from integrated phones/Web browsers to thin clients with full keyboards optimized for access to corporate networks and the Web. Devices are now appearing that will do everything—in a form factor that will work for many users. The iPhone's touch technology will launch a whole new class of access technologies and devices. Some will be touch screen-based, some will be hard keyboard-based—eventually—voice-activated. The biggest driver of this trend will be assessments about what employees actually need to do their jobs. Many employees will be given thin clients only, while a smaller number will be issued fatter clients. But the days when everyone gets a standard issue fat client loaded with bloatware are over. There is no reason for pay for unnecessary computing power, or support.

THE COMBINED EFFECT

The five trends discussed here are absolutely intertwined. MDM/BI will drive converged CRM with Web 2.0 technologies as accelerators, with all of this occurring within the changes occurring in the software development and delivery process and the thinner and thinner devices that will increasingly be used to access applications and data. The essence of the trends is their outward-facing nature. They shift business technology activity from within enterprises to beyond their firewalls. The real impact of the trends will be their contribution to the inward/ outward business technology shift (which began about 5 years ago), a shift that will transform the business technology relationship.

IMPLICATIONS

It is actually possible to completely re-architect around these trends. Imagine building a technology architecture and infrastructure from the ground up right now. Would you invest in large data centers that you would own and staff? Would you deploy hundreds or thousands of powerful laptop computers? Would you also give all of your employees PDAs? Would you continue to try to integrate lots of different data bases? Would you deploy large software applications or rent them?

Here is a list of things to consider given the trends discussed here.

Software Development and Delivery

- Evaluate your entire applications portfolio; use the process to decommission as many applications as you can.
- Inspect the licensing agreements on all of the applications, noting in particular clauses that require updates at specific intervals; search for off-ramps from the agreements; challenge punitive agreements.
- Identify the software that can be hosted by the vendor or a third-party: this is your target list for the new software development/delivery trend.
- Select several that make sense (given your licensing agreements) and pilot a hosting program; develop baseline metrics (against the traditional in-house deployment model) for assessing the effectiveness of the pilot, such as up-time, security, reliability, and so forth.
- Expand the pilot to include more hosted applications.
- Identify the application "extensions" (for your major enterprise applications) that will become "standard" for your company; pilot hosted versions of them as well; develop an interoperability quotient for all applications: applications that do not interoperate well should be discarded.

Web 2.0

- Understand the range of Web 2.0 collaborative tools and techniques that might enhance business processes; identify how they can be used to enhance performance.

Figure 4. The five intersecting trends

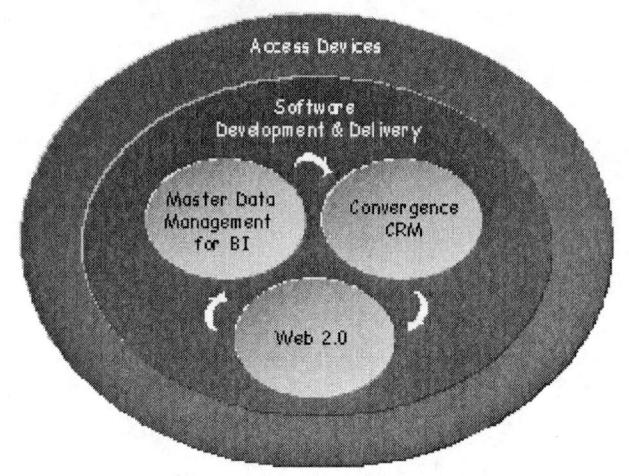

- Pilot wikis to build quick "course encyclopedias" and other forms of documentation.
- Pilot blogs to post and vet project assignments and othwerwise support collaborative discussions.
- Pilot podcasts to document content.
- Pilot folksonomies to organize content.
- Pilot RSS filters to create content streams.
- Pilot mashups to create tools, displays, and so forth.
- Pilot crowdsourcing to solve specific problems.

Master Data Management for Business Intelligence

- Assess your overall data/information/knowledge architecture for consistency, accuracy, scalability, and security.
- Develop and implement an MDM plan consisting of objectives, governance, and tools.
- Develop and implement an internal/external BI program consisting of objectives, governance, and tools.
- Integrate corporate strategy and tactics into the MDM/BI investment.
- Extend the analyses to monitor and optimize internal and external processes.

Convergence CRM

- Invest (invest, invest) in statistical modeling to develop a deeper understanding of the correlations among customers, sales, processes, products, and services; add additional variables to the models; identify customer life cycles that can be monetized.
- Integrate the models into formal sales and marketing campaigns.

Access Devices

- Segment users into classes within a computing hierarchy.
- Identify thin client options for consideration.
- Pilot a variety of thin client models ranging all the way from converged devices like iPhones to full function thin clients with traditional keyboards.
- Assess the effectiveness of alternative thin client models; expand the successful pilots.
- Decommission obsolete fat clients.

Together these trends present a snapshot of where business technology is going in 2007 on its way to 2010. These trends will drive lots of changes and trigger additional trends and changes that will continue the march toward integration, interoperability, collaboration, customization, and decentralization.

ENDNOTE

[1] Additional trends analyses can be found in Appendices A, B, and C at the end of the book.

Section II

Due Diligence Case Studies

Chapter IV
Venture Investing in Wireless Communications Technology:
The ThinAirApps Case

INTRODUCTION TO THE CASE[1]

This is a case study in venture investing. ThinAirApps (TAA) created, developed, and sold wireless middleware software products and services. Founded in July 1999, the company received its first and only round of venture financing in June 2000, raising $13 million in equity capital from Safeguard Scientifics, Inc., and a group of angel investors. The company's software products enabled enterprises of all types and sizes to rapidly and cost-effectively "wirelessly" enable their mobile workforces; giving professionals access to not only common data such as e-mail, calendars, and contacts (groupware), but also data from more complex desktop or server applications, databases, or other systems. All of this data resides "behind-the-firewall." Adding to TAA's uniqueness was its ability to access this data securely.

Let us remember that in 1999 voice-over-wireless was prevalent in the United States. Corporate America had largely embraced the administration of cell phones to workforces, and just one year before, AT&T Wireless had cracked the code with its One-Rate national business calling plan. However, the transmission of data (other than voice) over a terrestrial network to handheld devices such as cell phones and PDAs was not a broad commercial reality. In 1999, according to the Strategis Group, there were less than three million smartphone and wireless PDA units sold (by 2004 the number had exceeded 60 million). In addition to a host of players including Research in Motion, Palm introduced its seventh generation PDA device, the Palm VII. This first fully-contained wireless device for Palm exploited underused terrestrial data networks to provide users primitive (by comparison to today's technologies) access to a variety of wireless data, including Palm-sponsored e-mail and selected Internet destinations. The uniqueness of Palm and its open development platform, which gave rise to an industry that developed everything from games to productivity enhancing applications for use on the devices, created a watershed in the wireless data industry, and was largely the basis for the TAA business model.

TAA became an early leader in the rapidly developing yet immature wireless data industry. At the time of our investment, TAA's product, ThinAirMail, was the single most frequently downloaded add-on application for the Palm VII device—nearly 30,000 individuals were using it to access their e-mail wirelessly. While the company was just entering its revenue stage, it had already established significant market awareness by enabling tens of thousands of individual users to access their personal e-mail while on the move. The company's product was based on open standards—in other words it ran well regardless of which device operating system (Palm, BlackBerry, WAP-enabled cell phone, Java 2 Micro Edition, etc.), network (TDMA, CDPD, CDMA, etc.), or application (e-mail, or other enterprise application for which TAA wireless API's had been created) was being used. The company's revenue model was based primarily on enterprise sales of its flagship product, the ThinAir Server, which was a middleware server that connected to the corporate e-mail server (more often than not Microsoft Exchange) and allowed customization for connecting to other applications. Thus, the low-hanging fruit of instantaneous wireless data access could be realized—groupware including e-mail, calendar, and contact information was at a professional's fingertips while otherwise disconnected from the network. The TAA Software Development Toolkit (SDK) allowed organizations to build custom connections from their other enterprise applications directly into the ThinAir Server. The opportunities were significant, inventory pricing on a real time basis, instant ordering, real time accounting system updating, and all while away from the office and disconnected from the network.

THE ThinAirApps OPPORTUNITY

So what did the opportunity look like? The due diligence team was provided the following information about the company, the "space' and the financials.

The Company

ThinAirApps was founded in October of 1999 with the goal of developing the leading software platform for connecting enterprise data systems to any wireless device. The ThinAir brand has been established via ThinAir Mail™ (TAM) the first real-time e-mail solution for the Palm VII and most downloaded application from the Palm.Net Web site. That brand has been leveraged and extended with ThinAir Server™ (TAS), the leading wireless platform with over 1,000 installations in only 6 months. TAS was the first solution flexible enough to operate behind the firewall or hosted at a third party data center. TAS comes with provider software connecting common groupware/e-mail platforms like Microsoft Exchange, Lotus Domino/Notes, HTML/Web, and POP/IMAP to all leading wireless devices, including the Palm Pilot, WAP-enabled mobile phones, RIM Pagers, Microsoft PocketPCs, Handspring Visors, and every other standards-based device capable of connecting to the Internet. This real-time access is being extended beyond groupware to connect data from SQL databases, XML data sources, CRM, and other enterprise sources. With this robust offering TAS is the clear technology choice for enterprise customers to wirelessly enable their workforce, carrier-grade services targeting enterprise customers, and individual mobile professionals. With backing from Safeguard Scientifics, TAA is poised to quickly grow its market share in the huge wireless software industry.

The Market

Wireless device penetration is growing at a 66% CAGR domestically and 80% worldwide. There will be over one billion wireless devices worldwide by 2005. Increasingly, businesses are interested in enabling their workforces by extending the reach of their internal systems onto wireless devices. It is predicted that there will be 9 million enterprise wireless data users by 2003. Companies are expected to increase the number of enterprise connected mobile workers by 30%, spending approximately $10,000 per user.[2]

The Strategy

ThinAirApps is taking advantage of the explosion of wireless devices and their use as data access resources. Corporations are quickly realizing the efficiency gains and cost savings of wirelessly enabling their workforce which extends their existing systems outside the office. With the convergence of mobile phones and personal digital assistants (PDAs), the capabilities of devices are expanding greatly. In addition, with the focus by established players such as Nokia, Ericsson, Motorola, and Palm and the entry of Microsoft on wireless technology, the enterprise market is expanding quickly. However, the devices and carrier networks do not offer the complete solution. ThinAir Server is the missing component allowing for data center and behind the firewall access to corporate data on any device over any carrier network.

For a corporation's IT staff, ThinAir provides a single, centralized solution to manage all their devices and connect all their data sources in a secure, controlled manner. The platform is open with developer tools which allows for additional high-value applications to be added and customized for the enterprise. In this way, ThinAir Server is a strategic enterprise solution which grows with and adjusts to the enterprise's expanding wireless needs.

TAA is establishing ThinAir Server as the de facto standard for allowing all devices to access enterprise data. With over 1,000 installations of TAS, TAA has gained the market share and mind share of corporations as the leader in wireless software. Of the installed base, enterprises are in various stages of evaluation, pilot and approximately 100 are full paying customers. They range from small companies to large global 2000 firms across various industries. Customers include Bristol Myers Squibb, Johnson & Johnson, Morgan Stanley Dean Witter, and Andersen Consulting to name a few. And, through partnerships with VARs, ASPs, OEM/ISVs, and Integrators, TAA is quickly cascading its sales reach into the vast majority of corporations. Approximately 99% of enterprise leads are from inbound calls to ThinAir.

The international market is also ripe for TAS. With the opening of a sales and support office in Europe in Q1 2000 and reseller and integrator partnerships in Asia and Latin America, TAA is positioning itself to replicate the success of TAS in North America globally.

ThinAirApps is committed to being agnostic and open in the burgeoning and opaque wireless market regarding devices, protocols, and networks as technologies compete to establish themselves. In mitigating the risk and eliminating the uncertainty of depending on or combining with a failing technology, TAS offers a technology platform that is strategic and leverageable in the future for its enterprise customers. This way, an enterprise does not have to make a "bet" on any particular device or standard nor implement individual technologies for each data source. TAS will continue to support all new standards as they emerge, effectively removing the burden from the IT administrator.

ThinAirApps SOLUTIONS

ThinAirApps positions its products into three distinct but related lines of business targeting the enterprise customer.

ThinAir Server for the Enterprise

The current ThinAir Server offering is targeted at the enterprise customer as solution which can sit inside a company's firewall or at smaller hosted ASP data centers. The solution offers groupware access out of the box but will be expanded to offer applications for enterprise systems like CRM, ERP, time tracking, and others. The solution also has the ThinAir Software Development Kit and will be expanded with additional developer tools for internal IT developers and Integrator partners to customize ThinAir Server for other systems like custom SQL databases.

Carrier-Grade ThinAir Server

A highly scalable, robust, and reliable version of ThinAir Server targeted at carriers, wireless ASPs, and other high volume wireless service providers. The solution is focused on handling over 1 million users per installation in open, rapidly expanding messaging service environments. Carrier-Grade ThinAir Server will be available in Q3 2001 and rolled out to at least four services currently being signed to contracts.

ThinAir for the Mobile Professional

For the individual mobile user, ThinAir for the Mobile Professional is targeted at the individual whose enterprise has not deployed a full solution, but is a mobile professional and needs to leverage wireless services on a device they have purchased themselves. This solution includes a desktop redirector which allows for the forwarding of messaging services from desktop messaging software, a public server accessible on the Internet, and integration with other individual productivity tools. ThinAir for the Mobile Professional will be available in Q2 2001.

With this three-pronged approach, ThinAir offers a complete solution for all levels of enterprise data access from the individual mobile professional to the carrier. With this expansion of ThinAir Server for the enterprise, a ThinAir solution is available for all potential customers.

PRODUCT LINE

ThinAir Server™ is an open, secure, robust, data center scaleable enterprise platform. It can operate inside or outside of a corporate firewall, at an ASP's outsourced data center, and bundled as a piece of a larger solution. In addition, the ThinAir SDK™ allows enterprises and Integrators to extend connectivity to custom corporate systems and to integrate additional device types.

ThinAir Groupware Access™ is the first class of applications supported by ThinAir. Groupware Access gives instant, real-time access to common enterprise groupware/messaging technologies like Microsoft Exchange, Lotus Domino, POP, and IMAP servers. The product is bundled with the ThinAir Server.

ThinAirMail™ is a free e-mail client for Palm Pilot devices which uses the ThinAir Server to connect to e-mail and groupware systems. Since its release in October 1999, TAM has been the most downloaded application on Palm.Net and currently has over 30,000 active users.

ThinAirApps has a conservative but progressive attitude toward new technology. In continuing to embrace but not depend upon one particular technical standard or device, TAA is currently exploring emerging technologies to incorporate into its products. These include ongoing investigations into Bluetooth, 3G devices and networks, PDA/Mobile convergence, and data source integration.

PRICING

ThinAir Server is sold at a list price of $100 per user for enterprise licenses with volume discounts starting at 1,000 users. Discount schedules are offered for resellers and partners. Support is offered at a premium level for 50% of the license price and standard level at 20% the license price per year. Pricing for hosted solutions by ASPs is $3-4 per user per month which includes support. Upgrade fees are 50% of list price for major releases for customers on standard support (free for customers on premium support) and free to all customers on support for minor "dot" upgrades.

The Integrator Partnership Program, which allows Integrators to build custom applications with the SDK, is priced at $15,000 per year for 10 developer seats with additional seats costing $1,500. In addition, an associated Strategic Consulting Group to assist Integrators on their projects will bill their services per hour. ThinAir Mail is offered for free.

The Strategic Services group provides customized solutions and integration services for TAS customers. With this support, TAA is able to expand its range of solutions and create another high-margin revenue stream.

SALES

ThinAirApps is committed to direct and partner sales to rapidly gain market share for TAS. TAA already has in place a Director of West Coast Sales and a Director of East Coast Sales. In addition, TAA's four person business development department concentrates on ASP, Integrator, and OEM/ISV bundling partnerships. TAA currently is actively recruiting for International Sales positions as well as an experienced software channel salesperson to lead the department.

CUSTOMERS

ThinAir's target customers are domestic and international enterprises with large mobile workforces that require instant access to critical information. These are the companies who are driving the business adoption of wireless technology. Large companies with large investments in messaging and other communication and process technologies can leverage the ThinAir platform to extend them into the wireless realm.

TAA reaches its customers via five primary channels:

- Direct sales to larger enterprise customers
- Value-added resellers ("VARs") of TAS
- Enterprise ASPs: For enterprise customers looking for outsourced solutions
- Integrators: building wireless practices around TAS for customized enterprise solutions
- OEM/ISV bundling partners: To integrate as a larger offering wirelessly enabling other technologies

With this multitiered approach, TAA is gaining wide reach quickly and can leverage the sales forces and established relationships of partner companies. In addition, vertical partners in each space allow for deeper reach into early adopter industries without investment by TAA. ThinAir's goal is to become the de facto standard for real-time wireless connectivity to any enterprise data source.

ThinAirApps is also initiating an Integrator Partnership Program to stimulate development of custom enterprise solutions using TAS. This provides TAA with an alternative revenue stream from associated membership and consultation fees. In addition, it broadens TAS reach by leveraging the relationships and reputations of established Integrators.

MARKETING

ThinAirApps promotes its brand and products via eight primary platforms:

- Provide free offerings for individuals to access their e-mail
- Offer an evaluation version of ThinAir server
- Exhibit at prominent wireless and Internet trade shows
- Produce and maintain an effective Web site
- Advertise to business and technology audience
- Speaking at industry events
- Outreach to media and analysts
- Hosting technology seminars

TAA is committed to establishing its brand and products to enterprise customers through targeted but cost-effective means. Marketing costs are expected to scale proportionately with revenue projections and outsourced and partner opportunities will be leveraged when appropriate.

HISTORY AND ACCOMPLISHMENTS

ThinAirApps was founded in October of 1999 as a spinout of Creative Technologies New York (CTNY) an e-business integrator. ThinAir Mail and a precursor to the ThinAir Server were developed during that time released in October 1999 and February 20'00, respectively. ThinAir Server was the first behind the firewall solution for real-time wireless data access. ThinAir Mail is currently the most downloaded client from Palm.Net and has 30,000 active users.

MANAGEMENT TEAM

ThinAirApps is aggressively but intelligently expanding its workforce. Currently, about 50 full time employees span development, support, sales/business development, marketing, and management groups. TAA has a flat organization with experienced software and startup executives with proven track records of success. The TAA development team is composed of 15+ engineers with many years of combined wireless and enterprise software development experience.

Jonathan R. Oakes, Chief Executive Officer

Jon Oakes is CEO and cofounder of ThinAirApps. He is responsible for overseeing all the major functions of the company including strategic direction, marketing, sales, finance, and implementing day-to-day operations. Mr. Oakes is a frequent speaker at conferences and trade shows evangelizing the company's views on the current and future states of the wireless industry.

Mr. Oakes established a track record as a dynamic, entrepreneurial leader as a founder and partner of Creative Technologies New York (CTNY). In 5 years, he built the company from a small systems integration firm into a leading-edge e-business design and strategic consulting organization of over 45 professionals. At CTNY, Mr. Oakes spearheaded sales and marketing efforts, developed corporate strategy, and oversaw general operations. While with CTNY, the company was chosen as one of Deloitte & Touche's Fast 50, a ranking of the fastest growing technology firms in New York City. Mr. Oakes earned a Bachelor of Arts degree from Skidmore College.

Nathanial X. Freitas, Chief Technology Officer

Nathanial X. Freitas is the Chief Technology Officer and Cofounder of ThinAirApps, where he is responsible for overseeing the development of the technical architecture, as well as defining the technology roadmap to guide future strategic decisions and product development. He also works with existing and potential partners to strategize about and integrate the ThinAir Server platform into their wireless solutions. Mr. Freitas also often speaks at various conferences, tradeshows, and workshops, and participates in organizations such as the WAP Forum. His experience spans software development in academic, commercial, and governmental contexts.

Mr. Freitas received a Bachelor of Arts, Creative Studies, Cum Laude from the University of California, at Santa Barbara. His studies included classical and contemporary art music (composition and performance) and computer science. Mr. Freitas also worked and studied at the Center for Research in Electronic Art Technology (CREATE), a renowned lab focused on research and development of a new generation for software and hardware tools to aid in media-based composition.

After various independent consulting engagements, and being involved in two Los Angeles Internet startups, Mr. Freitas continued his research and studies at Santa Barbara from 1996 until 1998. During that time, he contributed to a number of important initiatives including the Alexandria Digital Library, a multimillion dollar project funded by the Department of Defense, NASA, and the National Science Foundation, where Mr. Freitas eventually found himself attaining security clearance to present his work to high ranking military officials. He also was CTO of Bodies, Inc., an online conceptual art "corporation," that won many accolades and whose work has been displayed around the world.

Seeking to branch out again from his experiences in academia, Mr. Freitas moved to New York, and began working as lead developer at Creative Technologies New York (CTNY). At CTNY is where Mr. Freitas lead the work on the dynamic enterprise application framework for thin devices (DEAFT), an early prototype of the ThinAir Server. That work lead to the development of ThinAirMail, the initial product of ThinAirApps.

Dr. Robert Yerex, Executive Vice President

Dr. Robert Yerex has 19 years of experience in the high-tech and software industry. Dr. Yerex earned a Bachelor of Science and Master of Science from the University of Minnesota and a PhD and MBA from Cornell University. Dr. Yerex pursued a range of academic studies, including statistics, numerical and quantitative analysis, economics, entrepreneurship, general management, finance, and computer science.

Dr. Yerex has been a consultant with American Management Systems and Rational Software, in addition to his own private practice. In 1993, he founded Objectshare Systems, Inc., and held the position of CEO until the merger of the firm with ParcPlace-Digitalk, Inc. In 1995, Objectshare was the eleventh fastest growing private technology company according to the San Jose Business Journal. From Objectshare, Dr. Yerex went on to be a Vice President at Aztec Software and a Vice President at Unity Software before coming to ThinAirApps.

RISKS

Technical: Wireless technology is in its infancy and while much is projected for bandwidth expansion and data usage, it is unknown what direction the technology will go.

Support/QA: There is a risk that testing new products and supporting existing ones will be a drag on the organization. Wireless software is more complex to support and QA then most other enterprise technologies because of the myriad of devices and networks it has to integrate with.

Competitive: While TAA is the leader in wireless data access for the enterprise, larger players could enter the market. Microsoft recently announced their intention to wirelessly enable their back office products but that release is at least 18 months away.

Market: While device penetration figures continue to grow globally and projections for the future are large the enterprise wireless software market may only end up being a niche market.

Financial: Scaling a horizontal enterprise software business is expensive especially in New York City.

FINANCIALS

ThinAir has a diversified and quality group of revenue streams. Direct and indirect licensing revenue provide the core of TAA sales and bundling efforts. In addition, partner Integrators provide licensing from custom projects they develop and also pay fees to be part of the TAA Integrator Partner Program. ASPs provide recurring monthly fees which scale with the number of users they sign on their service. Support contracts are annual and mandatory for all customers. In addition, TAA is building out its Strategic Services Group who bill on an hourly or project basis and support underlying sales of the products.

ThinAirApps raised an initial round from private investors totaling $1.2 million.

FORECAST

ThinAirApps projects 2001 revenues of approximately $8 million and $34 million for 2002. Monthly operational profitability is projected in late 2001/early 2002.

ADDITIONAL FUNDS

ThinAirApps is looking to raise a private equity round. TAA has already had a great deal of investment interest from strategic partners. A premium is put on strategic partners which will not only bring funds, but domain knowledge and relationships in the enterprise software industry. The funds will be used for expansion of staff, mostly in sales, and to invest in additional product development to broaden market reach.

INVESTMENT HIGHLIGHTS

Huge market. Wireless device sales are exploding: Wireless device penetration is growing at a 66% CAGR domestically and 80% worldwide. There will be over 1 billion wireless devices worldwide by 2005. By 2003 there will be 9 million enterprise wireless data users by 2003 with companies spending approximately $10,000 per user to mobile-enable them[3]

Early mover. TAA has over 1,000 installations of ThinAir server: TAA has established TAS as the best solution for connecting enterprise data to any device. With groupware support already in place and additional support for enterprise data sources coming, TAS will continue to be the most compelling wireless platform. And, with a flexible sales model allowing evaluation periods and pilot programs, TAS removes the hesitation to invest in new software.

Brand recognition. TAA has over 30,000 active users of ThinAir mail and the public ThinAir server: ThinAir Mail continues to be the best marketing tool for TAA. With over 20,000 active users, the ThinAir brand is recognized as the leader in wireless software. In addition, the public TAS supporting the free client demonstrates the scalability of the platform.

Strategic deal for ThinAir server to be backbone of a major device maker's wireless service offering: A major device maker has chosen the Carrier-Grade TAS solution to be the backbone of their next generation wireless service for connecting their devices to the enterprise and the Internet. The maker recognized TAS as the best solution for their service which is projected to be over 1 million users in the coming years. This relationship is a tremendous endorsement and customer reference for other enterprises looking at TAS. In addition, they use TAS for wirelessly enabling their own workforce.

ASP Deals with Aether, GoAmerica, Mobile Logic, Juno, E-Cal, Centerbeam, and UCSI: ASP relationships allow enterprise customers a hosted solution instead of managing TAS themselves. By partnering with the leading enterprise ASPs, TAA leverages their relationships and reputations in the enterprise.

OEM agreement with extended systems: TAA established its first OEM bundling deal in September 2000 with Extended Systems (ESI). ESI is bundling TAS with their existing XTNDConnect server. TAS was chosen because of its flexibility and provides ESI with the next generation of their product. ESI is the established leader in the enterprise connectivity market and the deal potentially provides annual seven figure revenue opportunities for TAA.

High-quality enterprise customers include Bristol-Meyers Squibb, Morgan Stanley Dean Witter, Anderson Consulting, and Johnson & Johnson. Many enterprises have already adopted TAS as their wireless solution and successful customer relationships provide excellent references for gaining new accounts. In addition, these large corporations are incrementally adding users and functionality to their TAS offering providing increased recurring revenue to TAA.

Agnostic. TAS is not tied to one device, protocol, or network: TAA has mitigated the risk in the confusing wireless market by not tying its software to a particular device, protocol, or network. This not only allows TAS to adapt to new technologies it assures and insures its enterprise customers investments.

Enterprise ready. TAS works openly, securely, and with corporate firewalls: TAS uses open standards for connectivity and uses the latest in encryption technology to make corporate customers comfortable with the product. TAS works inside and outside of existing corporate firewalls giving enterprises flexibility in how they implement the technology.

DUE DILIGENCE

The following recounts the most significant of the 15 due diligence criteria that were applied to the due diligence performed on this technology investment. Note again that the due diligence described here was performed by private equity venture capitalists—not by a technology manager in a specific company or a vendor. The due diligence lens adopted here was that of a venture investor who expected to see a substantial return on the investment in a relatively short period of time—the modus operandi of private equity venture capitalists.

The "Right" Technology/Few or No Infrastructure Requirements/Changes to Process and Culture

There is no question that TAA's technology qualified as the "right" technology. Wireless PDA and smartphone growth was established at a minimum of 50% annually for the foreseeable future. This trend proved to be only part of the story. Second generation wireless networks were already (in 1999) well built-out and technologies to enable third generation networks were being deployed in Europe and slated for the U.S. The infrastructure to support higher bandwidth data transmissions to devices was also becoming a reality. More importantly, the market for applications on handheld devices was without a standard. Many cell phones at the time were making use of wireless application protocol (WAP), which was created in 1997 by Ericsson, Motorola, and Nokia to enable online information and applications to be accessed by wireless devices. However, the two other predominant wireless device manufacturers, RIM (Research In Motion) and Palm, were pursuing their own protocols which required purchasing specialized middleware, and Microsoft Windows CE and J2ME were both showing signs of weaving their way into mainstream PDAs. Each of these standards required its own specialized middleware to access corporate data. Accordingly, the administration of these devices in a corporate setting posed an expensive and complicated problem unless users were to standardize on one device, like the ThinAir Server (TAS). Given the variety of networks and calling plans and the early stage of available pricing plans from mobile service providers, convergence within an organization on one device, let alone one carrier, was nearly impossible.

We were able to establish that ThinAir, on the other hand, enabled organizations running its ThinAir Server to manage multiple devices as well as diverse carriers off of one middleware platform. Further, the server product was both scalable and extensible to other applications used in the enterprise. The middleware platform required no significant additional capital expenditures by a user and even helped organizations avoid the cost of standardizing on one carrier plan or device. Connections to existing enterprise applications could be created using the SDK. The ThinAir server effectively represented a high-impact solution that required little infrastructure cost and little change to an existing business process or culture. The very existence of the TAA solution actually extended enterprise business processes outside of the boundaries of the office or wired connection. Because of these attributes, TAA was highly differentiated in the marketplace.

Our competitive analysis turned up several organizations that had specialized in one or more vertical application, yet no organization had set out to tackle the variety of competing standards or integration of varied devices. In fact, what we found was inertia in the industry—organizations were allying themselves with one standard or another—some producing applications for RIM Blackberries exclusively, others for Palm or Windows CE, and so forth. This fueled excitement for our investment that was shaping up as truly unique. However, it also created some anxiety relative to the task we were taking on along with TAA—to bring the industry together via a de facto standard for transmitting data to wireless devices. Was this risky? Absolutely. But the trends indicated that integration and interoperability would be requirements that just about every company (and prospective customer of the TAS) would highly value.

Horizontal and Vertical Strength/Quantitative Impact

TAA, by virtue of its technology, was disproportionately a horizontal play. The popularity of its horizontal applicability—wireless access to e-mail and groupware—served as a catapult to serve vertical specialties. Once TAA acquired "real estate" behind the firewall with its server, a platform was in place to develop vertically specialized wireless applications, such as in financial services or health care industry segments. However it did not possess the required domain expertise to create specific vertical solutions; so it essentially remained a horizontal opportunity (though with very real "vorizontal" capabilities).

The short term strategy to manage this deficiency was to sell its SDK into both the developer and end-user community, effectively enabling those with vertical domain expertise to develop applications on their own yet providing the company a share in that revenue. Over the long term, the company had developed a product roadmap that included a next-generation carrier-grade platform for application development on devices that would support all client-side code. The goal was to provide high level APIs to enterprise developers that could then leverage their existing domain-specific skill set. The company would continue to develop and support horizontal applications on its own.

Given the horizontal and groupware nature of the solution, TAA's value proposition provided only high level, "softer" quantifications of value. These included the net impacts of productivity enhancements from untethered access to corporate data, ease of device administration, and reduced costs in developing and implementing specialized applications. The company relied heavily on its channel partners to devise and sell on more quantitative value propositions within vertical specialties.

Solutions/Multiple Defaults

The company's product and distribution strategies were centered around a bundled product that often required integration and customization services, especially when a customer desired to provide wireless connectivity to an existing enterprise application other than typical groupware. By whom the integration and customization services would be provided depended on the distribution channel giving rise to product sale. For example, when a direct sale occurred, the company's internal service team would provide the integration and customization services. If a sale occurred through an indirect channel or was a direct sale to a very large organization, a channel partner would win the services contract along with the software sale. In an effort to push the ThinAirServer toward a vertical solution-type of sale, the company established a number of channel partnerships with wireless resellers such as GoAmerica, hosted service providers such as OmniSky, Juno, and Fiberlink, and wireless solution providers with vertical domain expertise such as Accenture and Vaultus. To facilitate the sale of an integrated vertical solution, TAA sold its SDK directly to its channel partners which would then resell customized applications developed using the SDK to end users on a vertically-specialized basis. This strategy effectively linked the horizontal and vertical strengths of the technology.

The wireless industry has a unique structure that is dependent on multiple segments, which include device manufacturers, carriers or network operators, and middleware applications developers. Around these segments are a variety of services firms focused on developing wireless applications for the enterprise market. But again, there had been no convergence on any one standard or protocol for the variety of devices, networks, and applications. We thus determined that our investment in TAA, which had the ability to mediate between and amongst the various standards and protocols, possessed multiple default exits. Given the appropriate level of execution and customer base growth, TAA and its intellectual property would prove valuable for a variety of large industry players across any of the multiple segments comprising the wireless industry. TAA had also already established itself as a leader among many of the players in these segments, particularly Palm, with which TAA had an extensive strategic alliance.

Experienced Management/Packaging and Communications/Industry Awareness

ThinAir was founded by a group of entrepreneurs in their late 20s. These individuals had built an internet integrator successfully, which is where they first developed the ThinAir technology. The company was energized with youth and creativity but was, despite previous experience, lacking a depth of business experience. We focused on identifying and assembling a roadmap in conjunction with management to address the deficiencies. TAA management was young and highly talented. Its product development capabilities and development infrastructure were a clear strength. We placed initial emphasis on sales and marketing because TAA had a completed product that had already been sold into the marketplace on a direct, yet case-by-case, basis. The primary task at hand for TAA was to build a distribution engine capable of scaling to the market opportunity at hand.

Early on in our diligence we had determined to make a commitment to the founder/CEO, which was based on his command over the technology and industry direction, as well as the loyalty he received from his employees. He had clearly established himself as the leader for the current stage of the company.

It was the company's creativity that resulted in significant market awareness for the TAS. Management's decision to give away its ThinAir Mail product to individual users of POP3 and IMAP e-mail accounts resulted in nearly 30,000 individual users at the time our investment was made, many cutting edge consumers who were promoting the company's innovative product. The "freeware" also landed ThinAirApps as the developer of the number one downloaded application for Palm devices for several years in a row, which, among others, certainly grabbed the attention of the top brass at Palm. In an effort to become the de facto standard for wireless access to enterprise data, the company adopted the tag line "Wide Open Wireless" or "WOW," which highlighted the uniqueness of the solution—enabling secure real-time wireless interaction with behind-the-firewall enterprise data. Beyond the sizzle of a market-leading young company run by young entrepreneurs riding the wireless wave in 1999, TAA boasted a roster of blue chip pilot clients, including Accenture, The McGraw-Hill Companies, Morgan Stanley, Textron, Bristol-Meyers Squibb Company, SBC, and Palm.

Partners and Allies

TAA had developed an extensive strategic relationship with Palm. This relationship, which was spurred by the popularity of the company's software among Palm VII users, spanned both consumer and enterprise strategies. At the time of the investment in 1999, Palm was predominantly a consumer-oriented company. It held greater than a 65% market share and boasted more than 12 million users, most of whom were individual consumers. Somewhat similar to Apple Computer, its devices were very popular among a niche market that had been surrounded by a developer community thriving on consumer demand for additional applications leveraging the Palm VII wireless capability—location-based programs, financial services applications, and personal entertainment, and so forth.

The enterprise market, however, represented the Holy Grail for Palm, yet the company ran a distant second in enterprise PDA market share to RIM and its BlackBerry devices. This market was less mature than the consumer market, though it was both larger and growing faster. The importance of the TAA server for Palm was obvious. If Palm could penetrate the enterprise market with any success, it had to do so initially with a scalable horizontal application area, such as e-mail and groupware. Through the partnership, Palm had licensed the TAA server as a core component of its next-generation messaging infrastructure. Palm had also actually chosen the ThinAir Server to satisfy its own internal mobile workforce requirements.

This relationship provided substantial validation of the significance of the ThinAir technology and market position. It would also define the company for its future direction.

Outcome

So what happened? Was the due diligence process "successful"? The investment was made and ThinAirApps was acquired by Palm in December of 2001 in a stock acquisition valued at approximately $20 million. The outcome for its investors was meaningful both in terms of absolute dollar return and return relative to average performance of other venture investments made in 2000 and in the wireless middleware industry. The outcome by venture investment standards, however, was less than spectacular. Remember that the investors invested $13 million in the company in 1999 and while a $7 million "profit" after 2 years may seem substantial, it might have been much higher had other events unfolded differently. For example, competition among the vendors—Microsoft, Palm, RIM, and so forth—might have reached a fever pitch triggering a much higher valuation for TAA's technology, instead of the relative "leveling" of the competitive space that occurred from 1999 to 2001.

All of that said, TAA's core technology remains today the basis for the Palm smartphone connectivity between a variety of applications, including its now seamless connectivity to the Microsoft Exchange (e-mail/groupware) server using the VersaMail e-mail client. This feat, by any measure, is a huge success certainly for the company and its founders.

But where did the investors fall short in its evaluation? Where did the company fall short in achieving its goals and objectives? What gave rise to it all? Let us begin with a discussion of the market dynamics that developed shortly after our investment was consummated.

In early 2001 Palm misstepped its launch of the Palm VII successor, the i705, by announcing its debut too far in advance of its availability. At the same time Palm had "stuffed" its distribution channels with Palm VIIs and demand for which, upon the announcement, fell sharply. This precipitated an inventory write down which caused Palm to substantially miss its earnings estimate. The stock and related value of the company took a nosedive. Around the same time an economic recession set in and projected sales of cell phones and PDA devices began to fall significantly. Nokia, Ericcson, Motorola, along with all the other major cell phone and PDA manufacturers entered a prolonged slump, from which they only began to emerge in 2004.

The economic decline snowballed many other events including the freezing of corporate technology spending. While research indicated many organizations had sights set on accessing corporate data and applications remotely through wireless devices, their propensity to spend immediately on this initiative evaporated. This was the most significant contributor to the outcome of our investment. Large capital expenditure on third generation wireless networks, which we believed would contribute to broader enterprise adoption, had also slowed dramatically. A "perfect storm"? Just about.

Considering the stubbornness of the industry regarding each player's own standards, we were counting on the enterprise to be the catalyst for convergence. However, 1 year into our investment we were faced with flagging enterprise demand for wireless applications—one default exit strategy arguably impaired. The company had made a significant commitment to one device provider through its strategic relationship with Palm, which itself was struggling. We had either poisoned the well on other device providers or had a long road to transitioning alliances—another default exit impaired. Finally, with shrinking corporate technology spending, many of our channel partners showed signs of distress and pilots and sales began to slow. The company that had otherwise built itself to be device agnostic suddenly found itself boxed in and codependent with only one device provider.

We had also put more effort into developing channel partners than we had identifying a strong internal sales professional. Combined with our horizontal strength and vertical weakness, we effectively had no industry specialization to fall back on for core revenue generation nor the internal sales and domain capability to generate vertical specializations.

The most significant flaw in our investment premise was the strength of the technology's agnosticism—its ability to work with a variety of devices, networks, and applications. While this premise essentially held true, it had been weakened by the company's lopsided actions with the device manufacturers. The strategic relationship with Palm effectively repelled other manufacturers: in the absence of a "pull" effect from the enterprise market for a multidevice platform product, the company was seen largely as a Palm-centric provider in the larger wireless marketplace. Other device providers, as with Palm, continued to promote their own middleware and closed standards. At this juncture there was no pressure from the enterprise market to deliver a ThinAir-type solution because the growth in deployments had failed to materialize. Even though TAA had been chosen as Palm's enterprise solution, Palm itself was embroiled in an internal debate about what type of company it was; in fact, it was struggling to find its consumer and business identities. It had failed to gain any traction in the enterprise market as it witnessed RIM gaining more and more market share.

The competitive landscape was showing signs of rapid change as well. Despite the economic situation, the private equity venture industry continued to invest into the wireless industry. As a result, numerous competitors had either entered the market or developed products that threatened to compete with TAA. The space was becoming crowded and there were early signs that TAA's market differentiation would erode. Many of the market entrants were pursuing vertical application strategies, currently a weakness of TAA.

At the end of the day with all the tumult in the marketplace, the company maintained its uniqueness yet risked losing its first mover advantage. Because TAA had no vertical specialty to fall back on, we moved expeditiously towards a sale to Palm, which had expressed interest in acquiring the company for some time. The company's first mover advantage—and its horizontal extensibility—would prove to be unsustainable over the long term. Today, we know that the horizontal play of behind-the-firewall access to groupware was only a near term "killer application" and a long-term commodity. This type of functionality is being incorporated into other platforms such as Microsoft Exchange. The longer term value in the wireless middleware and application segment is being realized through vertical application development and specialization. This vertical specialization is also giving rise to significant growth in service revenue, which TAA had determined to cede to its channel partners (and ultimately did not have the capability to deliver anyhow).

Where does the industry stand today? In 2005, there still exists a largely fragmented marketplace. If an enterprise wishes to deploy both Palm and RIM devices, it must at a minimum purchase a RIM-only middleware server. In mid-2004 Palm finally released its Treo-650 smartphone that had out-of-the-box groupware functionality with Microsoft Exchange. TAA is still the most widely used groupware and wireless application server on the Palm platform: Palm has chosen TAA core technology to be a critical element of its wireless solution for both its enterprise and consumer strategies (no other competitor has established this status with Palm or any other device vendor).

Palm has integrated the ThinAir Server to be the messaging platform and enterprise solution for its newest wireless devices. Palm currently private-labels TAA technology for its enterprise solution. Palm itself uses the TAA technology for all of its 2,000 employees. TAA technology has been developed to be device, protocol, carrier network, operating system, and data source independent—across Palm, Rim, Pocket PC—as well as any Java-enabled device (including cell phones) – enabling a unified solution for all mobile workforce needs. TAA enjoys strong industry recognition as a first mover in wireless middleware and groupware, particularly among both consumer and professional users of Palm and GoAmerica (RIM-based wireless Internet service provider)

But the Company's partnerships with any of the various large channel partners needed to proliferate its platform—Accenture, other Big 4, large integrators, and the like—have not materialized into sales. Enterprise demand has also lagged expectations. TAA is highly dependent on the Palm platform and its use is being driven primarily by Palm. The TAA technology provides Palm the needed solution to enable its own entrance into the enterprise market—widely believed to be the key to Palm's survival.

The competitive landscape in the overall wireless middleware and applications segment is becoming complicated and threats exist not only from direct competitors (primarily private entities such as Wireless Knowledge, Viafone, Everypath, etc.), but device producers (Palm, RIM, etc.) as well as established application and server companies like Microsoft, Oracle, and BEA Systems, among others.

CONCLUSION

$13 million yielded $20 million in 2 years. The due diligence process provided a green light to the investment. There was sufficient data to more than justify the investment: nearly all of the due diligence questions were favorably answered. Was the investment "successful"? Obviously, the answer is "yes." But the $13 million dollar investment could easily have returned ten times that amount had the events described above not occurred. But who can control the market or financial trends? Hence the inherent risk in venture—and all technology—investing, which is another way to say that due diligence can reduce but not eliminate uncertainty around technology investments. The due diligence process is disciplined but imperfect, since no one can predict the future market or financial trends that can easily and significantly impact outcomes.

ENDNOTES

[1] This case was prepared by Robert S. Adams and Stephen J. Andriole
[2] Data from the **Wireless Internet and Mobile E-Commerce Report**, Commonwealth Associates, October 2000.
[3] Data from the **Wireless Internet and Mobile E-Commerce Report,** Commonwealth Associates, October 2000.

Chapter V
Enterprise Investing in Remote Access Technology:
The Prudential Fox Roach/Trident Case

INTRODUCTION TO THE CASE[1]

This is a case of enterprise investing in a specific technology to improve the way its employees communicate and transact business with their customers. The company involved was Prudential Fox Roach/Trident (PFRT) realtors, an independently owned and operated member of the Prudential Real Estate Affiliate, Inc. PFRT is the fourth-largest provider of real estate services in the U.S., with nearly $16 billion in annual sales. The company operates 70 offices in the Pennsylvania, Delaware, and New Jersey area; there are 900 employees. Prudential Fox & Roach/Trident also contracts with more than 3,300 independent sales agents who work on commission. These agents were the focus of the technology investment.

The company is a full-service or real estate "solutions provider." The value proposition is wrapped around services, for buying, selling, renting, or relocating, and reach, through its offices and sales associates and the larger Prudential network with over 1,400 national offices and 40,000 sales associates throughout the United States and Canada. Prudential Fox Roach/Trident also provides financing, settlement, and insurance through The Trident Group.

Copyright © 2009, IGI Global, distributing in print or electronic forms without written permission of IGI Global is prohibited.

Up until the mid- to late 1990s, the processes that optimized production, sales and service—including real estate processes—were relatively well-bounded, with clear beginnings, middles, and ends. Today those processes are extended and continuous. One way to think about the change is to contrast discrete transactions with continuous ones. Discrete transactions, like selling insurance policies, buying disk drives to be included in PC manufacturing, or buying or selling stock, used to be transactions that one could begin in the morning and complete by afternoon or the next morning after running a transaction "batch." Today these transactions are continuous: one insurance policy is blended with another; cross- and up-selling—the desire to sell existing customers additional (and more expensive) products and services—are continuous goals; disk drive acquisition is integrated into PC manufacturing and buying and selling stock is simultaneously linked to tax calculators and estate planners. Real estate professionals simultaneously sell homes, mortgages, and insurance; some of them even sell (and re-sell) home decorating, nanny and shopping services; and they do all of this selling using digital technology at some point or another in the sales cycle. For example, so-called full-service real estate companies will list the house with the multiple listing service (MLS), advertise the house, offer open houses over weekends, help find financing, organize the settlement process, find vendors to perform whatever repairs are necessary, provide data about local and regional schools, provide names of approved babysitters, and names, addresses and references for licensed and insured electricians, plumbers, and landscapers. Much of this is done using data bases that the full service realtor has developed over time; much of the communication is through e-mail or through access to the company's Web site which has a password protected portal for its customers.

Connectivity among employees, suppliers, customers and partners—though far from complete—is enabling interactive customer relationships, integrated supply chains, and the business analytics that permit real-time adjustments to inventory, distribution, and pricing. The net effect is that time and distance have been compressed, and speed and agility have been accelerated. Perhaps no other market has experienced this compression more than real estate which in some parts of the United States has seen price increases of over 300% during the last 5 years. In addition, the average number of days that houses stay on the market before sale has decreased by 50% in many markets. Some markets see houses selling in a matter of days or weeks rather than months that was the norm in the 1980s and 1990s (though toward the end of 2005 the market began to slow).

In the real estate world, it is all about seller data, customer data, and matching this data with the specifics of each transaction. In order to do this, real estate brokers and agents must have immediate and reliable access to data, processes, required forms, and information about regulatory requirements and changes. The "life cycle" concept is especially relevant to real estate customers since a typical customer will buy and sell several homes, purchase several mortgages, and in the process obtain a great number of supporting services, like home inspection and insurance. Real estate buyers and sellers are near-perfect life-long customers.

Collaboration and data are especially relevant to the case described here: the challenge at Prudential Fox Roach/Trident was to keep everyone linked in a network of sellers, buyers, brokers, agents, forms, and regulations where no one has to worry about the management of the networks, the applications, or the communications devices. The challenge thus reduces to: how do we connect all of our stakeholders 24/7, reliably, securely, and cost-effectively?

The real estate world has been relatively conservative in its exploitation of the Web and the online processes that Web-based applications enable. Companies like Lending Tree were actually established in the mid- to late- 1990s, but their early revenues were slow to grow—until the roll out of broadband communications accelerated well into the general population. Real estate consumers (buyers and sellers) were also slow to embrace the Web until it became relatively easy to list or find homes via the Web sites of specific or general purpose realtors (such as Prufoxroach.com and Realtor.com, respectively). Now it is common practice to surf the Web for homes and then contact agents for visits and for companies to troll for buyers and sellers via the Web sites of related and unrelated sites.

The agents at Prudential Fox Roach/Trident were also relatively slow to adopt new technology, but since 2002 its adoption rate has skyrocketed. The challenge for the company was to make sure that their agents' use of the technology enhanced—not encumbered—all aspects of real estate transaction processing.

DESCRIPTION OF THE INVESTMENT OPPORTUNITY

Real estate companies and their sales agents face stiff competition for listings and clients. Finding new ways to expedite the complex property sales cycle offers an important advantage to a real estate company. Prudential Fox Roach/Trident sought a better way to give its mobile sales agents and branch office employees access to the myriad applications, forms, and databases, such as multiple listing services, needed to list and sell properties. The company wanted to deliver applications and the corporate intranet over the Internet, which offered wide availability and a familiar browser interface for agents, many of whom are technologically unsophisticated.

PFRT implemented the Citrix MetaFrame Secure Access Manager to augment its existing MetaFrame Presentation Server environment with secure access to applications and information over the Internet.[2] Independent agents and over 900 employees now have a single point of access to applications, databases, forms, and documents, and tools such as Web search engines from any standard Web browser.

The impact of the remote access implementation project shortened the real estate sales cycle, provided data security for remote and mobile users, and reduced support requirements by 60%.[3] The project also simplified computing for many nontechnical users. Perhaps just as importantly, the project has reduced the number of agents that have left the company for technology support reasons by more than 80%.

In the hot real estate market since the 2000, with interest rates at 30 year lows, real estate companies and their sales agents face stiff competition for listings and clients. Finding new ways to expedite the sales cycle offers an important advantage. Prudential Fox & Roach/Trident sought a better way to give its mobile sales agents and branch office employees access to the myriad applications, forms, and databases, such as multiple listing services, needed to list and sell a property.

The agents are always on the move and frequently work evenings and weekends so they can accommodate the schedules of buyers and sellers; the company needed a way to give these agents simple, real-time information access from any location, such as a home office, so they could make the most of their time with clients—and, if required, provide quotes or calculations on the spot. However, because these realtors are independent contractors and work on computers and network connections Prudential Fox & Roach/Trident does not control, it was vital to provide reliable and secure access to the company's—and the industry's—applications and data.

In terms of the requirements identified above, the real estate challenge proved formidable, since it touched just about every type of communications requirement. Figure 1 particularizes general collaborative communications requirements to the real estate field. The problems, however, at Prudential Fox & Roach/Trident were wide and deep. The new communications requirements were generating too many alternative solutions. Prior to the implementation of the Citrix-based solution, the company attempted to deploy and support thousands of agents and office employees with a conventional hardware/software implementation model. Soon after the roll out of this solution, however, major problems began to arise. Among the more troubling problems were:

- The inability to keep the applications that support the real estate transaction process current
- The inability to keep the process reliable or secure
- The inability to allow the process to rapidly scale
- The inability to keep ahead of the many "break and fix" requirements among computers that were "personal" or used for other than real estate business
- The inability to manage the process through distributed—but ungoverned—management policies and procedures
- A rising turnover rate among real estate agents which was, according to internal company surveys, increasingly driven by the nature and quality of technology

services provided (or denied) to agents by their brokers: in other words, without reliable, secure technology support agents will leave a real estate brokerage in search of better technology-enabled real estate transaction processing (the company reports that their normal agent turnover rate is approximately 20% per year and that approximately 40% of those agents identify "technology" as a primary driver of their decision to leave the company

In short, the traditional acquire/deploy/support technology model did not satisfy the collaborative communications requirements of real estate brokers or agents or mortgage brokers. This was largely due to the relationship that agents have with their brokers. Agents are essentially independent contractors. At the same time, brokers must treat them as full-time employees with a suite of benefits—like technology support. If brokers fail to provide them with adequate technology support (among other benefits), they will find brokerage homes elsewhere.

Given these pressures and problems, the traditional technology support approach had to be replaced by one that facilitated the secure connection of professionals, offices, branches, and partners. The Citrix® Access Infrastructure Solutions for Remote Office Connectivity enables organizations to securely deliver applications and information to remote offices and contact centers and maintain those applications and information from a central location.

With their diverse set of networks and devices, supporting and maintaining applications and information for remote users can be overwhelming. Integrating new branch offices and supporting mergers and acquisitions with disparate systems and resources brings additional complexity and challenges.

Providing a high level of support to these remote real estate professionals is critical to keep productivity and satisfaction high. Citrix Access Infrastructure Solutions simplified and streamlined the deployment and maintenance of applications and information to remote, mobile users and offered secure communications over any network and device, including the ability to:

- Configure, manage and enable application access from one centralized location, reducing the cost of provisioning branch offices individually
- Improve time-to-value for business expansion with accelerated delivery of enterprise resource planning (ERP), customer relationship management (CRM), sales force automation (SFA), and office productivity applications
- Eliminate the need to dispatch technology staff to service remote locations with centralized applications management
- Use the Internet to deploy applications securely with less bandwidth—at lower telecommunication and network costs
- Enhance customer service with centralized business-critical databases

- Protect data with a built-in disaster recovery system, and help eliminate costly downtime

Prudential Fox Roach/Trident concluded that the Citrix MetaFrame Access Suite was the easiest way for the company to provide secure, single points of access to its enterprise applications and information on demand, while ensuring a consistent user experience anywhere, anytime, using any device, over any connection.

The company implemented Citrix MetaFrame Secure Access Manager to augment its existing MetaFrame Presentation Server environment with secure access to applications and information over the Internet.[4] Over 3,000 independent agents and over 900 employees now use a single point of access to multiple applications, databases, forms, and documents, and tools, such as Web search engines, from any standard Web browser. Applications deployed via the Citrix solution range from older 16-bit software to "homegrown" solutions. They include Neighborhood Locator, School Report, and PC Forms that are required for house closings.

A real estate transaction can be a lengthy process. By giving agents flexible Web-based access to all the information resources they need, Citrix access infrastructure helps accelerate the sales cycle and deliver commissions faster. The MetaFrame Secure Access Manager helps the company's mobile sales agents optimize their time and shorten the selling cycle. With secure access over the Web, agents can obtain MLS listings, forms, loan information, school reports, and all of the other tools they need while on the road, at a property, or working from their home offices.

Supporting independent agents involves supporting many users logging in from devices that Prudential Fox & Roach/Trident does not own, over connections it does not control, which raises security concerns. The MetaFrame Secure Access Manager provides a number of security measures that helps the company protect corporate information. The approach delivers standards-based encryption of data over the network and allows Prudential Fox Roach/Trident to provide access based on user roles, so it can control who sees which information. With three main groups of users—corporate employees, employees of a subsidiary that provides mortgage and title services, and independent agents—Prudential Fox Roach/Trident can provide access tailored to each group's individual business needs.

One of the goals of the implementation was to simplify both the user experience and the technology administrator's job. Many of the agents are not technically savvy, so simplicity is critical; Citrix offers a consistent and simple interface no matter where the agent is logging on. In addition, it provides data backup for them. For administrators, Citrix enables efficient, centralized deployment of applications and updates. Under its former acquire/deploy/support model, Prudential Fox & Roach/Trident required three times as many field technical support staff to keep applications and systems up to date than it required after the Citrix implementation.

Figure 1. General collaborative communications and real estate trends

General Trends	Real Estate Trends
Changing Work Models & Processes, Including: - Telecommuting - Mobile Computing - Small Office/Home Office (SOHO) Computing	Real Estate Professionals are Increasing Working from Their Homes, from Their Cars, From Their Colleagues' Offices; Agents are "the" Mobile Professionals of the Early 21st Century
New Customer, Employee, Supplier & Marketplace Connections (B2B, B2C, B2E, Etc.)	Real Estate Professionals are Part of Larger Families of Product & Service providers, a "Marketplace" Unto Itself
Letters, Phones, Faxes, Face-to-Face; On & Off-Line, Synchronous & Asynchronous Communications	Real Estate Agents are More & More Technologically Dependent as They Expand Their Analog Communications
Near Real-Time Comparisons of Vendor Products & Services	Especially in the Mortgage Area, This is Now Happening
Disintermediation & Re-intermediation	Buyers & Sellers are Using the Web to Search & Pre-Qualify
"Whole Employee Management" & "Whole Customer Management"	Real Estate Professionals are Assembling Data on Customers & Suppliers for Long-Term Use
Multi-Purpose Access Points: Wired & Wireless PCs, Laptops Personal Digital Assistants (PDAs), Smart Cell Phones, Network Computers, Kiosks, Local Area Networks, Virtual Private Networks, Wide Area Networks & the World wide Web	Real Estate Brokers & Agents, as Well as Mortgage Brokers, are All Investing in Wired & Wireless Technology; Access Points are as Ubiquitous as Access to the World Wide Web; Desktops are Yielding to Laptops Which are Yielding to PDAs
Anytime, Anyplace Information Sharing	Buyers & Sellers of Real Estate are Insatiable for Information
More bandwidth (& Bandwidth Management), Access Points, Security, Reliability, Scalability & Distributed Systems Management	The Overall Trend in Real Estate is for More Bandwidth, Access, Reliability, Scalability - & Distributed Systems Management

The whole approach, with impact, can be summarized:

- Applications deployed
- Multiple listing service (MLS), a real estate database
- "School Report" by Homestore
- "PC Forms" by PCFORMATION
- "Know the Neighborhood" by eNeighborhoods
- "RealFA$T Forms"

Networking Environment

- Citrix® MetaFrame XP® Presentation Server, Feature Release 3 running on 7 HP DL380 servers
- Citrix MetaFrame secure access manager
- Microsoft® Windows® 2000 Servers
- Frame relay WAN, Internet

Key Benefits

- Shortens the sales cycle with Web-based access
- Provides data security for remote and mobile users
- Reduced support requirements by 60%
- Simplifies computing for nontechnical users
- Reduced agent turnover due to technology support problems by 80%

The solution was the result of 6 months of due diligence and then an almost equal number of months to pilot the technology and then roll it out to all of the agents and office professionals. Over the course of almost a year, the solution went from concept to reality, though there were problems. The company found itself between software releases of the primary vendor. This caused some delay in the piloting and implementation process. Some compatibility issues were also discovered during the piloting phase of the project. But once these issues were addressed (during the pilot phase of the project), the roll out began and met with almost instant success.

What happens when requirements outstrip our ability to satisfy them? We know from our experience that disaster soon follows. The initial agent technology support solution involved caring for each and every desktop and laptop computer for well over 3,000 agents. This approach failed on several levels. First, it was impossible to keep up with the demand. Each time an agent or employee had a problem, the company dispatched a technician. The number of problems quickly outstripped the company's ability to respond. Agents and employees were left waiting in an increasingly longer support queue. Second, the cost to maintain the approach was astronomical. Not only was the company responsible for "break and fix" for over 3,000 machines, it also had to keep all of the software applications current on all of the machines. This was difficult since it was impossible to control all of the versions of all of the applications on the agents' computers. Third, Prudential Fox & Roach/Trident wanted to begin to strategically leverage technology to support its agents, employees, and customers. The support model undermined this corporate objective.

In order to solve all of these problems, the Citrix solution was implemented. The initial results were more than just "promising." Cost savings in the support area are nearly 60% and the number of agents that have defected for technology reasons from the company since the solution was implemented has fallen by 80%. The access solution that the company implemented has solved several problems simultaneously. There is every reason to believe that the solution will endure and that the access portal that the company has created can be used for additional purposes over time.

DUE DILIGENCE

The following applies the due diligence criteria to this technology investment. The due diligence lens adopted here is that of an end-user who expected to see a substantial return on the investment defined primarily in terms of process improvement and cost-reduction.

The "Right" Technology

As suggested in Chapter I, the "right" technology assumes that the technology product or service is productive today—and likely to remain so. It assumes that the technology "works," and is capable of "scaling" (supporting growing numbers of users). It assumes that the technology is secure. It assumes that the technology is part of a larger trend, such as the development of wider and deeper enterprise applications, like enterprise resource planning (ERP) applications. But there is another dimension to "right." Technology does not develop in a vacuum. Those who create, buy, and invest in technology need to understand the relationship that specific technologies have with related technologies. For example, what is remote access (Citrix) technology? Is it a technology concept, a real technology, or a whole technology cluster?

Technologies can be mapped on to an impact chart which reveals that many of the technologies about which we are so optimistic have not yet crossed the technology/technology cluster chasm—indicated by the thick blue line that separates the two in Figure 2. Technologies in the red zone are without current major impact; those in the yellow zone have potential, while those in the green zone are bona fide. The chasm is what separates the yellow and green zones. Note the location of the Citrix technology on which the Prudential Fox Roach/Trident remote access solution was based.

The Citrix technology is well established and well supported. It is not "experimental" by any definition. In fact, it has been deployed for many years by a variety

Figure 2. Technologies, impact, and the chasm—and Citrix

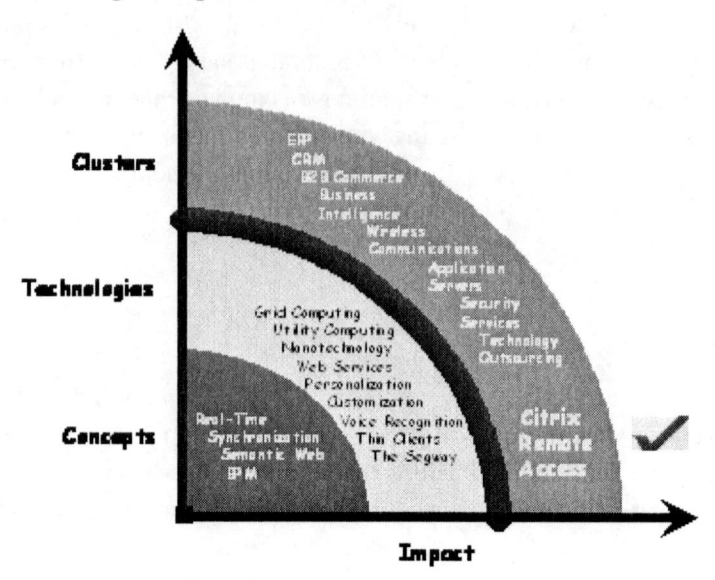

of companies in multiple vertical industries. There was virtually no risk in the adoption of this technology. It is a bona fide cluster.

Few or No Infrastructure Requirements

Technology solutions that require large investments in existing communications and computing infrastructures are more difficult to sell and deploy than those that ride on existing infrastructures. If technology managers have to spend lots of money to apply a company's product or service, they are less likely to do so—if the choice is another similar product or service that requires little or no additional investments. CIOs are incredibly sensitive to the law of unintended consequences: if an investment chain reaction is suspected as a result of a new technology investment the investment will not be made.

So what about Citrix? The Citrix technology was nearly plug-and-play (remembering that there is no such thing as pure plug-and-play). The Prudential Fox Roach/Trident team not only worked with Citrix in the form of pilot projects (to demonstrate how well the technology might work), but it also contacted Citrix customers to better understand their infrastructure experience with the technology. Overall, the team was convinced that the adoption of the technology would not require significant infrastructure investments (beyond the basic Citrix technology).

Budget Cycle Alignment

Let us talk about vitamin pills and pain killers, and buying Citrix technology. Regardless of where one sits, good timing is essential. It is important to understand market context, what is expected, and what is realistic. Capital markets fundamentally change the buying and selling climate; markets determine the popularity of pain killers vs. vitamin pills.

As Figure 3 suggests, the unusual aspect of the Citrix project was its ability to serve as both a vitamin pill *and* a pain killer. Remember that there was considerable break-and-fix pain connected with maintaining the personal computers of the real estate agents. But there were also vitamin pills surrounding the ability to connect 3000+ professionals 24/7—reliably and securely. Figure 3 indicates that the Citrix technology solution actually had infrastructure, tactical and strategic components. Projects like this, at least in terms of timing and impact, are very unusual. The team was happy to learn that the Citrix technology would achieve multiple goals.

It should also be mentioned that Prudential Fox & Roach/Trident was at the time of the Citrix technology investment enjoying one of the most aggressive growth periods of its entire history. The real estate boom (which some have referred to as a bubble) fueled much of this growth and profitability. The timing was excellent for such an investment: the senior management team was feeling particularly strong at the time. The senior technology managers were prudent to propose a pain killer/vitamin pill project during a period of unprecedented growth and profitability.

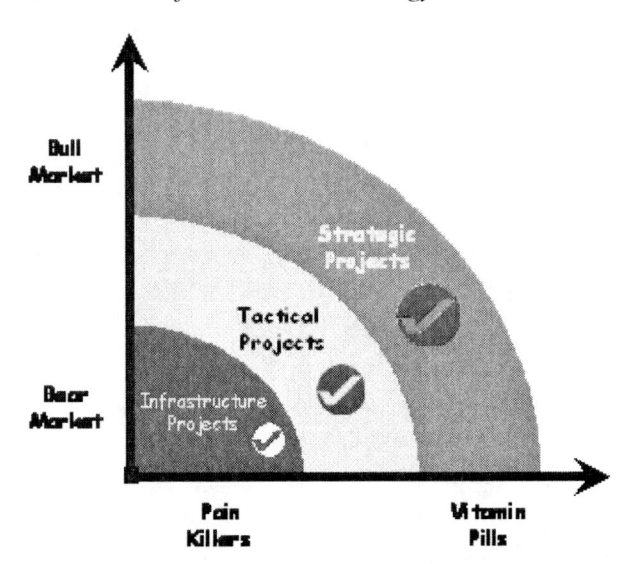

Figure 3. Investment drivers of the Citrix technology

Quantitative Impact

If a product's or service's impact cannot be quantified, then one has to rely upon anecdotes to persuade prospective customers that the product or service is worth buying. But if impact can be quantified, then it can be compared against some baseline or current performance level. Clearly, if quantitative impact is huge, for example, reducing distribution costs by 40% or increasing customer satisfaction by 30%, then it is easy to persuade customers about at least piloting a product or service. All technology investors like measurable quantitative impact.

Ideally, impact reduces some form of "pain," though obviously the impact of "vitamin pills" can be appealing. Quantitative impact also helps differentiate products and services. We already know that the Citrix technology solution has yielded a 60% reduction in support costs. We also know that the real estate agents using the new access procedures are happier and more productive (fewer and fewer are leaving the company because of technology problems). Interviews and surveys bear this out. But did the team expect enormous, or "average," quantitative impact? Expectations were more modest, in the 30% range, but customer (agent) satisfaction expectations were higher since so many agents had problems keeping their personal computers up and running.

The estimated and actual impact was significant, but remember that it is impossible to know precisely what the quantitative impact will be prior to a technology investment (though a structured demonstration pilot project can help reduce impact uncertainty a lot). To a great extent, we make estimated judgments about impact and then hope for—and try to influence—the best. On the other hand, good due diligence teams will talk extensively with others who have deployed a specific technology in an effort to reduce the uncertainty in the impact projections, and of course pilot projects are always a necessary part of the uncertainty reduction plan.

Changes to Processes and Culture

If a product or service requires organizations to dramatically change the way they solve problems or the corporate cultures in which they work, then the product or service will be relatively difficult to sell. Conversely, if a product or service can flourish within existing processes and cultures, it will be that much easier for organizations to adopt. The beauty of the Citrix technology-based solution was its ease of use and its ability to integrate almost seamlessly into the way real estate agents already do business. In effect, all that was required is for the agents to click on a new icon on their desktop to access the applications resident on a company server. There were no changes to processes or culture that made the due diligence team think twice about the investment.

Solutions

CIOs are always on the lookout for "solutions" that solve as many problems as possible. Vendors are aware of these requirements and try to provide just the right mix of services for the right price. The Citrix technology represented a partial solution to the technology requirements of the company. It represented a way to provide access to agents 24/7; it represented a way to connect agents to the applications they need to do their jobs; and it helped agents rethink the way they leverage industry and client data. While certainly not perfect, the technology represented a way to develop a remote access portal that simultaneously solved a lot of infrastructure, communications, data, and applications problems.

Multiple Exits

CIOs bundle their project outcomes within larger risk management frameworks. If a major application fails, they think about how to mitigate the impact; for example, smart CIOs will never cut over to a new application until the application has been thoroughly tested. This means that organizations frequently run two applications as they make sure that the new application does everything it is supposed to do. Prudential Fox Roach/Trident engaged Citrix and some independent consultants to run some pilot applications to determine how well the technology would work in their technology environment. These pilots reduced deployment uncertainty and relieved some pressure from the dreaded "what are the steps we should consider—if things go wrong?" question. The team at Prudential Fox Roach/Trident made sure that the technology worked before deploying it across the company. They even benchmarked the costs. All of this prework was intended to reduce the need for contingency planning and the need to specify "multiple exits."

Horizontal and Vertical Strength

The best products and services are those that have compelling horizontal *and* vertical stories, since customers want to hear about industry-specific solutions or solutions that worked under similar circumstances (like for a competitor). Without a good vertical story, it will become more and more difficult to make horizontal sales. CIOs expect their vendors to understand their business. Smart vendors organize themselves horizontally and vertically to appeal to their clients.

Citrix is primarily a horizontal vendor: its products and services are used in multiple vertical industries. It then customizes its offerings for specific clients in specific vertical industries. The pilot project tested the company's customized solution for Prudential Fox & Roach/Trident or, put another way, the pilot project tested the technology's horizontal and vertical strength. The piloted technology demonstrated that the technology would not "break" in the real estate world. In fact, once the new version was implemented, it distinguished itself.

Industry Awareness

CIOs have a tough time buying products or services with little or no name recognition. Most companies are unwilling to be technology "early adopters" simply because there is too much risk in the practice. Citrix is an extremely well known company with products that are industry standard. There was no danger of adopting a technology that no one had ever heard of or used extensively; Citrix was pretty much the standard for the application that Prudential Fox Roach/Trident had in mind. The industry research organizations have also consistently given Citrix high marks for their offerings.

Partners and Allies

CIOs expect a broad network of support. The Prudential Fox Roach/Trident CIO had the same expectations. Fortunately, Citrix has a network of consultants that can be used to help deploy and support Citrix products. Citrix itself has consultants available to help: as already noted, Citrix is a "cluster" of technologies, consultants, and support vendors.

"Politically Correct" Products and Services

Most technology managers will not risk their careers on what they perceive as risky ventures—even if the "risky" product or service might solve some tough problems. Buyers want products and services that will ease real pain. While "vitamin pills" are very nice to have, "pain killers" are essential. Reducing costs and staff, measurably improving processes, and improving poor service levels are pain killers that make buyers look smart. This is a good place to be—and invest.

"Politics" has a profound effect on business technology decision-making. Everyone relates to "politics" and the impact it has on corporate behavior. But politics is one aspect of the overall context that influences decisions. The others include the culture of the company, the quality and character of the leadership, the financial condition of the company, and the overall financial state of the industry and the national and global economies.

The three most obvious pieces of the puzzle include the pursuit of collaborative business models, technology integration and interoperability, and of course the management best practices around business technology acquisition, deployment, and support. Three of the other five—politics, leadership, and culture—are "softer"; two of them—the financial state of the company the global economy—are "hard" and round out the context in which all decisions are made.

Let us run through the variables. It is important to assess the political quotient of companies. Some companies are almost completely "political": a few people make decisions based only on what they think, who they like (and dislike), and based on what is good for them personally (which may or may not be good for the company). Other companies are obsessive-compulsive about data, evidence, and analysis; in the middle are most of the companies, with some balance between analysis and politics.

Corporate culture is a key decision-making driver. Is the culture adventurous? Conservative? Does a company take calculated risks? Crazy risks? It is an early—or late—technology adopter? Does the culture reward or punish risk takers? Technology investments must sync with the culture (as well as the rest of the variables that comprise the total decision-making context). As Figure 4 suggests, Prudential Fox & Roach/Trident's culture is essentially traditional but at times capable of morphing into a very entrepreneurial one.

What about the leadership? Is it smart? Is it old—nearing retirement? Is everyone already rich? Is the senior management team mature or adolescent? Is it committed to everyone's success or just its own? Is it compassionate or unforgiving? The key here is the overall leadership ability of the senior management team. There are some really smart, skilled, and honorable management teams out there and there are some really awful ones as well. Trying to sell a long-term technology-based solution to a self-centered team with only their personal wealth in mind simply will not work; trying to sell the same solution to a team that embraces long-term approaches to the creation of broad shareholder value usually works very well. As Figure 4 suggests, Prudential Fox Roach/Trident's senior management team is careful and selectively adventurous—good traits for optimizing technology investments.

How well is the company really doing? Is it making money? More money than last year? Is it tightening its belt? Has the CIO received yet another memorandum about reducing technology costs? Is the company growing profitably? Is there optimism or pessimism about the future? Is the industry sector doing well? Is the general economy looking good or are there regional, national or global red flags? What is the confidence level for the sector and the economy? Where is the smart money going? It is essential to position companies within the larger economic forces that define national and global bear and bull markets. Figure 4 suggests that Prudential Fox & Roach/Trident's financial situation is very strong and likely to remain relatively strong, though there are some signs that the general real estate market may be cooling.

Figure 4. The Prudential Fox & Roach/Trident /Trident context

- Traditional
- Selectively Entrepreneurial ...

Corporate Culture

Collaborative Business

- Strong Growth
- High & Growing Profitability ...

Financial State of the Company

- Careful
- Selectively Adventurous ...

"Leadership"

Business Technology Optimization

- Vertically Strong
- Cautious ...

State of the National/Global Economy

Technology Integration

Politics

Management Best Practices

- Clubby
- Friendly ...

Recruitment and Retention

Finding truly talented professionals to staff product and service companies is emerging as perhaps the most important challenge facing companies in all stages of development. Companies that have identified employee recruitment and retention as core competencies are more likely to survive and grow than those that still recruit and retain the old fashioned way. Creative solutions to this problem are no longer nice to have, but a necessity or, stated somewhat differently, creative recruitment and retention strategies are no longer vitamin pills. They are pain killers. CIOs expect their vendors to have lots of really smart, dedicated professionals. If there is evidence to the contrary, they are likely to not make the technology investment. Vendors understand this and try to surround their sales, marketing, and delivery efforts will the best and the brightest. The Citrix team was judged to be solid and professional. Their track record was investigated by the team by speaking with Citrix clients who had attempted to do essentially what Prudential Fox Roach/Trident was attempting to do.

Differentiation

If a technology company cannot clearly and articulately define its differentiation in the marketplace, then a large red flag should be raised about the company's ability to penetrate—let alone prosper in—a competitive market. Differentiation is critical to success and while not every differentiation argument is fully formed when a company is first organizing itself, the proverbial "elevator story" better be at least coherent from day one. CIOs need a ton of help here. In order to sell a technology investment, they need a business case that unambiguously describes how informed their choice is from all of the alternatives. Citrix is the market leader with relatively few viable competitors. The differentiation story was relatively easy to tell here. Alternatives were examined and rejected; none even got to the pilot phase.

Experienced Management

The key here is to see the right mix of technological prowess and management experience available to develop and deliver a successful product or service. There are other ideal prerequisites: experience in the target horizontal and/or vertical industry, the right channel connections, the ability to recruit and retain talented personnel, and the ability to work industry analysts, communicate and sell. The Citrix team is experienced and smart. The company is solid and has an enviable reputation in the industry. The specific individuals assigned to the Prudential Fox Roach/Trident project were more than qualified.

"Packaging" and Communications

While it may seem a little strange to acknowledge the primacy of "style" over "substance" and "form" over "content," the reality is that "style," "form," and "sizzle" sell. Product and service descriptions and promotional materials should read and look good, and those who present these materials should be professional, articulate, and sincere. Companies that fail to appreciate the importance of form, content, and sizzle will have harder climbs than those who embrace and exploit this reality. As a company, Citrix is mindful of the importance of packaging and communications. Their Web site (www.citrix.com) demonstrates the importance the company attaches to form and content—necessary in today's competitive business technology world.

CONCLUSION

The due diligence process yielded nearly all green lights. The investment paid solid dividends. Do all technology investment decisions end well? Obviously not, but having a filter through which to pass important questions helps reduce uncertainty about the outcome and impact.

ENDNOTES

[1] This case was prepared by Stephen J. Andriole and Charlton Monsanto.
[2] © 2004 Citrix Systems, Inc. All rights reserved. Citrix®, MetaFrame®, and MetaFrame XP® are registered trademarks or trademarks of Citrix Systems, Inc. in the United States and other countries. Microsoft®, Windows® and Windows Server™ are registered trademarks or trademarks of Microsoft Corporation in the United States and/or other countries. All other trademarks and registered trademarks are property of their respective owners.
[3] Based on data collected at the company in 2005.
[4] Citrix represents a solid solution to the problems at hand but certainly not the only solution available to Chief Information and Technology Officers (CIOs and CTOs). The key is to provide access to multiple applications and transaction processing capabilities in a way that minimizes support requirements for brokerage CIOs and CTOs.

Chapter VI
Venture Investing in Voice–Over–IP (VOIP):
The NexTone Communications Case

INTRODUCTION TO THE CASE[1]

NexTone Communications (NexTone), a voice over Internet protocol (VOIP) technology company, is an example of a venture-backed organization that has ended up virtually 180 degrees from where it started; and in some respects, the results are equally as different from the due diligence conclusions the investors drew upon making their first investments in the company. This example of our due diligence process is from the perspective of private equity venture capitalists looking for a return on their investment.

NexTone originally developed and sold software platforms that enabled advanced communications service delivery features over Internet protocol (IP) based networks. In other words, the NexTone service delivery platform enabled integrated communications providers (ICPs) of broadband services such as telecommunications companies, cable companies, Internet service providers (ISPs), and so forth, to offer their customers unique voice-related applications such as virtual PBX (in-office and multi-office communications systems), virtual private networks, find me-follow me (e.g., call portability to mobile, home, etc.), multimedia conferencing, and the like. Founded in 1998, the company raised its first round of venture financing of $2 million in equity capital from two early stage venture capital funds. After further developing its product and securing multiple pilot programs with a variety of integrated communications providers, the company raised a series B financing of $12 million from its existing investors and Safeguard Scientifics, Inc., as the lead. This chapter describes the due diligence that Safeguard conducted prior to its lead investment.

In 1999, the ICP/broadband services marketplace was explosive. The availability of capital to finance the build out of fiber optic networks, ISPs, and other high-speed communications networks was seemingly endless. Precipitated largely by the Federal Telecommunications Act of 1996, which allowed for new market entrants to compete with the incumbent telecommunications providers (such as Bell Atlantic, Bell South, U.S. West, etc.) using the incumbents' own infrastructure, the growth of new providers of communications services to commercial and residential customers was double, perhaps even triple, digit. Public equity and debt markets along with the private equity market had locked onto a capital-intensive quest to secure the increasingly valuable "pipe" into the home and/or business to provide an increasing array of services. Differentiation beyond simply provisioning data and voice services manifested in the offering of unique value-added services, and was becoming increasingly important as a competitive advantage to these ICPs. NexTone, along with many other VoIP businesses, was responding to this need that would consume a variety of newly available services due to competitive pressures developing in the services marketplace and the ubiquity of IP networks.

The company was on the cutting edge of developing new services for broadband IP communications. The company's iVANi system was designed to provide ICPs an end-to-end platform for developing and delivering value-added, differentiated services in a rapid, simple, and cost effective manner. At the heart of the iVANi system was the iServer, which was effectively a call control server, or call session manager. There was a platform management application called iView that let an administrator manage the provisioning of NexTone-enabled services. The iEdge 500 and 1000 were edge devices that were located on the end customers' premises. They would connect to analog telephones, PBXs, a cable modem, or to a LAN, and so forth, and enable the variety of value added services for the network or device to which they were connected. Finally, the iMobile was designed for mobile users traveling with a laptop or other mobile device. Some of the value-added applications enabled by using the NexTone platform included:

- Telecommuting, where ICPs could target business with large telecommuting programs, enabling users to install iEdge devices at home or remote locations and use the system to make themselves virtually present at the office—as if locally connected to the PBX or LAN—and receive and make calls using extension dialing, access network services, and so forth.

- PBX Extension, where companies with remote branch offices requiring voice and data connectivity to headquarters could, similar to the telecommuting application, be transparently present on the headquarters PBX or network.
- PBX Interconnect, which allowed two or more corporate locations to be connected using the public Internet as opposed to setting up a dedicated connection or private network, even when two different PBX brands are being used.
- IP Centrex, which eliminated the need for a business to purchase a standalone PBX and digital handsets by using the iServer to supply PBX-like functions (such as call hold, transfer, three-way calling, etc.) directly to analog handsets.

Among the more significant aspects of the investment premise was the shear market opportunity that existed for NexTone's products. There were so many new, well-financed ICPs—ISPs, CLECs, voice resellers, and so forth—competing on price that the need for differentiation and product bundling was very apparent. In due diligence, we substantiated that virtually all ICPs were searching for technology that would offer some competitive advantage—via a differentiated product offering—over the balance of ICP and incumbent competitors in the marketplace. It was this aspect of the investment premise, the market opportunity, that would result in NexTone dramatically changing direction in an effort to survive. Plainly put, the market opportunity evaporated. Beginning in 2000, recession, tightened capital markets, and excess supply converged to effectively render the majority of ICPs impotent and eventually bankrupt. However, NexTone remains in existence today and is thriving—based on a very different product set and target market.

BACKGROUND

NexTone was established in 1998. It has developed a system solution to address three critical issues faced by business customers:

- Eliminate current performance bottlenecks for high speed Internet access in a cost effective way by leveraging broadband technologies such as xDSL, fixed wireless, and Cable Modem.
- Consolidate multiple network lines for voice, fax, and data into a single high speed access link, thereby minimizing telecommunications costs,
- Increase mobility, productivity, and flexibility by enabling advanced services over the Internet such as Telecommuting, Remote Office Connectivity, and Virtual Office Suites.

NexTone's solution consists of iEdge devices that reside on customer premise, and a centralized system of management servers called the iServer. Together, the edge devices and the centralized management servers form a system which enables service providers to offer advanced voice data applications to their business customers. NexTone's products will be bundled as part of the services that the next generation service providers offer to their small and medium sized business customers.

PRIOR FINANCING

Prior financing was structured as follows: $2.4 Million in equity financing from Mid-Atlantic Ventures, Blue Rock Capital, the State of Maryland, and private investors, and $750,000 in debt financing from Silicon Valley Bank.

MANAGEMENT TEAM

NexTone's management team includes seasoned veterans of the telecommunications industry with over 70 years of total experience in product development, project management, marketing, and sales. Their backgrounds are a rich mix of experience with start-ups and large corporations. NexTone's management team has a proven track record of executing on the business plan. All projected milestones for 1999 are completed or on track.

Raj Sharma, President: MBA in finance and international business, MS in electrical engineering and computer science; 15 years of marketing and management experience in the communications industry with companies such as AT&T, IBM, Hughes Network Systems, and Newbridge Networks; experienced in marketing to the Fortune 1000 and International customers.

Ravi Narayan, Vice President, Finance and Operations: MBA in information systems & finance and MS in industrial engineering; more than 12 years of technical and business background in entrepreneurial companies and corporations such as Hughes Network Systems and IBM; manages Finance and Operations.

Sridhar Ramachandran, Vice President, Engineering: MS in computer engineering; more than 11 years of technical, business, and entrepreneurial background, encompassing different facets of the networking and telecommunications arena with companies such as GE, Bay Networks, and Hughes Network Systems; manages Product Development.

Steve Granek, Vice President, Sales: More than 20 years of sales, business, and entrepreneurial background, encompassing different facets of the networking and telecommunications with ADC Telecom, VPNet, Telet, and Microage; manages Sales and Business Development.

Dan Dearing, Director, Marketing: MS computer science and BS computer and information science; 16 years of sales, marketing, and support experience in the telecommunications and services industry both abroad and domestically with Sprint, GE, Hughes Network Systems, and Ciena; manages Marketing.

MARKET

NexTone has positioned its products to meet the increasing demand for value-added applications and services to be offered over high speed Internet access links. Emerging technologies such as xDSL, Wireless, and Cable Modem provide broadband connectivity to Internet at a very low price point. With the proliferation of these technologies, the "Next-Gen" service providers are increasingly motivated to bundle value added services for customer retention and differentiation. Frost & Sullivan expects the overall integrated communication market to grow at a rate of 149% annually through 2001, to about $1.89 billion. Within this market, NexTone is targeting large enterprises and service providers who cater to small and medium sized businesses.

COMPETITION

Competition within this market ranges from traditional router vendors to recently formed startups. To address this market, traditional router vendors are introducing voice over IP cards for their existing products and larger gateway products. NexTone's approach not only provides a key cost advantage but also focuses on the operational costs for the service providers. NexTone's Edge Devices have been designed from the start to be a low-density customer premise gateway. It is optimized in performance and cost for the small office environment. Vendors building carrier class IP telephony gateways provide solutions that are optimized for much higher port densities, making them suboptimal and expensive to be customer premise gateways. Vendors with existing customer premise products that are building cards to accommodate IP telephony features in these products must trade off efficiency to be compatible with their existing products and architectures.

NexTone's advantage over its competitors is its unique client-server architecture and its "whole solution." The architecture yields a system that is optimized to lower the total cost of operations for the service providers. Currently, the solution space is fragmented and it is up to the customer to piece different pieces of technologies from different vendors. By partnering with other vendors, NexTone offers a complete solution for the service providers to quickly enter the market and provide value added voice/data applications for increased revenue opportunities.

Strategic Partnerships

In order to bring a "whole solution" to its customers, NexTone is working extensively to forge strategic alliances with other key industry players. NexTone has formally initiated the "All for One" Partners Program, which is intended to bring together products and services from NexTone and all its partners to provide a total solution to NexTone's customers. Some of the partnerships in place and in progress are:

Technology Partnerships	Service Provider Partnerships	Business Partnerships
Copper Mountain (xDSL)	NexBell (Internet Telephony Service Provider)	Silicon Valley Bank (Equipment Financing)
Jetstream (PSTN Gateway)	GRIC (Clearing house)	
Mediagate (UMS)	Transnexus (Clearing house)	
MindCTI (Billing)		

Market Analysis

Currently most business customers access two different networks to carry out their daily business functions. One is the phone network and the other is the Internet or data network. Maintaining, managing, and paying for two separate networks is obviously inefficient. Perhaps more importantly, with two separate networks, it is difficult to leverage, combine, and utilize the intelligence that exists within the two networks. For instance, when a prospective shopper needs help and clicks on the "Talk" button of a business's Web site, the customer service agent should be able to view all the background information on this "Web shopper."

Convergence of voice and data is more than a great concept: it is revolutionizing the way people do business. Because of technological advancements, converged applications are starting to improve the way we work, sell, and communicate in today's competitive markets. And the wave of innovation has just begun.

The growth of the Internet and the success ISPs have gained by connecting businesses to the public Internet is now being applied to the corporate enterprise networks. Leading ISPs today offer solutions that allow corporations to run their wide area network over an ISP's TCP/IP network. This means using the same service, connections, and protocols for internal corporate networking and for external public Internet access.

Most small and midsize businesses need high speed, yet cost-effective way to connect to the Internet and other remote locations. They need to exchange e-mail, files, and other information with their employees, suppliers, and customers. They generally do not have a staff of in-house resources to install and manage a wide area network. The trend in small and midsize businesses is to outsource their data and telecommunications services.

The new breed of service providers: ISPs, CLECs, and cable companies are aggressively targeting these businesses to get their outsourcing business. Most small to midsize businesses want to treat networking as a utility over which they get all the services, that is, Web hosting, high speed Internet access, multimedia communication, and virtual private networking (VPNs). Business customers including telecommuters are seeking large numbers of additional access lines, mainly to put additional traffic over those lines. The same customers also increasingly want higher speed access to the Internet. And they want integrated services running over those access lines. The challenge for service providers is to figure out the right technologies to install so they can offer the kinds of services that the market demands today and in the future.

Being smaller, without the inertia associated with the larger telcos, the new service providers can take advantage of emerging technology to satisfy the growing need for bandwidth, better Internet access, and voice data integration. For instance, the service provider can leverage emerging technologies like xDSL in the local loop to offer high speed Internet access as well as voice data integration. In this case, voice has to be converted into IP packets at the customer premise. The service provider also needs to offer the same Centrex services that were available on the old circuit switched network from the ILEC. NexTone makes it easy for service providers to offer Centrex services to customers who switch to an integrated voice data service.

TECHNOLOGY DRIVERS

Over the next decade, telephone and fax traffic carried by IP networks are expected to slice away billions in revenues from the regular phone system. Voice over IP technology, which started out in 1995, has now matured to a point where it can coexist and compete with the traditional public switched telephone network (PSTN) for voice and fax traffic. We also observe a rise in product introductions that promise to make the interworking between IP based telephony and traditional PSTN telephony seamless. Newer technologies such as xDSL, wireless, and cable modems are already delivering high speed Internet access at very reasonable costs. With the availability of high speed Internet access at reasonable rates, it makes sense to integrate voice, fax, and data over a single connection to the Internet. This further lowers operational costs for service providers while becoming a one stop shop for subscribers. The service providers can attract new subscribers from the incumbents and also retain their existing base.

However, for advanced voice data applications to be widely acceptable and successful, existing telephony services such as PBX services will have to be preserved when telephony and fax migrate to Internet. Based on the acute interest and initial success of unified messaging of voice, fax, and e-mail, it is evident that there is a strong desire to consolidate messaging under one umbrella, that of the Internet. Internet Telephony, Unified voice, fax, and e-mail messaging and Internet based PBX services are ripe for integration and consolidation; and this represents a tremendous opportunity for NexTone. NexTone intends to capitalize on this shift by delivering Internet based PBX functions while rerouting voice and data traffic from the public switched telephone network (PSTN) to the Internet.

MARKET EVOLUTION

Small and midsized companies will enjoy the benefits of large bandwidth pipes that their supersized counterparts have taken for granted. Further, remote sites will finally get the bandwidth they need and deserve. That, in turn, will help to tightly link customers to corporations whether it is through remote learning to improve sales training or increased personal service to clients. It also means more sales and support data flowing toward HQ. Bumping up the bandwidth at branch offices will yield better customer service, boost sales prospects, and improve products and services.

Today, most headquarter sites are tied to satellite sites with private lines and frame relay, with management and financial apps consuming much of the capacity. In the near future, there will be hundreds or thousands of remote offices tied into the enterprise via high-speed DSL and wireless services. Those pipes will deliver voice and Internet access, along with virtual private networks (VPNs) and PBX extension services. Small and midsized businesses will be up there with larger enterprises, participating in the "net economy." It allows network mangers to dedicate big bandwidth not just to big sites but to small ones.

NexTone expects the business voice traffic volume to migrate to the Internet gradually. Business users will start putting a small percentage of their voice traffic over the Internet initially. In the short term, businesses will primarily be driven by price arbitrage that exists between the PSTN and IP telephony, both for domestic as well as international minutes. NexTone expects that domestic price arbitrage will only exist in the short term as shown in the following graph. This is because the telcos have significant room to lower the price per minute of use (MOU) and be competitive with IP telephony MOU. For international calls, IP telephony will continue to play a big role over the long term. This is because the price arbitrage is huge for international calls and the settlement charges levied by the foreign PTTs, which lead to huge price arbitrage, are likely to continue for a long time.

NexTone believes that the real driver for IP telephony in the U.S. will not be price arbitrage. Instead, it will be productivity enhancing applications that rely on

Figure 1.

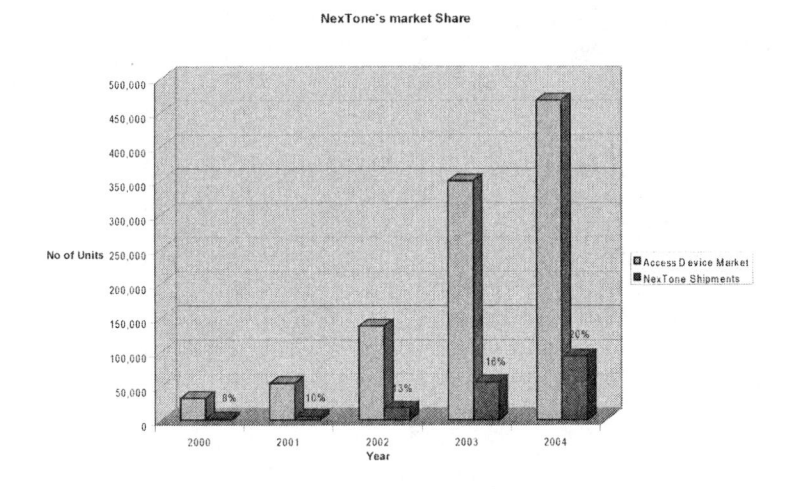

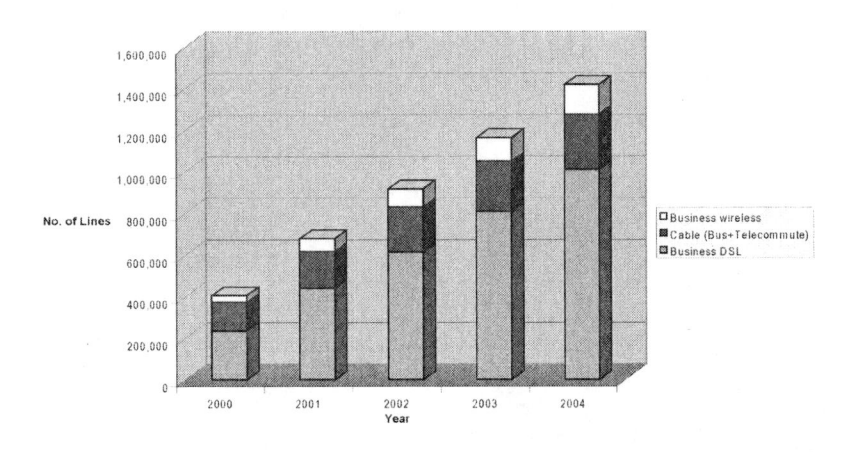

IP telephony. Applications such as Internet call centers, unified messaging, workflow, and collaboration will increasingly depend on IP telephony and will play a key role in saving time, enhancing communications, and making businesses more efficient.

NexTone will acquire 8%, 10%, 13%, 16%, and 20% of the integrated access device subscribers over the next 5 years.

Marketing Strategy

NexTone's marketing strategy will be unfolded in phases as shown in the following.

Phases	Phase 1: Market Entry	Phase 2: Market Penetration	Phase 3: Market Dominance
Marketing Plan	Direct sales to Enterprises Indirect Sales through service providers Bundling of NexTone products with other services Coexistence with PBX/Key Systems Interoperability/Alliance with other industry vendors Proving the technology	Indirect sales through service providers Direct sales to Fortune 1000 Technology and concept acceptance	"Can't get fired for going with NexTone" Service providers and enterprises seek out NexTone solutions
Products	iEdge 500 Family Applications: Telecommuting, Retail Industry, VPNs	iEdge 1000 Family Moving up the food chain Communications center - Internet access + Voice + Fax Market Entry -> Market Penetration	iEdge 5000 Family Complete PBX replacement Market Penetration -> Market Dominance

Product Strategy

The NexTone solution consists of the iServer and the NexTone clients: iEdge500, iEdge1000, and iMobile. The system provides voice, data, and fax convergence right at the customer premise utilizing broadband access to the Internet. The system delivers value added services such as Virtual Corporate PBX extensions, Centrex services, roaming, VPN, and unified messaging services (UMS). The NexTone solution operates over the Internet and leverages any type of Internet access mechanism including xDSL, wireless, cable modem, and ISDN. The client, which resides on an edge device or a PC, interfaces with existing fax, phones, and PCs and redirects traffic over the Internet. In cases where the destination is not connected to the Internet, the client, transparent to the user, forwards the call to alternate discount carriers or conventional PSTN to direct fax and phone traffic.

INTELLECTUAL PROPERTY (IP)

The biggest barrier to entry for NexTone's competitors is to duplicate the scalability that NexTone's architecture provides. NexTone has designed a system that can handle thousands of registrations and address resolutions. Using a unique way to scale the system to handle hundreds of customers, NexTone has implemented an Internet PBX architecture that is hard to duplicate.

Virtual partitioning of the server allows the service providers to offer different levels of service to different customers in way that is easy to provision and manage. NexTone has a minimum of 9 months lead over its closest competitors.

Also, the bandwidth management algorithms implemented in the client software provide superior voice over IP quality compared to competitive solutions.

DUE DILIGENCE

The following reviews the most significant of the 15 criteria that were applied to due diligence performed on this technology investment.

The "Right" Technology/Few or No Infrastructure Requirements

Our due diligence yielded sufficient evidence that NexTone was pursuing a technology that was applicable in the present and future marketplace. The growth of broadband services and the preponderance of IP based communications was evidence enough to substantiate the NexTone value proposition. Cost savings associated with deploying virtual PBX or IP Centrex as compared to the alternative of spending tens of thousands, even hundreds of thousands of dollars in the case of larger organizations, for PBX-type services was very compelling. The icing on the cake was the existence of a burgeoning distribution channel in ICPs being financed by Wall Street and venture capitalists across the globe; and these ICPs were demanding the types of products NexTone could provide in order to compete more effectively themselves. In short, the technology trend that NexTone was riding was well-financed and founded on an already established communications services industry that was undergoing dramatic technological disruption. NexTone played right into this disruption. Of significance on the balance, however, was the fact that the NexTone product set had yet to prove scalable in a tier 1 ICP (such as Sprint, ATT, etc.) setting. This was deemed a technology problem and not insurmountable. In fact, a portion of the capital being raised in the round in which Safeguard was participating was to finance product development on a highly scalable basis.

While the NexTone product set was reliant on a costly digital communication infrastructure, its sale did not necessarily trigger such an endeavor. The disruption occurring in the communications marketplace had already given rise to massive spending on broadband infrastructure. Again, NexTone's product was largely software based and dovetailed with its target customers' strategy with regard to products they would offer their own customers. Further, NexTone products were based on an open architecture that effectively worked over any ICP network.

Horizontal and Vertical Strength/Quantitative Impact

We viewed NexTone as a horizontal play since its products did not carry any particular vertical benefit or any propensity to be sold into any one vertical over another. However, given that NexTone sold only to ICPs, one could view it strictly as a vertical play. From that perspective, the key for NexTone was that its products enabled the provisioning of typical voice features by ICPs that otherwise could not provision such services. We viewed the strength in this as a "vertical specialization that could be sold-thru on a horizontal basis." In other words, we would not have been supportive of a model whereby NexTone itself undertook to sell products directly to end users, regardless of what vertical specialty.

The quantitative aspect of NexTone's story was very compelling. We assessed this on two levels. First was the quantitative benefit for the company's direct customers—ICPs. As previously mentioned, the iVANi system enabled ICPs to bundle a variety of services to its end user customers. We were able to quantify at a high level the benefit ICPs derived from the NexTone set. In our diligence calls with numerous ICPs, we compared price points to those of incumbent communications providers and other competing ICPs. For example, we determined that, consistent with what the ICPs were telling us, with a product offering that included PBX-type services, ICPs could be able to increase their average revenue per line (ARPU) by anywhere from 3% to 7%. With other features that NexTone provided, the ARPU could be further increased.

On the second level, where the ICPs were selling to end users, we were able to establish an even more compelling quantitative impact. Specifically, in the case of a new PBX sale, an end user could purchase a system including digital handset for upwards of $30,000. Depending on how large the organization, costs could run as high as $100,000 and then some. Using the NexTone-enabled virtual PBX or IP Centrex over an ICPs network, the same services could be delivered for very little marginal cost to the ICP, leaving a great deal of room for pricing and savings to the end user.

Changes to Process or Culture/Compelling Differentiation Story/End-To-End Solutions

As previously mentioned, we were determined to think of the NexTone product set as a sell-thru to end users by the ICPs. This vantage point served us well in reviewing the ultimate salability of NexTone's wares. To that end, we emphasized in our diligence the ease with which NexTone products could be deployed by an end user and any disruption that may be caused. This was both a technology analysis and customer behavior assessment.

Our technical review of the NexTone product set turned up no concerns regarding the quality of service in a NexTone deployment. Specifically, end users of voice expect "five nines," or 99.999% reliability. Not yet a reality, we gained sufficient comfort that it was achievable in the near future through continued product development efforts. From a pure customer behavior perspective, we gained insight thru end user diligence that, as expected, the NexTone-enabled services were effectively identical to those already being provided by incumbent communications providers. In short, we were able to conclude that the use of NexTone technology in its designed applications would not produce any unwanted change in process or behavior.

We have already discussed at length the importance of NexTone as a differentiation story for its customers. We also focused diligence efforts on understanding how many other direct competitors existed and how NexTone was able to differentiate itself from them in the eyes of its own customers—the ICPs.

The company's primary differentiator was its end-to-end solution set. Compared to much of its competition, which was focused on individual aspects of the VoIP market and its applications, NexTone had developed a solution for its customer base that worked "out-of-the-box" for ICPs and their customers. For example, in several cases ISPs using NexTone products were able to offer feature-rich voice services to their customer base. Think of it from the perspective of "1999"—your ISP offering voice services!?

Multiple Defaults

This aspect of our diligence proved to be a blind spot. We concluded that while multiple exit opportunities existed for our investment, at least in terms of other VoIP technology providers, NexTone had only one vertical specialty—ICPs. If something unpredictable were to happen in this marketplace, NexTone, along with its potential acquirers, would be in trouble. As mentioned, something did occur and our exits, while not altogether vanished, at least appeared frozen for some period of time.

Ultimately, the risk of not having multiple defaults will prove to not be a significant factor in the NexTone investment. We will discuss this further in the conclusion to this example.

Experienced Management/Industry Awareness

NexTone was founded by three first-time entrepreneurs with a complimentary set of experiences in the communications industry spanning tech R&D, general management, and marketing. These entrepreneurs had exhibited an ability to add to the team where holes existed, particularly in sales.

However, our risk-taking on a first-time entrepreneur team proved costly when the market turned on the company. Decisions to control spending and shift strategy were slow to occur and the founder team was divisive in achieving the changes that needed to occur. A significant factor in NexTone's survival was the ultimate addition of a more experienced executive to lead the company thru the painful changes that were required. However, it should not go unmentioned that the founder team, particularly its technology expertise, was instrumental in the strategic shift that has led to the company's success today.

Because of the founding team's experience in the communications technology marketplace and their expertise in marketing, NexTone did possess an industry awareness that gave the company a competitive edge. NexTone and its products were well-socialized in the VoIP marketplace from thoughtful promotion within the industry itself to market analysts, such as Yankee Group, which had written on the potential of the company's technology. Further, and of significance to our diligence, the company had assembled an advisory board of thought leaders and ICP executives who also lent credibility to the company's efforts.

CONCLUSION

NexTone today remains a provider of VoIP technology to the communications marketplace. Its products and target market, however, are entirely different from our original investment premise.

During the 2000-2002 timeframe when the ICP marketplace was imploding, NexTone was effectively left with no buyers for its technology. The economic problems impacted the adoption incidence of VoIP and the NexTone technology was not scalable to serve those tier one service providers that had decided to enter the market. So with a great deal of pain and struggling, some downsizing, management augmentation, and additional capital infusions, NexTone adapted its core technology to produce a solution for the one marketplace where VoIP was being used on a large scale—internationally. Outside of the U.S., VoIP was been embraced as a low cost alternative to long distance communications. Quality of service aside, VoIP was being used by both developed and underdeveloped countries to connect people at a low cost alternative to wire line infrastructure. Large international voice service providers were "shunting" traffic over IP networks and avoiding costly long haul handoffs to other carriers. However, there were a number of problems that international VoIP service providers faced.

The most significant of these problems was interoperability of equipment used in the handoffs across disparate IP networks. Two VoIP protocol standards had been widely deployed—H.323 and session initiation protocol (SIP). These two standards were incompatible and as such service providers using opposing standards were unable to handoff VoIP traffic in an effort to find the lowest cost route for any given voice call. NexTone adapted its core session controller technology to provide H.323 and SIP interworking among networks. At on point it was the only company doing so. Accordingly, when a VoIP service provider used the NexTone interworking session controller, they could shave a great deal of cost from their network services. Further, the technology had a network externality effect—the more service providers using it created a greater opportunity for cost savings alternatives for all users. Finally, once the technology was installed, its untimely removal could bring down an entire ICP network. As such, NexTone had both pricing power and good receivables turns from customers. This is an important point because many of the company's new customers were tier two or three carriers that were not good credits.

In 2002, NexTone quickly began to address this market for its new products and continues today to grow revenue and product offerings. In November of 2005 the Company announced a $35 million venture financing. Of the original founder team, only the CTO—the creator of the NexTone technology—remains with the company.

NexTone ended up a good example of how critical just a few of the 15 evaluation criteria can be in any given investment. While many of the criteria discussed above held true—few or no infrastructure requirements, quantitative impact, changes to process or culture, compelling differentiation strategy, market awareness, and so forth—the success in the investment came down to just a few, right technology trend and multiple defaults, that created the problems we experienced. Remember, NexTone today is a success, but it is not because the early investors were accurate in assessing these few criteria. Additional capital, new leadership, and a lot of luck has rendered NexTone successful.

In short, NexTone proved a good bet on the VoIP technology and market trend. However, the development of the market took much longer than the investors ever anticipated. Our reliance on the company's applicability to one marketplace—the "Right Technology Trend"—was very risky and almost resulted in a failed investment. While largely market-related risk and unforeseeable, we were effectively relying on a new and burgeoning market of ICPs as the sole outlet for NexTone. And we never developed scenarios for the direction of the company in the event this market slowed or never materialized. Ultimately, we took on far greater "market risk" than was acceptable. This issue also highlights the risks we took in not identifying multiple defaults for NexTone. Perhaps, in collaboration with management, we could have identified earlier the interworking market opportunity. That would have saved a great deal of the investors' money.

ENDNOTE

[1] This case was prepared by Robert S. Adams and Stephen J. Andriole

Chapter VII
Enterprise Investing in Radio Frequency Identification (RFID):
The Oracle Case

INTRODUCTION TO THE CASE[1]

This is a case of enterprise investing in a specific technology—Radio frequency identification (RFID)—in an effort to add another technology-based service to, in this case, Oracle Corporation's, repertoire of products and services.

Oracle USA is one of the top software companies in the world today, with over 50,000 employees, $14 billion in revenues, and 150,000 customers. Oracle USA spends over $1 billion a year in research and development (R&D). Deciding upon a technological course requires a substantial investment in market research to ensure that R&D funding leads to products whose marketability goes from intuition to fruition. One of Oracle's latest investments has been in the area commonly referred to as sensor-based computing (SBC) centered around a technology known as Radio Frequency Identification, or "RFID."

RFID stands for radio frequency identification and is an automated data collection technology that enables equipment to read tags without contact or line of sight. RFID uses radio frequency (RF) waves to transfer data between a reader and an item in order to identify, track, or locate that item. A typical RFID system is made up of three hardware components:

- An antenna
- RFID tags (transponders) that are electronically programmed with unique information
- An RF module with a decoder (transceiver)

According to the Gartner Group (a technology industry research and consulting organization; see www.gartner.com), RFID may become the major technology enabler for the transportation sector in the next decade.[2] They go on to say that it would be dangerous for technology vendors to sit behind the adoption curve as RFID takes off, probably between 2007 and 2009.

RFID will unfold in waves throughout state and local government, according to the market research firm Input Inc., of Reston, Virginia. The first wave might involve commercial trucking, the second wave might target borders and ports, and the third wave might address traffic management. As the demand for RFID grows, state and local governments will use federal funding for transportation and homeland security to help them implement RFID projects and initiatives.

State and local customers now are more receptive to RFID, because they view it as part of the larger wave of wireless technology projects sweeping the government market. ACS, Inc., for example, has a firm grip on RFID for revenue collection and regulatory compliance in the transportation sector. Nine states now use the company's EZPass electronic toll program, and 24 states use its PrePass program to weigh and monitor the credentials of trucks crossing state lines. RFID could be deployed more broadly not only as an integral part of intelligent transportation initiatives, but also for better transportation security; RFID could be used for real-time tracking of commuters' patterns as part of managing traffic congestion, or it could be used to track hazardous materials on interstate highways.

Unisys is trying to expand its work with major ports in the wake of its participation in the Homeland Security Department's Operation Safe Commerce pilot. The company has an RFID project with the Port of Seattle and is hoping to expand the solution to other ports in the region. Unisys also is pursuing a pilot project with the Texas Animal Health Commission that could serve as a national model for identifying and tracking livestock. The National Animal Identification System is a federally funded project that, if rolled out nationally, would let states track livestock through various marketing venues. Using RFID tags and readers, it also would compile data that could be used in the event of disease outbreaks and other health issues.

RFID tags are classified as either active or passive, each with its own characteristics, as Figure 1 suggests.

With RFID, companies can more accurately track assets and monitor key indicators, gain greater visibility into their operations, and make decisions based on real-time information. RFID is just one type of sensor-based technology; others

Figure 1. Types of RFID tags

	Passive Tags	Active Tags
Power	No battery; obtains power	Battery on board
Use case	Retail, disposable	Expensive items, no network presence
Data	Unique identifier on tag is used to lookup details about item stored on the network in real-time	Written to the device
Storage	Less space than active tags	More space than passive
Shelf life	Shorter than active tags	Longer than passive tags

include moisture, light, temperature, and vibration sensors. Increasingly, RFID tags are being combined with sensors and tracking technologies like global positioning systems (GPS) to give companies greater visibility into their supply chains for reduced risk and optimized business processes.

ORACLE'S RFID INVESTMENT STRATEGY

RFID is an important emerging data-capture technology that further extends the value of Oracle with new (mobile) business solutions. At first, Business Information systems acquire more accurate, better, and faster data, enabling major cost savings and performance improvements. Ultimately, electronic product sensing offers unique ubiquitous item-level information, enabling new business processes and supply chain transparency. "Product sensing" is an important step in closing the control loop of operational, tactical as well as strategic business processes, hence enabling real-time Corporate Performance Management and actual "Pervasive Intelligence"—a key to Oracle's future success and that of Oracle's customers.

Oracle's sensor-based services are a comprehensive set of capabilities to capture, manage, analyze, access, and respond to data from sensors such as RFID, location, and temperature. In most RFID applications, the data volume is extremely high. The higher the volume, the more the application matches Oracle's forte. Based on Oracle Database, Oracle Application Server, Oracle Enterprise Manager, and the Oracle E-Business Suite, Oracle's sensor-based services enable companies to quickly and easily integrate sensor-based information into their enterprise systems. Oracle's solution includes a compliance package, an RFID pilot kit, and integrated support in Oracle's E-Business Suite and Oracle's Application Server.

Some initial pilot applications of this technology bundle include:

- NASA **Dryden Flight Research Center** has successfully deployed a program that integrates Oracle's RFID and sensor-based technology to improve the management of hazardous materials and enhance security and safety, while significantly reducing
- **McCarran (Las Vegas) Airport** sites Oracle RFID technology as having a main role in resolving baggage-handling issues: RFID-tagged luggage will reduce baggage misdirections to less than 1%

As RFID standards evolve, Oracle will push solutions down into the technology such that customers do not need to do it themselves. The ultimate vision is to offer sensor-based computing and predefined services such that customers can cost effectively and efficiently benefit from RFID technology.

Oracle offers enterprises the ability to incorporate RFID (and other sensors) technology into their application and business processes with a complete solution comprising of an RFID platform as well as prebuilt RFID-enabled applications. Oracle's RFID platform is based on Oracle Database (version 10g) and Oracle Application Server (10g) and can be utilized to RFID-enable any existing or new application. The Oracle E-Business Suite, an ERP suite built on this technology platform, will offer out-of-the-box support for transactions that are automatically triggered by an RFID read.

Oracle provides one of the most complete RFID solutions in the market, including a platform for developing RFID solutions and applications that are RFID-enabled. Oracle is actively working with partners to deliver an end-to-end solution ranging from RFID reader integration to application processing, by leveraging the technology and scalability strengths of the Oracle Database 10g and Oracle Application Server 10g coupled with the business processing capabilities of Oracle Applications.

Some key features of the Oracle RFID solution include:

- **Database and J2EE developers:** Oracle developers can begin integrating data from RFID and other sensor technology into their applications today; no need to learn a new development environment or have intimate knowledge about the sensor technology.
- **Robust technology and proven scalability:** RFID requirements are integrated into existing business process and data management technology, namely Oracle Database and Oracle Application Server.
- **Business process expertise:** Oracle Applications
- **Integration of event and data management:** Historically event and data management has been separated, leading to complex interaction and complex processing for non basic functionality such as composite events and event ag-

gregation. Integrated data and event management simplifies many functions and enables new features, such as supervision of "missing" events, which are very important to supervise perishable goods.
- **Policy management:** Describe policies declaratively whenever possible and procedurally whenever necessary.
- **RFID hardware agnostic:** Extensible framework to build new drivers as well as use prebuilt drivers for leading RFID reader vendors.
- **Roadmap for sensor-based computing:** Oracle's RFID platform cannot only support RFID's but other types of sensors as well.

The Oracle RFID Platform enables a distributed infrastructure that can acquire, filter, and deliver massive amounts of real-time data for use by any enterprise application. The advantages to using the Oracle RFID solution include:

- **Fastest adoption of RFID technology:** Application developers do not need specialized knowledge about RFID hardware. Furthermore, any Java developer can access (RFID) data and events via simple standard interfaces.
- **Easiest integration with reliable Real-Time data:** Reliable-messaging framework offers filtering and mapping of sensory events to business events and real time communication with sensors, edge computing, as well as business partners.
- **Investment protection:** Different hardware vendors offer different API's. Plug and play architecture for RFID readers isolate business applications from shifts in technology and standards and enables companies to refresh tag technology without affecting the applications.

ORACLE'S RFID DEVELOPMENT ROADMAP

RFID technology is revolutionizing the entire business process. The center of the business process is now the item, which can be tracked in real time. By definition, all the business processes have to be integrated to track an item as, for example, it passes through the supply chain to the consumer. Existing applications will need to be retooled to support this concept and the underlying platform has to support an event driven model as the item passes from one business step to another. Oracle is positioning itself to provide both the applications and the technological platform for RFID computing.

RFID HARDWARE CONSIDERATIONS

Oracle's RFID Platform is hardware agnostic, that is, the equipment that read data from the RFID tags, referred to simply as readers, can be from any manufacturer such as those made by Intermec, Alien Technology, EMS, and Matrics.

Oracle Enterprise Manager (OEM) provides system administrators with tools to monitor the complete RFID Platform, including the Oracle Database, all the components of the Oracle Application Server and Oracle Application Server Wireless for edge processing. OEM provides a central console for monitoring and managing RFID readers. Support for "heartbeat detection" ensures that the RFID readers are live. System administrators and technicians can be alerted via e-mail, pager, or voice when a reader goes down. Flexibility to support other alert triggers is also available.

RFID STANDARDS

OracleAS Wireless provides an extensive driver framework to enable integration with any new drivers that are available. These drivers enable any RFID reader to plug and play into OracleAS Wireless, thus isolating RFID applications built on this platform from the underlying hardware infrastructure. Oracle offers complete integration with Oracle Applications Warehousing/Logistics products for RFID triggered transactions for Shipments and Receipts—with more to follow. More than 150 adapters are available to integrate with foreign data sources and applications. Additionally, Oracle is actively recruiting partners to integrate their applications with the Oracle RFID Platform.

Oracle's RFID platform will support the industry's RFID standards as they are finalized. Oracle already has a lot of components that align with the AutoID/EPC architecture. For example, the Oracle Database 10g already supports XML and XQuery. The same infrastructure can be used to support PML. Oracle's directory service is suitable for supporting the Object Name Server.

In addition, to further understand the requirements of RFID enabled applications, Oracle is already working with a number of partners who are AutoID Center members to integrate their products with the Oracle platform. Oracle will track EPCglobal and AutoID Labs as it defines the standards for RFID computing both for the platform and application and ensure that our RFID solution meets operation requirements and delivers the most scalable and cost effective solution. Oracle is also in the process of joining EPCglobal.

INTEGRATION

Together, Oracle Database 10g and Oracle Application Server 10g provide the basis for sensor-based computing. Oracle's solution is an event-driven architecture that is deeply integrated with state of the art data management, and hence can capture, store, filter, process, and analyze events from any source or sensor. Oracle supports bar code readers today and Oracle's object technology supports complex and large data types. It allows applications support RFID to define information provided by other sensors such as temperature, laser, radiation, and so forth. Only Oracle provides the crucial integration of event and data management.

Oracle's RFID platform does not impose any constraints on the size and specification of the identifier. Oracle Database 10g's object technology supports complex and large data types as a fundamental object in the database. Oracle has successfully tested up to 96 bits and will support advances in industry technology and requirement. Oracle offers complete integration with Oracle Applications Warehousing/Logistics products for RFID triggered transactions for Shipments and Receipts—with more to follow. More than 150 adapters are available to integrate with foreign data sources and applications. Additionally, Oracle is actively recruiting partners to integrate their applications with the Oracle RFID Platform.

MANAGEABILITY, SCALABILITY, SECURITY

The wireless component of the Oracle RFID Platform integrates with leading RFID readers/writers and can support both scenarios described. The platform manages, processes, and intelligently filters data from RFID readers before passing it on to the database server and the data warehouse for business intelligence and mining. RFID tags can contain a unique identifier (e.g., EPC number), which is then used to lookup information about the item that it is attached to in the backend (e.g., database). Information about the item can also be directly written to the RFID tag itself if supported (e.g., Intermec's passive tags support read and write), thus eliminating the need to have a lookup service to obtain information about the device. The Oracle RFID Platform also supports using a combination of storing/retrieving information directly off of the RFID tag and using the unique identifier to lookup information stored on the backend or Internet.

Oracle provides one single interface to access all information (e.g., tag data and PML service) about a tag at any particular point in time. This provides application developers significant advantages over systems where the tag data interface is different from the lookup service interface.

The Oracle RFID platform enables event filtering at several different levels. The wireless server integrated with RFID readers at the edge performs first level filtering. Oracle Application Server Wireless provides built-in support for redundancy filters, collision detection, cross reader filters, and high level filters (e.g., pallet level vs. item level). The application server in the middle tier can do further filtering before passing the data up to the database server.

Oracle provides customers with the flexibility to further define filtering rules. Oracle9i Release 2 provides a rule engine where arbitrary data can be filtered based on interests. This provides users with the ability not only to react immediately to "raw" events from the edge, but also react to "derived" events coming from operational and warehouse data. This allows users to "stay on top pf things" with little or no programming.

A key feature to Oracle's RFID Platform is the support for a new data type—EPC data type—in Oracle Database 10g. Based upon this, identity management is enabled via EPC lookups. Queries of the current state of data, rules evaluations, continuous queries (CQ), and flashback queries (these are queries to previous states of the database) provide the most comprehensive tracking of items. Flashback data effectively provide a full history of every (current and historical) status of all items in the supply chain. Further analysis is accomplished and delivered with Business Intelligence and Portal.

Lifecycle management requires a journal of states of data, which is a very difficult task. This requirement leads to complex data structures and significant additional development cost. Experience shows that the access to historical data is even more difficult. To provide a solution to this problem, Oracle provides users not only access the current state of data but also access to previous states. Users can ask any query in reference to any point in time, assuming the underlying data are still available. This functionality allows users to find any state that ever existed, or, looking from an auditing perspective, to access any historical state. No programming is required to. Additionally, Oracle allows users to access data in a time period, giving user access to the evolution of individual records and record set (or collections).

Oracle's RFID solution provides a solid opportunity to achieve scalability. Oracle Database 10g, Oracle Application Server 10g, and Oracle Application Server Wireless 10g are based on the latest grid technology and designed to scale on demand. The Oracle technology is known for its high level of scalability. The database provides a unique feature call Real Application Cluster (RAC).

Oracle's RFID Platform offers fine grain security and a framework for device and data security policies. As the industry develops best practices, Oracle will offer prebuilt policies that support industry recommendations and legal practices. Companies can also define their own policies using the security framework.

Security of information associated with a tag and hence with an item can be set at several levels. The Edge Server can be set to only monitor certain types of tags and items. Fine grain security of data stored in the backend allows the flexibility to define access rules both upstream and downstream. For example, as an item travels through the supply chain, each supply chain participant can enable future participants to see the history of the item. At the end of the supply chain, a store has the option of giving the customer access to the history of data on the item. On the flip side, each participant in the supply chain can prevent future participants from accessing the history of the data on the item.

The Oracle RFID Platform supports data federation via a distributed architecture, which includes a central server that can connect to multiple Edge Servers, each of which can manage multiple readers. Data can be federated at the Edge Servers by summing together reads from multiple readers, and additionally, low-level events filtered from the Edge Server can be federated with other data on the backend and then passed on to the application.

ORACLE RFID PARTNERS

Oracle's RFID solution partners are essential for Oracle to deliver a complete RFID solution to new and existing customers. These partners will leverage Oracle's expertise in data management, business process management, and application development while providing essential value added services or capabilities at critical points within the RFID value chain. Together, Oracle and its partners will continue to build and integrate RFID components to give customers an increasingly complete and integrated set of RFID services and solutions.

Below is a partial list of partners with whom Oracle has currently achieved a level of integration or interoperability. These partners are particularly important because many customers plan to conduct pilots or trials in the near term and they often have constrained budgets.

The near-term opportunity is for Oracle to leverage these particular partners to minimize Oracle' cost of sales as well as investment risk on part of the customer. Over time, the number and depth of RFID partners will increase significantly, providing Oracle with greater flexibility and Oracle customers greater value. The upcoming RFID Compliance Package, RFID Pilot Kit, and availability of Oracle's new edge services in Oracle Application Server 10g will be a key enabler for growing this population of partners, which includes:

- Alien Technology is a leading supplier of radio frequency identification (RFID) hardware that enables consumer packaged goods companies, retailers, and

other industries to improve their operating efficiency throughout their supply chains. Alien Technology is a leading supplier of radio frequency identification (RFID) hardware that enables consumer packaged goods companies, retailers, and other industries to improve their operating efficiency throughout their supply chains.
- Intel is widely recognized as one of the world's most advanced manufacturers, and is reviewing best-known methods from around the world, as well as developing pilots to test the usage and value of RFID technologies and processes within our own high-volume manufacturing facilities. Intel is evaluating current practices and exploring RFID's use to speed the flow of materials, enhance tracking and tracing, and enable breakthrough improvements in manufacturing productivity.
- Intermec RFID systems are well suited for a wide range of applications that include inventory and asset management, access control, compliance tracking, and personnel and vehicle identification. Intermec offers the broadest range of RFID hardware, software, and implementation services, all compatible with adopted or emerging standards, ensuring interoperability and providing a clear technology migration path. Oracle has partnered with Intermec and together we have developed drivers to Intermec's RFID readers. The Oracle RFID Platform provides an abstraction layer above the readers, such that application developers do not have to integrate directly with the readers and manage, filter, and process the raw data reads.
- For more than 18 years, Loftware has been providing advanced barcode and label printing software technology, including RFID. Loftware's global marking solutions are hardware and printer independent, and extremely robust within high-demand environments. Loftware provides both server and client side technology, allowing mission critical applications to meet the barcode compliance needs within the global supply chain.
- Pragmatyxs is an e-business firm specializing in "last mile" integration of labeling, data collection, and RFID solutions. Pragmatyxs works closely with its partners, Loftware, and Intermec, for enterprise label and smart tag printing, data collection, and RFID devices.
- Printronix is a leading provider of industrial and back-office enterprise printing solutions for customers the world over. The long-term market leader in line matrix printers, Printronix has earned an outstanding reputation for its full selection of thermal, laser, and network solutions technologies, all supported by unsurpassed service.
- Tata consulting services (TCS) is Oracle Worldwide Certified Advantage Partner with more than a decade of experience with Oracle's products. TCS offers the following RFID services: consulting, application development, system integration, training, and support.

THE HORIZONTAL AND VERTICAL MARKET STRATEGY

Successful technologies provide solutions to real business problems. The path to success is a logical progression[3]:

- Determine the industry context of opportunities and necessities
- Determine the organization's context with the industry to form a business plan

This is the process followed by McCarran Airport. They did not simply say, "RFID for Baggage Handling is the latest in technology that should be deployed." Instead they determined that baggage handling was a part of a larger business strategy within an industry where baggage handling has been a historical problem looking for a solution.

Airports have been looking at a theme entitled Common Use Self Service (CUSS) whereby certain functions traditionally managed by the airlines are outsourced to the hosting airports who in turn bill the airlines on an as facilities are used. This airport-airline relationship is not new as runways have been a shared resource since the beginning of the modern flight. What is new are the services being discussed—kiosks, gates, clubs, and baggage handling.

Baggage handling, therefore, is but one aspect of the CUSS industry-wide theme being embraced by airports and airlines all over the world in an evolving manner. At a typical airport, and McCarran was no exception, 10% of all bags have been misdirected. McCarran averaged about 65,000 bags per day so 6,500 were misdirected. The cost of reuniting a bag with its rightful owner can be as much as $150 per customer. Rounding the cost estimate, misdirected bags could easily be a $1 million per day problem.

McCarran decided to pursue CUSS. Today you can find evidence of success with shared kiosks (icons of all airlines appear on the opening screen), shared gates, and shared baggage handling facility. Each of these were part of the business strategy. The IT Plan for baggage handling required a $125 million project to transform to an RFID system which, after testing, shown a less than 1% misdirection rate. The savings for the airport and the airlines can be measured in real return on investment—the project payback period is about 6 months. The intangible history comes in enhanced customer service and satisfaction.

Oracle's investment in FRID technology is formidable. What do the due diligence criteria say about the decision to add RFID technology to the company's repertoire of products and services?

DUE DILIGENCE

The following applies the due diligence criteria to the RFID technology investment. The due diligence lens adopted here is that of a company trying to generate additional revenue and profit through investments in new technology-based products and services. In this case, the technology is RFID.

The "Right" Technology

As suggested in Chapter I, the "right" technology assumes that the technology investment target is part of a larger trend. What is RFID technology? Is it a technology concept, a prototype technology, or a whole technology cluster?

Technologies can be mapped on to an impact chart which reveals that many of the technologies about which we are so optimistic have not yet crossed the technology/technology cluster chasm—indicated by the thick blue line that separates the two in Figure 2.

Technologies in the red zone are without current major impact; those in the yellow zone have potential, while those in the green zone are bona fide. The chasm is what separates the yellow and green zones. Note the location of RFID technology.

RFID technology has not crossed the chasm, yet. Is it well on its way? Absolutely, but there is still uncertainty about standards, the technology itself, cost, and support, especially as it involves the storage and analysis of all of the data that active and passive RFID tags will generate. Oracle's investment in the technology represents a controlled gamble. It is controlled because Oracle is in a position to actually influence the direction the technology takes. But make no mistake that investments in RFID technology, involve some risk.

Few or No Infrastructure Requirements

Technology solutions that require large investments in existing communications and computing infrastructures are more difficult to sell and deploy than those that ride on existing infrastructures. If technology managers have to spend lots of money to apply a company's product or service, they are less likely to do so, if the choice is another similar product or service that requires little or no additional investments. Vendors, like Oracle, are well aware of the caution that surrounds new investments.

Is the light green here? It is yellow (moving to green).

RFID comes with a price-tag. The data base infrastructure must be capable of receiving and processing the data generated by the tags. The communications infrastructure must be capable of securely receiving data. Investments in RFID

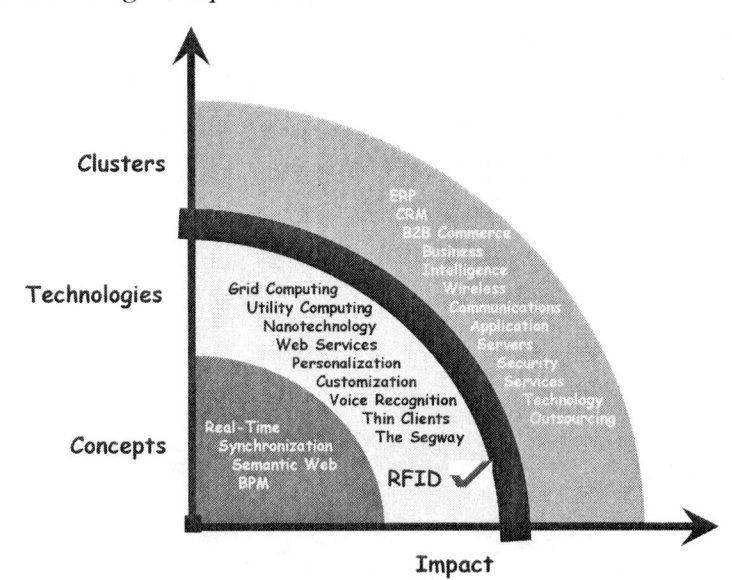

Figure 2. Technologies, impact and the chasm—and RFID

standards must be made to make sure that there is communications and analysis compatibility. In short, RFID technology comes with some strings attached, strings that are changing as the technology moves toward standardization. Companies like Oracle that choose to invest in RFID technology-based products and services are gambling just a little on the future of the RFID market size and on the technology standards and standard processes that will emerge over time. In order to ease the concerns that technology managers might have connected with the adoption of RFID technology, Oracle must make sure it provides paths to infrastructure readiness to prospective customers of its RFID products and services.

Budget Cycle Alignment

As Figure 3 suggests, the primary focus of RFID is initially strategic. This is how Oracle would like to position the technology. Their investments are positioned to help their prospective customers exploit their supply chains, logistical operations and, ultimately, their market share. RFID is initially strategic but over time will become more tactical and part of technology infrastructures. The competitive advantages that Oracle expects its customers to achieve via the adoption of FRID technology will initially be strong, but over time, as more and more companies adopt essentially the same technology, the competitive advantages will diminish.

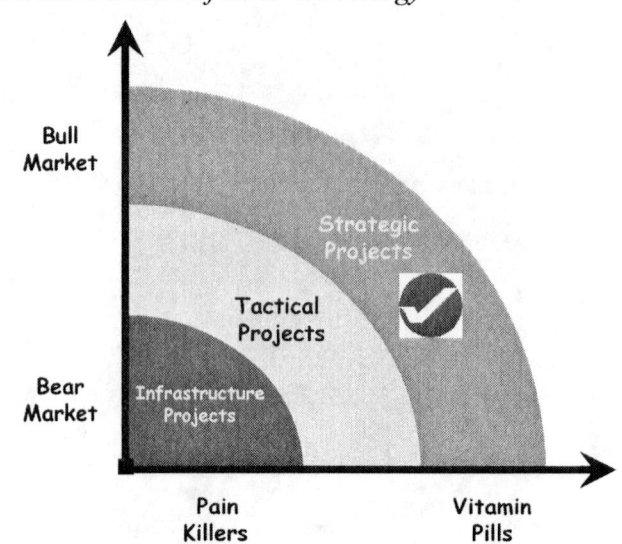

Figure 3. Investment drivers of RFID technology

During the course of its investments in RFID technology, Oracle remained financially strong and while the global market for technology suffered during the 2000 to 2004 period, there was still lots of money available for technology investments, especially if there were expected to yield downstream dividends. Clearly, RFID technology qualified for continued investment during what was otherwise a relatively bear technology market.

Quantitative Impact

When standards settle and industry deploys RFID technology across its manufacturing, distribution, and retail systems there is every reason to believe that cost savings will be substantial. At the same time, impact is still hypothetical, save for early data from some pilots. Oracle is "promising" cost savings as well as process efficiencies, that RFID technology will not only solve some old problems but will generate more business.

It is impossible to know precisely what the quantitative impact will be prior to a technology investment. To a great extent, we make estimated judgments about impact and then hope for, and try to influence, the best. On the other hand, good due diligence teams will talk extensively with others who have deployed a specific technology in an effort to reduce the uncertainty in the impact projections, and of course pilot projects are always a necessary part of the uncertainty reduction plan. Oracle has already deployed the technology in several locations, with excellent results. It is Oracle's job to demonstrate that the technology can save money, make money, and improve customer and supplier relationships, among other quantitative results. They are proceeding well toward this objective. But individual customers will have to conduct their own pilot projects to determine the quantitative impact RFID technology might have on their operations.

Changes to Processes and Culture

If a product or service requires organizations to dramatically change the way they solve problems or the corporate cultures in which they work, then the product or service will be relatively difficult to sell. I have said this before and it applies to the adoption of RFID technology as well. Oracle needs to make sure that its customers can adapt their processes to RFID technology in ways that do not disrupt current manufacturing, distribution, and retail processes. The RFID technology suite that Oracle has developed is intended to do just this, though there will inevitably be processes they have to define and integrate into the new technology. Overall, the assessment here is mixed: industry (Oracle's customers) may have to redesign some key processes to fully exploit the capabilities of RFID; and because RFID technology is part of collaborative supply chain planning and management, the processes will have to be at least semijointly redefined with partners and suppliers. This is always challenging and may well require Oracle and its partners to step up—as they have in the past—to helping their customers exploit their new RFID products and services.

Solutions

Companies are always on the lookout for "solutions" that solve as many problems as possible. Vendors, like Oracle, are aware of these requirements and try to provide just the right mix of services for the right price. RFID technology-based products and services are not end-to-end solutions for the technology infrastructure or architecture, but they do represent—as bundled by Oracle—a segmented RFID "solution." This means that the integration and interoperability requirements of the technology are largely solved by Oracle, especially if the customer is already an Oracle data base management customer. While the nature of RFID technology is not "stand alone" or "end to end," Oracle has bundled the elements of the technology enough for it to qualify as an integrate-able partial solution to a whole host of larger requirements.

Multiple Exits

The Oracle team is betting heavily on RFID technology. Is this a good bet? Probably. Regardless of where the technology settles, there will be a large market for FRID technology. The only risks that need to be mitigated are the size and speed of the investment: too much money too fast could create an imbalance in the supply of and demand for RFID technology.

Horizontal and Vertical Strength

The best products and services are those that have compelling horizontal *and* vertical stories, since customers want to hear about industry-specific solutions or solutions that worked under similar circumstances (like for a competitor). Without a good vertical story, it will become more and more difficult to make horizontal sales. CIOs expect their vendors to understand their business. Smart vendors, like Oracle, organize themselves horizontally and vertically to appeal to their clients.

Oracle is primarily a horizontal vendor: its products and services are used in multiple vertical industries. At the same time, it organizes and customizes its products and services for specific clients in specific vertical industries. RFID is initially horizontal but it also has immediate vertical applications. RFID is thus a true horizontal/vertical enabling technology.

Industry Awareness

There is no problem here: just about everyone on the planet is aware of RFID technology. Some would say that the technology industry itself is hyping it as one of the next "killer apps." Industry research organizations, standards setting organizations, and pundits are all talking about just how important the technology may be to many vertical industries.

Partners and Allies

CIOs expect a broad network of support for the products and services they purchase. Oracle has a long history of partnering with lots of third party vendors and consultants. The Oracle channel of partners is wide and deep—and they cultivate it. There is no danger of becoming orphaned with an Oracle product or service. All lights green here.

"Politically Correct" Products and Services

Technology investors will not risk their careers on what they perceive as risky ventures—even if the "risky" product or service might solve some tough problems. RFID is generally perceived as a hot technology—the "right" technology for a lot of applications. Oracle has a long history of risk-taking and innovation. The corporate culture, leadership, and politics all align with investments in new technologies expected to increase market share and margins. In short, RFID is easily "politically correct" for Oracle.

Recruitment and Retention

Oracle is a blue chip technology company that has little trouble attracting the best and the brightest. Buyers of Oracle's technology products and services can expect lots of available talent directly from Oracle and also from its substantial network of partners.

Figure 4. Oracle context

Differentiation

Differentiation is critical to success and while not every differentiation argument is fully formed when a technology is emerging, technology buyers need to understand why the product or service they are about to buy justifies their investment. Oracle has done a good job describing why it is RFID technology-based products and services are solid, secure, scalable, interoperable, and cost-effective. Industry analysts compare and contrast Oracle's RFID offerings with other vendors. CIOs, CTOs, and technology managers need to assess all of the differentiation data. That said, the largest vendors will have relatively complete stories about how their products and services are "better" than their competitors'. RFID vendors should offer pilots to demonstrate just how different their products and services are; buyers should demand that the largest RFID vendors subsidize these pilots.

Experienced Management

The key here is to see the right mix of technological prowess and management experience available to develop and deliver a successful product or service. There are other ideal prerequisites: experience in the target horizontal and/or vertical industry, the right channel connections, the ability to recruit and retain talented personnel, and the ability to work industry analysts, communicate, and sell. The Oracle RFID team is experienced and smart. The company is solid and has an enviable reputation in the industry.

"Packaging" and Communications

As a company, Oracle is mindful of the importance of packaging and communications. Their Web site, www.oracle.com/technologies/rfid/, demonstrates the importance the company attaches to RFID form and content—necessary in today's competitive business technology world. Oracle sponsors RFID conferences and meetings and already has a substantial number of white papers that describe its approach to the application and support of the new technology.

CONCLUSION

The due diligence process yields nearly all green lights. But there are some key uncertainties that Oracle and its prospective customers must address. Oracle, like many vendors, is making a bet on the viability of RFID technology. As suggested above, the bet is almost certainly a good one, but there are still standards and other technology issues—like the cost, security, and reliability of RFID tags—that must be resolved before widespread deployments can take place. Oracle has a vested financial interest in the early adoption of RFID technology. Its investment in RFID makes perfect sense for a company of Oracle's stature. The buyers of RFID technology, on the other hand, should proceed somewhat cautiously, and push Oracle (and the other RFID vendors) to complete the RFID story—which they will.

ENDNOTES

[1] This example was prepared by Ralph Menzano with Stephen J. Andriole
[2] "4 New Technologies Will Improve Cargo Handling and Logistics," 2004 Gartner, Inc., May 20, 2004
[3] Making IT Happen, Ralph Menzano, Publications Unbound, 2001

Chapter VIII
Technology Product and Service Development in an Enterprise Architecture Service Capability:
The LiquidHub Case

INTRODUCTION TO THE CASE[1]

This is an example of an investment in a new technology service. LiquidHub, Inc., assessed the market and its own service capabilities and decided to make an investment in enterprise architecture to better serve their clients. A significant investment was necessary to understand the market, develop an overall go-to-market enterprise architecture strategy, and retool professionals to be able to provide the very best services available. Would the investment pay?

BACKGROUND

When LiquidHub was founded as a systems integrator and technology consultancy in early 2001, it arrived at its name based on a commitment to address the enterprise integration needs of their clients. They needed some way of tying their technology together, a "hub." But such a hub is formed from a complex set of solutions and processes, always changing, so a solution to this problem needs to be adaptable, to be "liquid." In early 2001, the potential of e-business had not been fulfilled and many organizations felt frustrated and feared that the promises of the Internet would be as fleeting at those that fueled the dot.com craze. While they shared the sense that the fundamental issues surrounding enterprise systems integration remained unsolved, they were sure that the integration of the Internet into businesses was not a fad. At issue was determining—indeed defining—how that integration would evolve. LiquidHub's service model always focused on its core competencies in IT strategy, applications architecture, and applications development. But when the company went to do IT strategy or to discuss architecture with a client, it too often seemed like they were being asked to devise that strategy only in the context of immediate needs, with the IT decision divorced from the business needs today or tomorrow. This realization began a 4 year effort to develop a powerful alternative approach, one that could address the needs of their clients, structure their engagements with them, provide long lasting value for them, and differentiate LiquidHub in the technology services marketplace.

THE CHALLENGE

The company's clients face some difficult challenges in their efforts to expand their business capabilities, including those enabled by information technology. As they completed multiyear, cross divisional initiatives such as ERP, customer relationship management (CRM), and supply chain management (SCM) implementations, both technology and business management professionals realized that it had become increasingly difficult to envision a rational systems- and process-integration approach. While each of the leading vendors in these spaces touts the ease of integration their solutions affords, integration has actually grown more vexing as the sheer number of "enterprise" systems has increased. Today, one of the largest obstacles to enterprise technology initiatives is that they are inconsistently defined, not only by the technology providers, but also by the enterprises themselves. Clients face a number of challenges in this increasingly critical and increasingly complex IT environment.

Among other things, they need:

- **Agility:** The accelerating rate of change in business cycles and the difficulty in maintaining an equivalent rate of innovation within an enterprise's technology architecture cause many organizations to lose their edge, resulting in missed opportunities, inflexible processes, decreasing share-holder value, and

poor customer satisfaction. Enterprises that do not develop agile enterprise architectures will be unprepared to deal with a consumer-driven economy that is based on shifting alliances, quick product life cycles and global competitiveness.
- **Efficiency:** Organizations now realize that the Internet is not just a low-cost distribution channel, but also a low-cost technology platform—enabling an increasing number and diversity of businesses to share information and to coordinate commerce processes. The Internet and its associated technologies provide a powerful, flexible, and extremely cost-effective competitive advantage, with legacy investments and processes given equal weight. The challenge is figuring out how to take advantage of the efficiency gains the Internet might give them.
- **Competitive capability:** Organizations need to enter new global markets quickly and gain competitive position, while simultaneously growing in existing markets, connecting the enterprise with local partners, suppliers, and government agencies. In addition, many organizations have achieved their growth via aggressive merger and acquisition strategies, so rapid deployment of common business processes and integrated technical infrastructures is necessary to keep competitive advantage. Finally, competition also necessitates close collaboration, particularly as the organization relies on ever increasingly complex interactions with a greater number of suppliers, customers, employees, and trading partners.

In short, the company's clients need to integrate existing technology investments with evolving applications architectures while addressing their changing business needs.

THE APPROACH

The journey to address clients' needs followed two parallel paths, which eventually converged into the service offering now called The enterprise services transformation roadmapSM (ESTR). The first (or "top down") path followed the discipline of enterprise architecture (EA) and resulted in a process for doing pragmatic enterprise architecture planning (EAP) based on our simplified EA Reference Model. The second ("bottom up") path followed the evolution of integration technologies, through EAI, then adopting service oriented architectures, and finally adopting a broader approach through enterprise services architecture planning.

At the confluence of these two paths, we arrived at an integrated model that helps clients understand how to plan for the future, even when that future seems

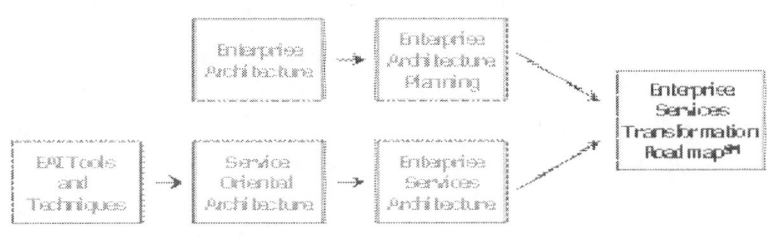

Figure 1. Paths to enterprise services transformation

uncertain from a business and/or technology perspective. At the same time, this approach helps firms begin to simplify their often-chaotic technology architectures. This unified planning approach, resulting in a dynamic roadmap for our clients to follow, has brought clarity to our clients' technology planning efforts as well as our implementation engagements.

The Enterprise Architecture Path

We first realized that a comprehensive understanding of the whole environment, one encompassing both IT and business considerations, was critical for success. The company began by adopting (then adapting) some of the best practices in enterprise architecture. Since their clients' business cycles continue to shrink, often from several years to just a few months, both the business units and the information technology (IT) organization experience enormous pressure. Adding to this pressure is the fact that these two communities historically "speak different languages" and are infrequently synchronized in their intent or their methods. Once an enterprise addresses its two distinct but related architectures, one for business and one for technology, it can align and synchronize them. That new framework, or enterprise architecture, can enable a highly responsive and efficient organization by synchronizing business and IT. For organizations competing in a rapidly changing economy, one that is competitive, global, and unpredictable, survival depends on agility and responsiveness to changing industry and economic forces, addressing product commoditization, government regulations, emerging technologies, and globalization.

Over the last decade, many organizations designed technology architecture roadmaps as part of their annual information technology strategic plans. These technology roadmaps and associated technical blueprints represented paths toward reaching a common technology vision across an enterprise. However, IT planning was typically done following the completion of the business strategy, making it a tool to drive cost reduction via operational efficiencies rather than a well-considered method to further existing business aims or shape new ones. In addition, the joint pressures of Year 2000 remediation and the rise of the Internet prompted many organizations to react, rather than act, making decisions out of a sense of urgency that was not integrated into a longer term vision for technology in the enterprise. And historically, IT organizations have been closely aligned with finance, reporting up through the CFO, positioning technology as subservient to an overall business strategy. It comes as little surprise that the traditional architecture models developed in such an environment are not able to support the flexibility that business process outsourcing, new business alliances, and the introduction of Web services, portals, and collaboration demand.

While the concepts behind enterprise technology architecture have been available for years, many organizations have not successfully implemented sustainable architectures that bridge the gap between business and technology. Industry best practices, combined with the experience of our own clients, has led us to bridge that gap with enterprise architecture, a framework that allows an organization to move beyond the historical, technology-oriented focus of traditional architectures to one that encompasses business architectures (including business process models, operations management, organizational models), technology architecture, and governance.

Depending on the IT and business cultures of a given organization, an enterprise architecture can be used in a variety of ways. As a reference model, it can define and inventory the relationships between an organization's business processes, IT assets, and project portfolio. As a method or process, it is a strategic planning tool that also functions as a central repository for all standards, guidelines, and policies that support consistent application integration and maintenance at the enterprise level.

As a reference model, enterprise architecture not only forms the foundation for a holistic analysis of an organization's current capabilities and future needs, but also provides the means to better scrutinize areas of focus for necessary technology investments. And, as a process, enterprise architecture helps to optimize the portfolio of projects that represent the current-state to future-state transition on an ongoing basis.

The most successful organizations are those who are able and willing to integrate their business and technology priorities, linking them through a common set of architectural principles and standards, with enterprise governance supporting the enterprise architecture. Together, business architecture, technology architecture, and enterprise governance can define a balanced approach to planning and executing on a strategic direction.

Business architecture specifies these supporting processes within the organization as well as those processes that are shared with the organization's suppliers, trading partners, and customers. Because they often contain the business logic that details these processes, enterprise application suites are also included in the business architecture. Whether they are labeled as ERP, as supply chain management, financials, back office, HR systems, or CRM, these software suites are usually configured to hold an organization's business process knowledge and workflow. In parallel, technology architecture is a logical model of the technology principles, models, and standards that are defined to support the organization's business model and architecture. It is used as a technology reference blueprint that supports ongoing planning and design of the enabling information systems critical to a business model and strategy.

LiquidHub's EA reference model helps clients articulate the current and future state of the enterprise's processes and technology platforms in a single diagrammatic statement. Having made visible these resources and their potential connections to one another, the diagram helps organizations to frame relationships between business and IT architecture, strategic planning, and ongoing governance.

From a planning perspective, enterprise architecture is an effective technique for planning, designing, implementing, evolving and managing the business and technology architecture. The major difference between a theoretical architecture, one that is used as an idealized future state model, and a practical architecture involves the process of its implementation. As a process, enterprise architecture connects the business drivers, the technology investments, and the realities of project priori-

Figure 2. Enterprise architecture reference model

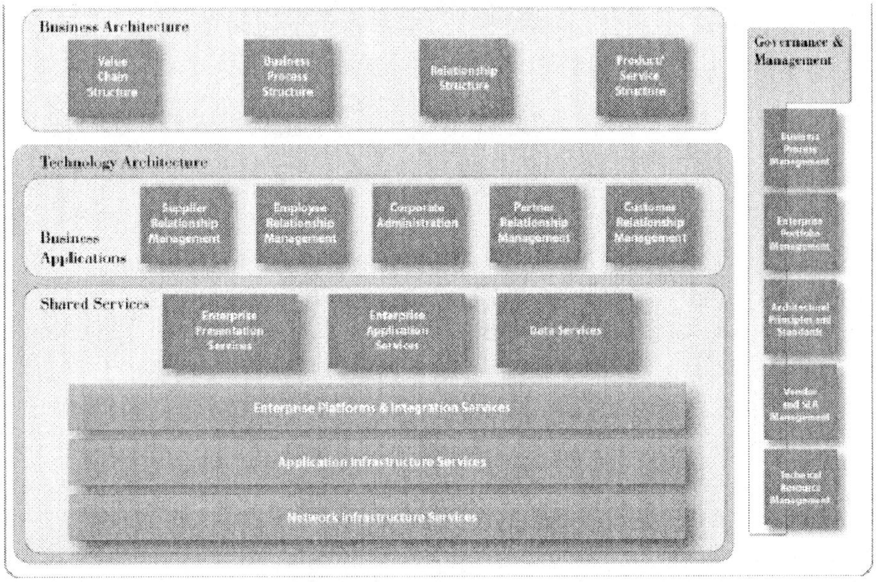

tization within an enterprise. Portfolio management—the ongoing prioritization, supervision, and evaluation of a series of technology and business projects—actually ensures the successful implementation of an enterprise architecture.

But even with these powerful "top down" tools, clients continued to face the day to day questions of IT management:

- How should I build new applications?
- How do I reduce costs in technology selection and implementation?
- How do I get ahead of the complexity of my applications architecture?
- For answers to these questions, clients looked to the applications integration space, to find the "glue" that would tie applications together, hopefully melding them into a unified technical framework.

The Enterprise Services Path

LiquidHub's pedigree is as a systems integrator, skilled at applications architecture and development. So while EA provided a disciplined process to work with clients in business and technical strategic planning, clients also sought help solving day-to-day technology implementation and integration issues. In short, clients still needed solutions to the problems generated by the growing complexity of systems and the need to integrate them.

EAI

During the Internet boom, the difficult problems of systems integration remained behind the scenes. From the mid-1990s onward, enterprise application integration (EAI) provided an infrastructure to connect information sources, acting as a go-between for applications and their business processes. Early applications running on monolithic systems served individual target departments well, but information exchange between systems was limited. The evolution of client-server technology and improved relational data management technologies expanded the tools available for integration.

Enterprise application integration software suites evolved to tackle the complex requirements of application integration, providing key areas of functionality, including:

- An integration broker (providing a set of services for message transformation and intelligent routing)
- Development tools for specifying transformation and routing rules and for building adapters into applications

- Off-the-shelf adapters for popular enterprise packaged applications (e.g., SAP R/3)
- Monitoring, administration and security facilities
- Message oriented middleware (MOM)
- Business process managers, e-commerce features, and portal services

EAI-enabled business process integration throughout the value chain can produce huge benefits for companies, but even with these products, the path to get to these benefits was often arduous. Three major problems that companies ran into while utilizing an EAI system are:

- A divergence of modern application architectures, with developers following Microsoft's .NET architecture as well as Sun's Java 2 Enterprise Edition (J2EE) on a variety of platforms (IBM, BEA, Open Source, etc.)
- A trend within IT to purchase "packaged" applications to satisfy niche business requirements
- The development of a wrapper for legacy applications, which may have no integration interface at all, is required to facilitate an integrated architecture

Implementing EAI tools was still more challenging because of their heterogeneity: each product required users to interface with systems and the workflows in its own way. The challenge faced in EAI is in the very separation of its elements. While these point solutions, such as message brokers or application servers, provided desperately needed functionality to the IT organization, the diversity of vendor solutions coupled with constantly changing standards have made true organization-wide EAI seem an unattainable dream. Numerous vendors supply application servers, message brokers, and business process management tools. Others have entered the pure play EAI market providing integration solutions that combine other integration technology concepts through proprietary implementations. Given these challenges, while we continued to work to implement EAI technologies, we did not see it as a strategic technology that would address our clients' long term business and technical challenges.

Service Oriented Architectures (SOAs)

EAI is focused on allowing multiple disparate applications to work together, sharing data, processes, and other critical functionality across the enterprise. But the disparity of EAI platforms makes it difficult to address the core problem—increasing systems complexity. In contrast, the service oriented architecture model does away with middle-tier message brokers and instead exposes application functionality as services. These services are made available for synchronous invocation, similar to API calls, but in a standard format. Fundamentally, SOAs are focused on wrapping existing applications with a well-defined interface that transforms the application into a set of services accessible by other applications. SOA is enabled by the emergence of widely accepted interface standards, encoding standards such as XML, and a new class of applications known as Web services which make SOA available via discoverable Web-accessible interfaces. By enabling the loose coupling of applications, organizations can pursue a previously risky "best-of-breed" approach when selecting business logic solutions. In addition, transforming existing architectures to a services model drives more rapid and cost-efficient development by removing a layer of complexity.

SOA should not be considered solely from an IT perspective. It is important to remember that the business side of an organization plays an integral part both in defining and realizing the benefits of SOA. SOA, then, can be defined as a set of design and organizational principles that expose business logic and data as independent reusable services. The overall objective of SOA is to structure the organization such that business vision and requirements can drive IT towards the same goals without technology impeding achievement of objectives. Exposing IT assets as reusable services gives business managers the ability to capitalize on previously unrealized integration points. For example, exposing customer-buying patterns to an order fulfillment model could enable more direct target marketing or the up sell of complementary products.

From an IT perspective, SOA is a method of interfacing two or more disparate applications via what is known as "loose coupling." In a non-SOA approach, applications are typically tightly coupled, that is, each application is directly aware (connections are predefined) of the applications it integrates with and the interfaces between applications rely on business logic for successful processing. Loose-coupling, in contrast, separates the interface components from the business logic of the underlying applications. This is an advantage, in that new connections can be made between applications by changes to only the interface layer. This process of "wrapping" existing applications into a business service removes a layer of complexity that has traditionally been present between disparate systems.

This model allows business managers and IT organizations to treat applications and processes as defined components that can be mixed and matched at will. The SOA model also supports strong IT governance in that the business services layer relies on proven interface standards. Encapsulating business logic as described above allows IT organizations to tackle business requirements in a more focused manner and with shorter development timeframes.

Enterprise Services Architectures

Expanding the SOA concept beyond particular business processes and IT development is known as enterprise services architecture (ESA). ESA takes a holistic view of an entire organization, across disciplines, and identifies the high-level components that provide specific business/process functions. The focus of an ESA is to create a flexible IT architecture where IT assets are aligned with an agile business architecture, allowing adaptability to current and future business needs and thereby reducing the cost of transactions and overall analysis.

ESA is profoundly changing the way that software vendors build and use applications. Legacy applications are systematically being broken apart and redesigned as layers that can be leveraged as reusable services and components. This is important, as organizations today face redundancy in many areas of their architecture including systems, processes, and data. A common goal among IT organizations is to reduce this redundancy in order to remove complexities. Rationalizing systems, processes, and data into a superset of capabilities allows organizations to simplify their existing architectures. Removing complexities increases business agility by simplifying architectural decisions (fewer components to consider) and the loose coupling of a SOA allows for more rapid development. The ability to reuse existing services and components decreases project durations and thus lowers overall development costs.

In this context, it is clear why enterprise services architectures are getting serious consideration as business and IT leaders look to address upcoming development projects. The loose coupling and wrapper architecture of SOA has the potential to address the elusive goal of treating applications as services that can be dynamically combined to address real business problems, especially when addressed enterprise-wide.

ESA can add value to an organization due to its ability to align the IT organization with the business drivers of an organization. The process of abstraction that is completed as part of SOA removes application-specific complexities. This allows IT personnel to focus on how to best use their enterprise portfolio of applications and data stores to meet the business requirements. Prior to SOA, IT organizations' main concerns were related to the underlying technology and how to make disparate systems work in unison. The business problem to be solved frequently fell outside of their purview.

It is important to keep in mind that SOA is not a short-term efficiency program. Organizations that invest in SOA are thinking long-term, willing to invest in a foundation that initially may not impact ROI (or even decrease ROI on particular development projects), because they recognize SOA will help them meet future business requirements in a more efficient and less resource intensive way. The pace of change is likely to increase going forward and business managers will have to continually adjust business models to meet customer and competitive pressures. To be effective, these adjustments will have to be implemented rapidly and without the interference of IT limitations. Although an SOA will not remove all IT obstacles, a well-designed SOA does give IT managers a new method for creating solutions based on proven business applications.

But implementing ESA should be a strategic investment, not merely a single project. To realize the benefits of a services model, organizations have to approach SOA with a combination of a tactical "bottom-up" approach and a strategic "top-down" approach.

Tactically, some organizations have begun by identifying legacy applications with data sets (or features) that, if exposed, could benefit other systems or processes (then creating services around the access of this data). Strategically, the process begins at the enterprise level with the identification of services that drive the organization technically and from a business aspect. This process gives rise to the high-level roadmap of all the technical and business services that define an organization; with roadmap in hand, business and IT units are able to evaluate subsequent tactical projects to ensure compliance with the overall direction of the organization, and can adjust the map to engage unexpected changes in circumstance.

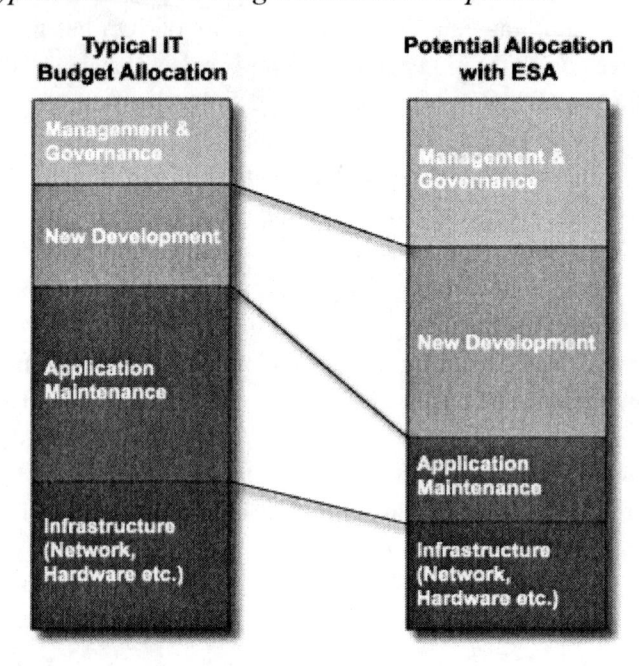

Figure 3. Typical and ESA IT budget allocation comparison

THE LIQUIDHUB ENTERPRISE SERVICES TRANSFORMATION ROADMAP

The combination of "top-down" and "bottom-up" approaches makes sense for organizations that must deal, simultaneously, with changing business environments and technical architectures. IT and business leaders need to be able to see the big picture and understand how technology enables it. IT decision makers, architects, and developers need to ensure that each technical implementation fits into a larger whole, and is reusable and scalable.

As beneficial as the principles of EA are, and as obviously valuable as SOA is, independently they do not give clients the confidence and clarity to answer the simple question "how should I proceed?" That is still another reason the roadmap is so useful. It gives direction, shows the stakeholders what steps to follow in planning for their enterprise business needs, implementing service oriented architectures, and building on those investments as their organization grows and changes. Intuiting how valuable this map would be—both in itself and as a reminder that the process involves charting both trajectories—led us to the planning and implementation methodology we now call The enterprise services transformation roadmap, or ESTR. The business survival of many organizations depends on their ability to have the information they need in order to respond quickly and appropriately to rapidly changing market conditions. Acquiring and using that information often depends upon having equally agile enterprise technologies. Changing marketing conditions demand innovation, in business strategy and in the information technology that supports and enables it. The successful organization must have the confidence and agility that is based on a flexible and responsive technology infrastructure. Such an organization is an agile enterprise, one that effectively utilizes current assets and incorporates new technologies as the market and the business require. LiquidHub's enterprise services transformation roadmap helps organizations plan for technology simplicity and reusability, providing a roadmap to the agile enterprise.

Figure 4. The need for the agile enterprise

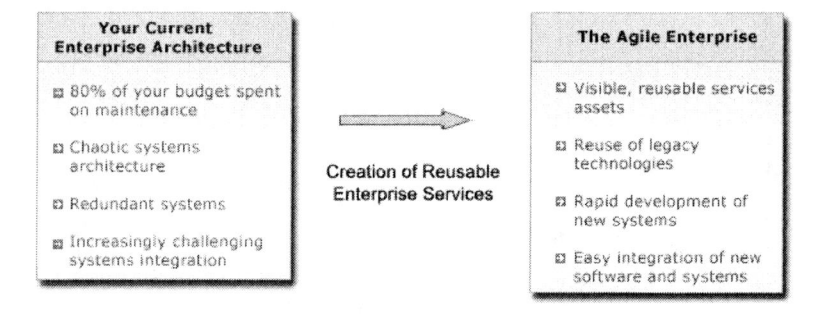

In response to organizations' need to maximize their agility through their information technology capabilities, ESTR delivers an incremental strategy and planning process that identifies and leverages the value of existing IT assets while delivering a flexible technology architecture that will work long-term. Based on the principles of enterprise architecture and service oriented architecture, ESTR provides LiquidHub's clients with a clear process for evaluating business needs, identifying existing technology and process assets, and planning the implementation and integration of new technologies in a way that ensures technology reuse and lower total cost of ownership.

ESTR is defined by four "streams" which define the technology and process domains. At each point along the streams, specific tasks and deliverables are identified that help an enterprise incrementally improve reusability and agility in technology implementation and integration.

The four streams are:

- **Program management:** How an enterprise manages its priorities and resources.
- **Enterprise business services:** How an enterprise documents and describes the basic business processes that underlie its operation.
- **Information management:** How an enterprise manages and utilizes the data, content, and information at its disposal.
- **Technology shared services:** How an enterprise creates reusable enterprise technology assets and scalable frameworks.

LiquidHub has worked with clients in life sciences, health care, financial services, pharmaceutical, insurance, and other key industries to successfully integrate new technology frameworks with enterprise and legacy systems. The enterprise services transformation roadmap helps our clients meet unique and changing needs, simplify new technology development, and become more agile. ESTR serves as LiquidHub's strategic implementation methodology, encompassing both planning and strategy expertise as well as our implementation experience and approach. Our clients continue to benefit from this unified approach, blending the best of enterprise architecture, service oriented architecture, and best practices in implementing them in real organizations.

DUE DILIGENCE

The following applies the due diligence criteria to LiquidHub's investment in a new service offering—Enterprise Services Transformation. The due diligence lens

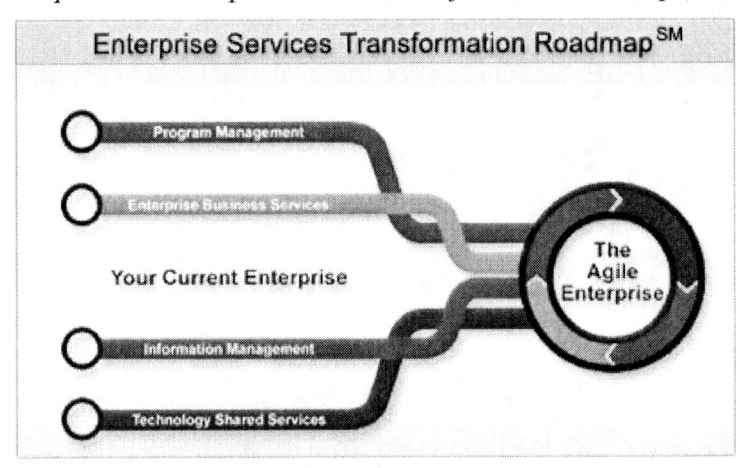

Figure 5. LiquidHub's enterprise services transformation roadmap (ESTR)

adopted here is that of a company trying to generate additional revenue and profit through investments in new technology-based service.

The "Right" Technology

As suggested in Chapter I, the "right" technology assumes that the technology investment target is part of a larger trend.

Technologies can be mapped on to an impact chart which reveals that many of the technologies about which we are so optimistic have not yet crossed the technology/technology cluster chasm—indicated by the thick blue line that separates the two in Figure 6. Technologies in the red zone are without current major impact; those in the yellow zone have potential, while those in the green zone are bona fide. The chasm is what separates the yellow and green zones. Note the location of SOA-EST technology services.

SOA-EST technology has not crossed the chasm, yet. Is it well on its way? Yes, but there is still uncertainty about standards, the technology itself, cost, and support, and LiquidHub is not large enough to actually determine the direction in which the standards or technology will move.

Is the light green here? It is yellow (moving to green). Is there risk here? Some but it is manageable.

Few or No Infrastructure Requirements

Technology solutions that require large investments in existing communications and computing infrastructures are more difficult to sell and deploy than those that ride

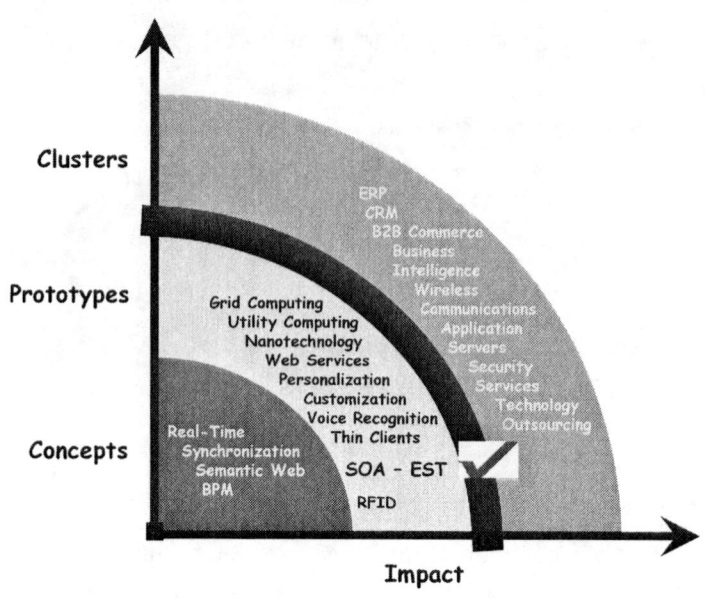

Figure 6. Technologies, impact, and the chasm—& SOA - EST

on existing infrastructures. If technology managers have to spend lots of money to apply a company's product or service, they are less likely to do so if the choice is another similar product or service that requires little or no additional investments. Vendors—like LiquidHub—are well aware of the caution that surrounds new investments. EST requires an investment. Companies that want to take full advantage of Web services and service oriented architectures will have to invest both in process and infrastructure changes. This might give some companies pause. LiquidHub is fully capable of providing strong business cases to their clients, but when companies are in cost-cutting modes they will resist expensive new initiatives. That said, EST pay back is substantial and far-sighted companies will be happy to make the necessary investments to achieve significant results.

Budget Cycle Alignment

As Figure 7 suggests, the primary focus of EST is initially strategic. This is how LiquidHub would like to position the technology. EST is initially strategic but over time will become more tactical. The competitive advantages that LiquidHub expects its customers to achieve via the adoption of EST technology will initially be strong, but over time, as more and more companies adopt essentially the same processes and technology, the competitive advantages will diminish.

Part of LiquidHub's challenge will be to convince clients that EST will add long-term business value. This can be done, and while the initial impact will be slow to gather, over time impact will be enormous. Smart companies will understand this early.

Quantitative Impact

It is impossible to know precisely what the quantitative impact will be prior to a technology investment. To a great extent, we make estimated judgments about impact and then hope for—and try to influence—the best. On the other hand, good due diligence teams will talk extensively with others who have deployed a specific technology in an effort to reduce the uncertainty in the impact projections, and of course pilot projects are always a necessary part of the uncertainty reduction plan.

LiquidHub's business case for EST contains quantitative impact data. Clients need to see that their investments in EST will pay quantitative dividends. At the same time, there are no guarantees in technology life so quantitative impact is by no means assured. Improved service and reduced cost—not to mention to ability to integrate new processes and technologies—are some of the impact metrics that are tracked by EST clients. Depending on how "chaotic" an infrastructure is, cost savings, improved services, and dramatically increased integration and interoper-

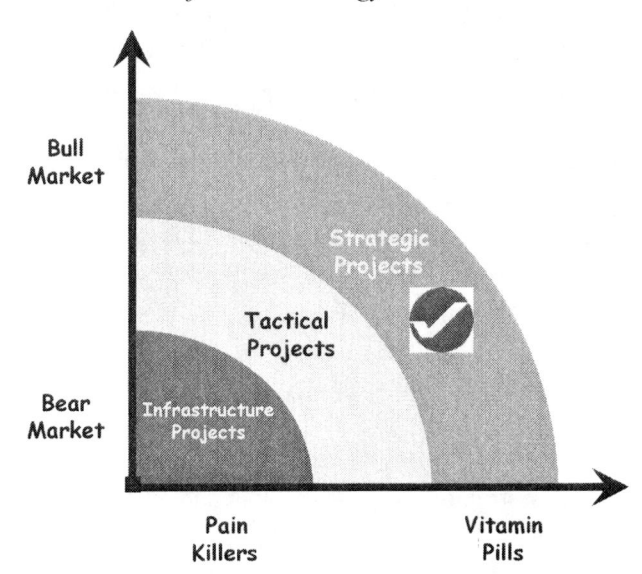

Figure 7. Investment drivers of EST technology

ability could all be in the 50% - 75% range (though companies not so chaotic would enjoy a 25% - 50% improvement range).

Changes to Processes and Culture

Perhaps the most significant change is what EST implies about a company's approach to business technology governance. Specific decisions need to be made to optimize EST investments. Some companies may balk at changes to their governance policies, but if changes are not made investments will fail. Beyond governance there are changes to applications, data and communications architectures that are also necessary to optimize a company's commitment to EST. LiquidHub will have to assist their clients with business cases to justify these changes.

Solutions

EST is a set of technologies, processes, models, and software. It is also a state of mind and a business technology acquisition and deployment philosophy. As such, EST is a solution, especially since it can affect the entire business technology relationship. That said, EST is an objective from which companies must migrate from their current infrastructure and architecture.

Multiple Exits

LiquidHub is betting heavily on the EST service. Is this a good bet? Very likely. There should be a large market for the service so long as LiquidHub (and others) raise everyone's consciousness about the importance of EST. Much of the business technology community has been relatively quiet about "killer apps" and "disruptive technologies" since 2000 and the collapse of the dot.coms. Major investments in new service offerings are risky. They are primarily based on trends analyses and plans to be ready to satisfy new client requests for help. Web services and Service Oriented Architectures represent two major trends that LiquidHub correctly anticipated in time to develop their EST service offering. If the company is wrong about these trends and their service offering is ill- or mistimed, they will lose a great deal of money. There are so alternative exits here: they need to be right or the investment is lost.

Horizontal and Vertical Strength

EST is a horizontal service offering that can be customized for specific vertical industries. EST is a very strong horizontal offering and a high potential vertical one. Green lights here.

Industry Awareness

The world is still abuzz about the Web services, SOA and now—through LiquidHub's efforts—EST. The company has produced White Papers, done industry interviews, presented at industry conferences, and communicated with the leading industry analyst organizations.

Partners and Allies

LiquidHub has partnerships with several leading vendors who provide access to software, support services, and even hardware. The company is well represented with especially the major technology vendors.

"Politically Correct" Products and Services

LiquidHub has gown dramatically since 2000 when it was cofounded by three entrepreneurs. The company is not "conservative"; rather, it is has an aggressive strategy to continue to grow regardless of the capital markets or the overall state of business technology spending. In fact, the company actually grew over 50% a year during one of the toughest times in the industry—the period from 2000 to 2003. Management is aggressive and smart. Timing is excellent as we move from the Web services and SOA pilot phase to where the technologies begin to define who new infrastructures and architectures.

Recruitment and Retention

LiquidHub has solid management and has continuously recruited new talent as they have grown and honed their service offerings. Perhaps just as importantly, the company has invested in its existing staff through in-house and external training.

Differentiation

Differentiation is always tough—especially when industry awareness about new technologies and approaches to business technology optimization is high. LiquidHub is not the only company to offer enterprise architecture services based on Web services and SOA. In fact, all of the major services vendors offer similar services. So what is the difference? First, LiquidHub is offering a turnkey service in its EST offering. Migration is a key component of LiquidHub's offering, a component that is sometimes lacking in competitors' offerings. LiquidHub also priced its EST services lower than the major technology service vendors adding another layer of differentiation. All of that said, ultimately differentiation will be determined by their performance and the performance of the competition.

Experienced Management

The senior management team at LiquidHub is solid and experienced. They have been at the helm of the company since its founding. They are the architects of the EST service. Clients can feel comfortable about the company's leadership.

"Packaging" and Communications

LiquidHub has packaged their EST offering well. As discussed above, the company has generated significant industry awareness of its EST offering and has participated in a variety on horizontal and vertical industry conferences. All lights are green here.

CONCLUSION

The LiquidHub EST offering is still new. Does the due diligence analysis suggest that their investment in this new service made sense? Yes, clearly. All lights are green. The service should do extremely well in the market place—so long as the company continues to deliver excellent service to its clients. But given its record of growth over the years it is likely that LiquidHub will continue to be a highly regarded partner by its clients.

ENDNOTE

[1] This example was prepared by Robert Kelley and Jonathan Brassington with Stephen J. Andriole

Chapter IX
Venture Investing in E-Mail Trust Solutions:
The Postiva Case

INTRODUCTION TO THE CASE[1]

This is an example of venture investing in an e-mail trust solution developed by Postiva, a company that provided "e-mail trust solutions" and also had a complementary line of business in privacy education and consulting services. "E-mail trust solutions" are best-of-breed information technology products and services that uniquely combine aspects of e-mail content security (e.g., antispam, antivirus), authentication, and customer relationship management technologies. In short, the company's products enabled e-mail sender authentication and accountability, more efficient e-mail processing, and filtering capabilities for ISPs and recipients and heightened privacy, security, confidence, and control to consumers as they wade through increasingly cluttered and danger-filled e-mail in-boxes.

This example discusses the company's business plan and then filters the investment opportunity—from the perspective of a venture investor—with the 15 due diligence criteria.

BACKGROUND

Postiva, Inc., was formed in December 2000 to assure privacy and trust and enable optimization of messaging platforms, including e-mail, instant messaging, two-way paging, short messaging service, and other existing and future forms of interactive communications ("messaging"). The company developed a patent-pending message-based trust infrastructure that enables various forms of business transaction processing (the "POSTiva Platform") upon which its software products will be built. The company believed that with one proprietary architecture, it can solve three of the most significant issues facing the Internet and messaging—privacy, legal compliance, and productivity—and enable smart messaging solutions that help enterprises optimize existing business processes and conduct novel messaging-based transactions. The POSTiva Platform could also be used to integrate and provide interoperability for many applications, serving as secure "lite" transport for data between applications, architectures, operating systems, and devices.

The company initially hoped to generate revenue by offering trust business services (an e-mail/messaging trust seal and privacy consulting services) and software and infrastructure relating to privacy and legal, regulatory, and best practices compliance—the RightPathtm compliance system. The company then expected to leverage and build upon its installed base and technology of the POSTiva Platform to offer a smart messaging suite of solutions, EM-Agents.tm. EM-Agents.tm intended to provide novel and light-transport productivity and transactional solutions for direct marketing and customer relationship management (CRM) messaging, message-based billing and payment solutions, streamlined and asynchronous information gathering and retrieval and supply chain management, workflow and productivity processes.

The company believed that significant value and competitive advantage derived from the architecture of the POSTiva Platform that uniquely is intended to provide a common platform for both legal and regulatory compliance as well as productivity and transactional smart messaging solutions. The company designed the POSTiva Platform to provide companies with a number of benefits, including, but not limited to (1) automated/anonymous legal, regulatory, and best practices compliance, (2) new business opportunities, (3) heightened productivity, (4) a simplified means of integration and interoperability, and (5) as a result of the foregoing, an increase in the return on investment on installed technology and marketing practices.

The POSTiva Platform and the related trust business services were intended to provide a novel and sustainable solution to problems posed in an approximately $60 billion market that encompasses privacy, security management, application integration, e-mail marketing, and other services. The privacy and security management planning market is projected to reach $5 billion by 2004, while the market for application integration services is estimated to reach $17 billion in 2005. With regard to the specific or general markets related to the functionality of the company's smart messaging solutions, the company will compete directly in or tap into the following markets: the overall markets for outsourced CRM and e-mail marketing services are projected to respectively grow to $4.8 billion in 2004 and $7.3 billion by 2005; bill payment services were estimated to be an $11 billion market in 2000; the aggregate corporate online procurement marketing sector will reach $10.1 billion by 2005; and all of mobile- and location-based commerce is estimated to be a $5 billion market in total by 2004.

The company sought to raise up to $750,000 for seed stage funding. Proceeds from the offering would be used to launch the trust seal and privacy consulting businesses, to continue the development of RightPathtm software and EM-Agent™ suite, and to establish sales and marketing initiatives, hire additional key personnel, launch brand-building activities, develop strategic alliances/partnerships, and for general corporate purposes.

THE MARKET AND THE OPPORTUNITY

Privacy and Legal Compliance

Privacy and building consumer confidence are essential for the increased consumer adoption and use of the Internet and interactive communications. Consumers often refrain from making purchases online and do not trust online companies with safeguarding their personal information. For example, nearly two-thirds (65%) of 779 consumers in an IDC survey on privacy said that—more than once— they had not made a purchase in the past 6 months primarily due to privacy reasons. A recent Forrester Research report found that only 6% of consumers show a high degree of confidence in trusting a Web site with personal data.

There is an evolving labyrinth of legislation and regulation. Forrester Research also reported that seven of eight Americans support the introduction of Internet privacy legislation. In response to strong constituent sentiment, more than 1,000 privacy and antispam laws have been introduced into Congress over the past 2 years. While all of this activity has not resulted in the enactment of broad privacy legislation, a growing thicket of sector-by-sector regulations already exists; for example, privacy of children, i.e., the Children's Online Privacy Protection Act of 1998 (COPPA), privacy of financial information, i.e., Gramm-Leach-Bliley Act of 1999 (GLB), and privacy of medical information, i.e., Health Insurance Portability and Accountability Act of 1996 (HIPPA). Forrester examined the Fortune 100 and concluded that 73 must already comply with at least one of four sets of federal privacy regulations that became law in the last 2 years. Additionally, there are numerous and inconsistent state, local, and international laws and regulations, and more are being proposed. For example, 18 states have varying laws on unsolicited commercial e-mail; and as of mid-February, 2001, there were 314 privacy-related bills pending in 42 states, 36 Internet-specific privacy bills pending in 11 states, and

14 identity-theft bills and 51 financial privacy bills also pending in different states. Additionally, strong privacy laws have recently been enacted in the EU, Canada, and parts of Asia. As a result of this evolving legal and regulatory labyrinth, business and marketers are increasingly bewildered by the existing and proposed patchwork of federal, state, local, international, and industry-specific laws and regulations and best practices. Moreover, given the evolving patchwork of legislation and regulations, businesses are often unaware of all applicable laws and regulations or are unable to comply with such across all business units and/or within all jurisdictions where customers may be located.

Compliance with privacy-related legislation can have significant costs and is often ignored. For example, industry studies concluded by Ernst & Young LLP and the Information Services Executive Council concluded that proposals to limit companies from sharing or selling customer information without permission would cost 90 of the largest financial institutions $17 billion a year of added expenses, and would result in a $1 billion "information tax" on consumers through costs tacked on to products from catalogs and Internet retailers. Also, 1 year after the passage of the Children's Online Privacy Protection Act, researchers at the University of Pennsylvania's Annenberg Public Policy Center found that most Web sites geared for children had not followed federal requirements for privacy.

Enforcement efforts and liabilities are beginning. While businesses and marketers can potentially be the subject of civil litigation from and liabilities to individuals and business for breaches of privacy, violations of laws and regulations, including COPPA, GLB, and HIPPA, can subject the offending party to financial penalties and/or even criminal penalties, including jail sentences. For example, the Federal Trade Commission (FTC) recently levied $100,000 fines against three Internet companies for violating COPPA for allegedly unlawfully collecting names, addresses, telephone numbers, and other information from children under 13. Additionally, the FTC also brought enforcement proceedings against three businesses for violating privacy provisions of GLB where the FTC is seeking fines and the disgorgement of the profits related to certain improperly obtained personal financial information.

Legislation and regulation is not efficient and may be harmful. Businesses, marketers, Internet service providers, e-mail service providers, and other interested parties have been lobbying legislatures and policymaking bodies worldwide on policy issues given (1) the potential costs with complying with enacted legal and regulatory requirements, (2) the potential liabilities and brand exposure that may result from noncompliance, and (3) the adverse business ramifications, whether intentional or not, that legislation and regulation may have on long-standing and/or innovative business practices. In order to prevent preemptive and less efficient governmental regulation and its unintended consequences and adverse business ramifications, such lobbying efforts have largely called for trade association sponsored and/or supported self-regulation standards, seal programs, and new technologies as superior alternatives to legislation and regulation.

With respect to the promise of new technologies, to date, privacy tools have been mostly Web-based, complicated, lack compliance/accountability, user control features and have been marketed to individuals who undertake the installation and use thereof. A recent study from The Privacy Leadership Initiative determined that as few as 15% of consumers are installing and utilizing available privacy-protecting technologies. Additionally, corporate enterprises are increasingly realizing that they proactively must provide their customers with privacy and security solutions, and accordingly a recent study by Forrester predicts that corporations will triple the amount they spend on security such that security spending in the U.S. will top $19 billion by 2004, including more than $5 billion of which is related to security management and planning (including privacy audits, policy setting, and implementation and privacy-related tools).

PRODUCTIVITY

As the volume of e-mail and other forms of messaging proliferate, e-mail and messaging increasingly is viewed as a tool both enhancing as well as detracting from productivity. A recent report from United Messaging shows that at the end of 2000, there were 891.1 million electronic mailboxes worldwide, representing a 67% increase in the number from the previous year. Also, the report estimated that there were 345 million corporate messaging seats worldwide, representing a 34% growth for the corporate e-mail sector. The report also showed that at year-end 2000, 45% of U.S. consumers and 75% of workers in the U.S. use e-mail regularly, representing a total of 58% of the population. These findings match well against a Gartner Group research estimate that corporate e-mail now carries up to 75% of a company's communications.

While businesses have adopted e-mail as a cost- and time-efficient means of communication, the volume of personal e-mail comingled with business messages may detrimentally affect productivity. Gartner Group research estimates that the average American worker spends 4 hours of each day reading, writing, and forwarding e-mails, and IDC reported that of the 6.1 billion e-mails sent daily in North America, 3.4 billion are work-related (56%) and 2.7 billion are personal e-mails (44%). Based on these reports, it can be approximated that the average American worker spends roughly between 1.5 and 2.0 hours of each day (potentially, up to 25% of a business day) reading, writing, and forwarding personal e-mails.

The situation will exacerbate. By 2006, Jupiter Media Metrix predicted that the volume of e-mail messages per user will increase forty fold from today's levels with the amount of spam representing about one-third of all e-mails. Notwithstanding the benefits of e-mail to an enterprise, spam and personal use of corporate e-mail systems are undercutting and will consume an increasing amount of the cost-savings and/or productivity gains facilitated by e-mail and messaging. The company believed that as the volume of e-mail and messaging rises, businesses increasingly will desire and implement tools that can optimize the efficiency of messaging by, in part, segregating personal e-mail from business-related communications, prioritizing and categorizing business-related messages, and minimizing spam.

BUSINESS-TO-BUSINESS INTEGRATION AND MESSAGING OPTIMIZATION

With the advent of enterprise applications, complications and the need for coordination have arisen. Over the last decade, companies have invested heavily in enterprise applications to automate and improve the efficiency of their internal business processes. As competition has increased and markets have become more dynamic, companies have begun to recognize that they must coordinate more closely with their customers, suppliers, and business partners. Traditional enterprise applications, however, do not readily support business processes beyond the borders of an enterprise. As a result, companies relying only on their enterprise applications have not been able to easily integrate their business processes with those of their customers, suppliers, and other business partners to achieve productivity gains.

In the midst of this environment, the Internet has emerged as a crucial medium for electronic commerce, communication, and sharing information. The widespread adoption of the Internet as a business communications platform has created a foundation for business-to-business relationships that has already enabled organizations to tap new revenue streams, streamline cumbersome processes, lower costs and improve productivity. Businesses are seeking business-to-business integration solutions that allow them to utilize their existing portfolio of enterprise applications to exchange information and transact business with customers, suppliers, and other business partners over the Internet.

The variety of computing environments and the inability to share information across those environments have been a major impediment to business-to-business integration. Current business-to-business integration (B2Bi) approaches are costly, problematic, and ineffective. Additionally, since most B2Bi approaches are Web-centric, they require persistent connections to the Internet by users. The company believed part of the next wave of B2Bi innovation would focus on enhancing and extending messaging to make it smarter and an appropriate platform for B2Bi. Factors driving this trend include (1) B2Bi increasing penetration internationally and in other markets where Web access is limited and/or costly and/or e-mail penetration is materially greater, (2) the proliferation of mobile platforms, devices, and applications, and (3) the increasing concerns of business about the security and privacy risks related to applications that require persistent Internet connections. This trend is underscored by the prediction that although e-mail and messaging will evolve into various forms (e.g., IP voice-enabled e-mail, ubiquitous instant messaging, intelligent-agent associated e-mail, embedded e-mail), it will remain the dominant application of the Internet as it increasingly becomes smarter and proactive. It is the company's view that e-mail and messaging will travel not just from network to network and between varying devices, but e-mail and messaging will also transport data and connect applications with applications. The company believed that e-mail and messaging infrastructure that affords trust and supports standardized and optimized functionality of e-mail and messaging is necessary for e-mail and messaging to become a reliable transport upon which there can be interoperability between heterogeneous (open/proprietary) messaging architectures, disparate enterprise applications (messaging, business, transactional or otherwise), emerging security and privacy technologies, and business process systems.

Based on the growth of the Internet and the expectations for growth in communications, e-commerce, and mobile devices, the company believed the market opportunity for B2Bi software, including messaging-based forms, is substantial. For example, a recent study by IDC focusing on the market for online exchanges predicted that the market for B2Bi will reach $17 billion in 2005.

In addition to being the cornerstone of its B2Bi technology and services, the company intended to leverage the installed base and technology of the POSTiva Platform to extend and enhance messaging by enabling it to be more intelligent and transactional.

Specifically, the company intended to develop a smart messaging suite of solutions, EM-Agentstm, that principally includes message-based productivity and transactional solutions for (1) direct marketing and customer relationship management messaging, (2) message-based billing and payment solutions, (3) streamlined and asynchronous information gathering and retrieval, and (4) supply chain management, workflow, and productivity processes. The company believed the markets for the functionality enabled by the EM-Agentstm suite to be substantial. The company anticipated competing directly in or taping into the following markets: the overall market for outsourced e-mail marketing services ($7.3 billion by 2005—Jupiter Media Metrix) and CRM outsourcing ($4.8 billion in 2004—IDC); billing systems ($11 billion in 2000—Industry Standard) and bill payment ($40.4 billion in transactions by 2005—ZonaResearch); m- and l-commerce ($5 billion in total by 2004—IDC); and corporate online procurement marketing ($10.1 billion in the aggregate by 2005—Forrester Research).

THE POSTIVA SOLUTION

Positioning, Strategy, and Marketing

A cornerstone of the company's solution was its patent-pending message-based trust infrastructure that enabled various forms of business transaction processing (the "POSTiva Platform") upon which its software products were built. The company believed that with one proprietary architecture, it can (1) solve three of the most significant issues facing the Internet and messaging—privacy, legal compliance, and productivity—and (2) enable smart messaging solutions that help enterprises optimize and integrate existing business processes and applications and conduct novel messaging-based transactions. Additionally, the company believed that the architecture of the POSTiva Platform integrated with existing privacy and security applications and infrastructures. The POSTiva Platform also could be used to integrate and provide interoperability for many applications, serving as secure "lite" transport for data between applications, architectures, operating systems, and devices. The POSTiva Platform was intended to become an enabling layer in today's and especially tomorrow's distributed architectures where all forms of messaging occur.

As illustrated in Figure 1, the company planed to execute its strategy in integrated phases: (1), E-Mail Trust Services and Products and (2) E-Mail/Messaging Optimization (Productivity & Transactions). The company positioned itself to have multiple revenue sources. In the first phase, the company expected to generate revenue by (1) providing privacy consulting and chief privacy officer support services, including assessments, monitoring, education, and/or outsourced chief privacy officer functions, (2) developing, marketing, and licensing an e-mail/messaging trust seal program, and (3) developing, marketing, and licensing RightPathtm automated legal, regulatory, and best practices compliance tools. The company believed that through its privacy consulting and trust product offerings it would help companies (1) comply with an evolving thicket of domestic, international, and industry-specific laws regulations and best practices, (2) reduce costs by managing and automating portions of the compliance process, (3) increase long-term customer value by enhancing trust with and protecting the privacy of prospective and existing customers, and (4) increase the return on investment with respect to customer acquisition and retention efforts and existing marketing and CRM practices and technologies.

In the second phase—the e-mail/messaging optimization phase—the company expected to leverage and build upon the installed base and technology of the POSTiva Platform to generate revenue by developing, marketing, and licensing its messaging agent product suite, EM-Agentstm, that would include message-based productivity and transactional solutions for (1) direct marketing and CRM messaging, (2) message-based billing and payment solutions, (3) streamlined and asynchronous information gathering and retrieval, and (4) supply chain management, workflow and productivity processes. The company believed that the POSTiva Platform and its productivity and transactional smart messaging solutions would provide companies with a number of benefits, including, but not limited to, (1) new business opportunities, (2) heightened productivity, (3) a simplified means of integration, and (4) as a result of the foregoing, an increase in the return on investment on installed technology and business and marketing practices.

As described in a later section, the first phase is designed to build awareness, credibility, and trust among companies, marketers, intermediaries, and consumers. The company views its balanced approach to e-mail privacy standards among its key competitive advantages and an essential basis to maintaining its relationships with both privacy advocates and businesses/marketers. Specifically, the company believed its initial privacy-centric, nontechnical efforts (offering the e-mail/messaging trust seal and privacy consulting services) and the derived relationships would provide a cost-effective platform to (1) introduce and market RightPathtm, the company's legal, regulatory, and best practices automated/anonymous compliance application and (2) develop over time e-mail optimization tools and products. The company's strategy in the first phase was to use both direct and indirect methods to reach marketers, chief privacy officers, chief security officers, lawyers, online advertisers and support services, Internet and e-mail service providers, and other professionals that need to be aware of, respectful of, and/or compliant with privacy issues, laws, regulations, and best practices. As such, the company intended to

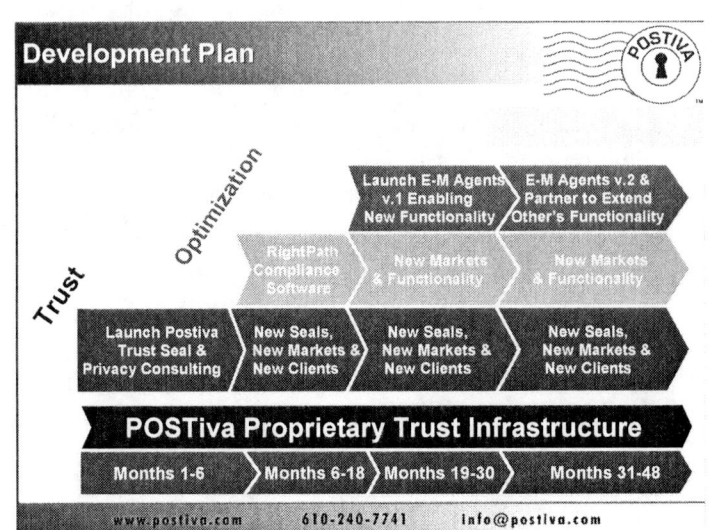

Figure 1. Postiva development plan

utilize direct efforts that include sales agents and staff, trade shows, sponsorships, print advertising, direct mail and e-mail, telemarketing, and public relations; and indirect efforts involving strategic alliances with trade associations and consumer interest and privacy groups.

The company's phased approach allowed it efficiently to tailor its product development, marketing efforts, and use of financial and human resources based on the learning, success, and feedback obtained during each phase as well as to flexibly respond to existing conditions and emerging trends in its target markets and in the financial community.

REVENUE AND BUSINESS MODEL

The company's business model incorporated multiple revenue sources. Let us begin with a discussion of its products.

Products

Phase I: E-Mail Trust Services and Products

Privacy Consulting

The company began offering privacy consulting through its E-Privacy Group to help companies design, develop, and validate their privacy policies, practices, and procedures. The consulting services offered include privacy workshops and training programs, expert consulting services with business, and cross-discipline expertise, assessments, chief privacy officer outsourcing, chief privacy officer support services, self assessments and privacy monitoring, notification, and alert and response services. The company expected that its initial clients would be direct marketers and larger companies with multiple and international business operations, each having to operate in varying legal environments and specific industry best practices. While the fee structure for consulting services generally was on an hourly basis, there were fixed fee rates for certain introductory workshops and educational services.

The Postiva "Trusted Message" Seal Program

The Postiva "Trusted Message" Seal Program was an initiative dedicated to building trust and confidence between e-mail senders and receivers. The company developed a standards-based, third-party "seal" program that both addresses users' concerns about e-mail privacy and attempts to meet the specific business needs of business and marketers alike. The Postiva Trusted Message Seal Program was composed of a multifaceted assurance process that established e-mail sender accountability, thereby making e-mail receivers more comfortable when communicating online or providing personal information in connection therewith. The program and its standards were developed in connection with direct marketing trade associations, privacy advocates, policymakers, and technology futurists. It was the company's intent to leverage the relationships that the company had with these individuals and associations to concentrate its initial marketing efforts on their respective constituencies and members. Seal licenses would be priced based on a formula including company size and messaging volume. Over time, the company intended to introduce additional seals, including, but not limited to, seals regarding heightened security and commerce involving messaging.

RightPathtm Compliance System

The company's first and flagship software product was the RightPathtm privacy, legal, and regulatory compliance system. The RightPathtm compliance system was based on the POSTiva Platform and was designed to utilize a real-time rules database that incorporated compliance procedures for applicable federal, state, local, international, and industry-specific (i.e., financial, healthcare, children, and direct marketing) laws, regulations, industry trade group guidelines, and best practices. The company anticipated a significant demand for its RightPathtm compliance software because it expected it to be the first software and/or infrastructure solution that enabled compliance by companies with certain laws with differing applicability and consequences based on specific personal information of the intended recipients (e.g., age and/or geographical location of the recipient) without recipients exposing, delegating, and/or sharing such personal information, i.e., automated and anonymous compliance. Moreover, the company believed that demand for the RightPathtm compliance system and its other trust products and services (i.e., trust seal program and privacy consulting services) would be driven by the evolving legal and regulatory environment, increased concerns among businesses and marketers regarding potential liabilities and/or their inability to comply with or keep track of the existing and proposed patchwork of federal, state, local, international, and industry-specific laws, regulations, and best practices. The company intended to structure its RightPathtm compliance system software licenses with a base license fee based, in part, annual number of mailings and/or sales. Additionally, the company intended to charge additional and ongoing fees for support, installation, training, and/or information services (e.g., updated legal and regulatory rules databases).

Phase II: Messaging Optimization (Productivity and Transactions)

The company expected to leverage and build upon its installed base and technology of the POSTiva Platform to offer smart messaging solutions that help enterprises optimize business processes. Specifically, the company intended to develop a messaging agent product suite, EM-Agentstm, that provided novel and light-transport productivity and transactional solutions for (1) direct marketing and customer relationship management (CRM) messaging, (2) message-based billing and payment solutions, (3) streamlined and asynchronous information gathering and retrieval, and (4) supply chain management, workflow, and productivity processes.

The company believed several of the EM-Agentstm proprietary product features to be novel and to have significant commercial application. Although the related patents covering such features were filed, some are still pending.

Several of the key product features that the company intended to build into EM-Agentstm over time include:

- **Marketing and CRM agents:** Based on the POSTiva Platform architecture, the patent-pending smart messaging marketing and CRM solutions were designed, in part, to resolve the "privacy paradox" of the Internet—balancing the conflicting desire of consumers on one hand to generally keep private personal information and the practice on the other hand of businesses and marketers to leverage such information online and off-line in customer acquisition, cross- and up-selling, and customer retention efforts. The company anticipated a significant demand for these solutions because it believed the solutions would be the first software and/or infrastructure solution that enabled personalized services, targeted, and dynamic offers and customer acquisition without users exposing, delegating, and/or sharing personal information—i.e., true confidential, anonymous targeting, and profiling. License fees associated with the marketing and CRM agents typically were set on a per mail/campaign basis, consistent and competitive with rates for enhanced e-mail marketing functionality. Additional installation, development, analysis, and training fees were also contemplated.
- **Transactional, billing, and payment agents:** The company also filed patent applications relating to a system and method for conducting predefined and/or reply-based transactions via an e-mail/messaging infrastructure. Based on such patent applications, the transactional, billing, and payment solutions would enable a number of single-click/reply-based and other personal and business transactions, and payments and purchases via e-mail and messaging. License fees associated with the transactional billing and payment agents typically

would be on a per transaction/bill basis, consistent, and competitive with rates related to credit card processing enhanced services or those of other payment systems/technologies. Additional installation, development, accounting, and training fees were also contemplated.

- **Streamlined and asynchronous information gathering and retrieval agents:** Recent studies by both Accenture and Jupiter-Media Metrix have suggested that the wireless industry must overcome several significant problems before wireless Internet access can grow in popularity. Both reports identify the difficulty of data entry and the number of keystrokes required to request and receive information on a wireless device as an impediment to growth. The company believed that much of the commonly requested information sought on the Web could more quickly, safely, and cost-effectively be obtained via e-mail and messaging. Accordingly, the company developed streamlined and asynchronous information gathering and retrieval agents. License fees associated with the information gathering and retrieval agents typically would be on a per transaction/inquiry basis, varying based on value/costs of information database accessed (alternatively, there could be an overall license with a cellular or wireless network provider covering a set of queries or functions). Additional installation, development, and training fees also were contemplated.

- **Supply chain management, workflow, and productivity processes:** The company filed patent applications and intended to develop related solutions regarding various supply chain management, workflow and productivity processes, including (1) automated e-mail/messaging requests for quotes, B-2-B and group procurement, and auction bidding, (2) segregation and rerouting of personal e-mail/messages from business e-mail boxes and messaging systems, (3) corporate messaging bandwidth and flow prioritization, management, and spam squelching, and (4) categorization and prioritization of business and personal e-mail/messages. The company anticipated a significant demand for these solutions because several of these solutions are novel and given the potential cost savings, efficiencies, and productivity gains for large enterprises related to solutions of this type. License fees associated with the supply chain management, workflow, and productivity process agents typically would be on a per transaction/purchase basis, consistent and competitive with rates related to auction services, credit card processing enhanced services, and business-to-business e-mail procurement exchanges/technologies. Additional installation, development, and training fees were also contemplated.

COMPETITIVE ADVANTAGES

The company believed that it had several additional competitive advantages over potential competitors. These advantages included:

- **First end-to-end range of privacy and trust products and services:** The company believed that it is in a strong competitive position due to its strategy of offering multiple and complementary privacy and trust products and services. The company's strategy of creating a full-service company focused on a model that included standard-setting, consulting services, and development of related compliance software and enhancement creates significant cross-selling opportunities, given the complementary service offerings and customer bases.
- **Automated and anonymous legal compliance:** Also, the company anticipated a significant demand for its RightPathtm compliance system because it expected it would be the first software and/or infrastructure solution that enables automated and anonymous compliance. Also, while the company's seal program focused on e-mail and messaging best practices, the company also believed its design of a technology-driven platform to automate legal and best practices compliance would afford it a sustainable competitive advantage.
- **Integration and interoperability with disparate platforms and applications; partner rather than compete:** While the company would compete with many of the companies discussed below in the Competition Section hereof on a product-by-product basis, the company was unaware of any company similar to Postiva concentrating on trust/privacy as well as optimization services and products. Also, as illustrated in Figure 2, while all such potential competitors may have their respective business strategy and product lines, Postiva believed that the interoperability and integration features of its proprietary POSTiva Platform provided it with a competitive advantage as well as created opportunities for the company to partner—instead of compete—with such companies to enable new and extend the functionality of their products.
- **Credibility and balance; relationships with privacy advocates and marketing organizations:** The company believed that its balanced approach to privacy standards and relationships with both privacy advocates and businesses/marketers (including trade associations) were among its key competitive advantages. Also, the company believed the composition of its Advisory Board and their active participation would further support its commitment to balance and credibility. The company also believed its initial privacy-centric, nontechnical efforts (offering the e-mail/messaging trust seal and privacy consulting services) and the relationships derived would provide a cost-ef-

fective platform to (1) introduce and market RightPathtm compliance system and (2) develop over time, e-mail optimization tools, and products.
- **Superior and scalable technology:** The company believed that the technology and architecture underlying the Postiva Platform were based on a foundation and algorithms that provide superior and scalable technology that posts a barrier for competitors. The company believed that significant value and competitive advantage derived from the architecture of the POSTiva Platform that was intended to provide a common, open platform for both legal and regulatory compliance as well as productivity and transactional smart messaging solutions.
- **Patent portfolio, intellectual property, and proprietary methodology:** In an effort to obtain a sustainable competitive advantage related to the POSTiva Platform, the RightPathtm and EM-Agenttm software and functionality, the company filed five patent applications and one provisional application with the U.S. Patent and Trademark office. The company also sought trademark protection for key trademarks and servicemarks.
- **Management; experienced managers, marketers, technologists, lawyers, and academics:** The company believed it assembled an unparalleled management team, particularly for an early-stage company, with extensive experience in business, marketing, technology, security, privacy, law, finance, medicine, and e-commerce (see "Management" and "Advisory Board"). Also, the company was positioned to develop proprietary, original, and balanced privacy products and services because of its dedicated staff, consultants, and advisors.

COMPETITION

Due to its strategy of offering multiple products and services, the company believed that it was in a strong competitive position. While the company may compete with many of the companies discussed below on a product-by-product basis, it believed that the interoperability and integration features of its proprietary POSTiva Platform would provide it with a competitive advantage as well as create opportunities for the company to partner, instead of compete, with such companies to enable new and extend the functionality of their products; see Figure 2.

Generally, the company's competition included:

- **Trusted message seal program:** There are several companies and organizations offering privacy seal programs, such as Trust-e, BBBOnline, and CPA WebTrust. Currently, these programs are Web-centric seal programs that offer

Figure 2. Competitive map

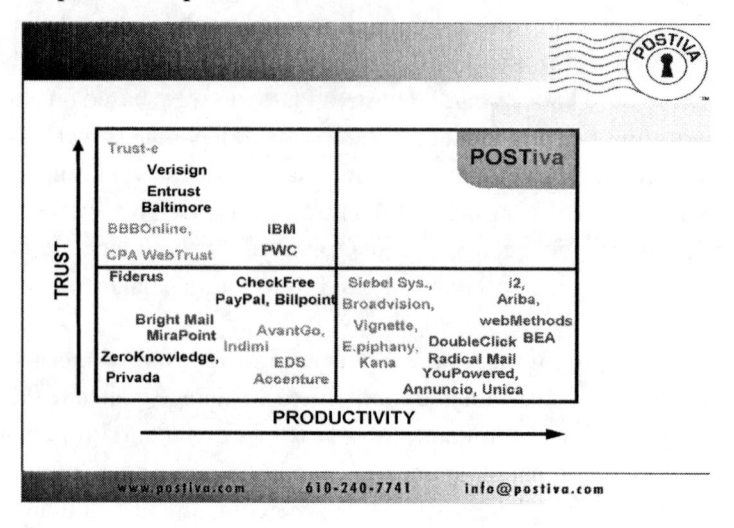

seals in connection with a voluntary pledge to abide by and include certain provisions in the privacy policy included in a particular Web site. While the company's seal program focuses on e-mail and messaging, the company also believed its design of a technology-driven platform to automate compliance would afford it a sustainable competitive advantage.

- **Privacy consulting:** Consulting and accounting firms, including PriceWatehouseCoopers, IBM, Fiderus, and Privacy Clue, that provide, at least in part, privacy-related consulting, auditing, assessment, and/or management services are many and varied. The company expected increased competition in this market sector and would keep a constant vigil to stay in a leadership position. Also, the company believed its relationship with both marketing trade organizations and privacy advocates provided a competitive advantage in reaching the appropriate balance between business objective/marketing techniques and privacy concerns.
- **RightPathtm compliance system:** There are a number of companies that offer privacy or security related software and services, such as Verisign, Entrust, Baltimore, Microsoft, Privada, YouPowered, and ZeroKnowledge. Additionally, with respect to the application of the POSTiva Platform as a platform for B2Bi, business-to-business integrators and enablers include WebMethods, BEA, i2, and EDS. While the company potentially would compete with certain companies with respect to potential customers, the company's products and services were generally intended to be complementary to the products and services offered by each of the privacy, security B2Bi, and enablement

companies. By partnering with these companies, the company believed that it could integrate with certain of their products and services and extend and/or enhance the related functionality.
- **EM-Agentstm:** There are various potential competitors depending on the nature of the particular EM-Agent solution. Potential competitors for the company's Marketing and e-CRM Agents included e-mail enhanced functionality companies (Radical Mail, DoubleClick, and Unica) and e-CRM firms (Siebel Systems, Broadvision, Vignette, and E.piphany). The company's transactional, billing, & payment agents competed with bill payment and payment system companies (Check Free, PayPal, and Billpoint). The company's supply chain management, workflow, and productivity processes solutions competed with antispam companies (BrightMail and MiraPoint) and productivity and e-procurement companies (i2, Ariba, and CommerceOne). Potential competitors with the company's streamlined and asynchronous information gathering and retrieval agents may include various mobile information and commerce information management, content, and enabling companies (including Avant Go, Indimi, and Multex).

Despite the potential competition outlined in this chapter, the company believed that it could succeed in these market sectors by offering differentiable and higher quality services and end-to-end solutions for all aspects of privacy and legal, regulatory, and best practices compliance. Also, the company believed its diversified strategy created significant cross-selling opportunities, given the complementary service offerings and customer bases. Accordingly, the company believed its strategy placed it in a very strong long-term competitive position.

TECHNOLOGY

To the extent the company will engage in software and application development related to legal, regulatory, and best practices compliance, it would utilize proven state of the art technology and partners to quickly and efficiently develop the underlying technologies and components. To date, the company selected technical components and infrastructure, keeping in mind, ease of implementation, cost, reliability, performance, and scalability. The company's goal was not to be on the bleeding edge of technology but the leading edge of proven technology. This would insure that its products would meet the needs of its customers and provide a stable infrastructure that it could be built upon easily and efficiently in the future.

Also, the company retained Synnestvedt & Lechner, LLP to begin the development of a patent application portfolio. To date the company filed five patent applications and one provisional application related to the POSTiva Platform, RightPath,^{tm} and *EM-Agents^{tm}* software products and the related functionalities and methodologies.

SUMMARY OF PROJECTED FINANCIAL PERFORMANCE

Table 1 presents the company's forecasted financial performance post funding. These projections were developed based upon certain assumptions, and were of course subject to risks and uncertainties. The company made no representation or warranty as to the accuracy or completeness of this information.

MANAGEMENT: OFFICERS, DIRECTORS, AND KEY CONSULTANTS

The management team of the company represented a spectrum of experience and accomplishment in Internet, marketing, law, academics, business, and computer technologies. The company's management team included:

Name	Age	Position
Vincent J. Schiavone	42	Chairman, President & CEO
James H. Koenig, JD	36	Executive Vice President; Chief Development & Legal Officer; Director
Simson Garfinkel, CISSP	35	Chief Privacy Officer
Michael Miles	43	Chief Financial Officer
Michael Miora, CISSP	48	Vice President, ePrivacy Group
Ram Mohan	32	Chief Technical Officer
Geoff Mulligan	44	Senior Architect

Vincent J. Schiavone, President & Chief Executive Officer

Mr. Schiavone has been deeply involved in Internet commerce, privacy, and security since 1986. He is the founder and former CEO and Chairman of the Board of the 4Anything Network, a venture-funded, first of its kind nonhierarchical Internet search engine that includes 4,500 vertical topic sites including 2,000 topical and 2,000 city and country sites. With Schiavone in charge from June 1996 through April 2000, the company grew from concept to 120 employees and over 8 million unique visitors per month, making it one of the top 100 trafficked Web sites according to Media Metrix. In 1999, as part of the 4Anything Network, Mr. Schiavone

Table 1.

($000s)	2001		2002		2003		2004	
Revenue								
Phase I – Trust	$	662	$	5,396	$	16,696	$	46,836
Phase II - Transactions & Optimization		-		293		3,381		23,559
Total Revenue	$	662	$	5,688	$	20,077	$	70,395
Cost of Sales								
Phase I – Trust	$	211	$	1,395	$	3,252	$	6,032
Phase II - Transactions & Optimization		-		47		439		2,356
Total Cost of Sales	$	211	$	1,441	$	3,691	$	8,388
Gross Margin								
Phase I – Trust	$	454	$	4,004	$	13,448	$	40,508
Phase II - Transactions & Optimization		-		246		2,941		21,204
Gross Margin	$	454	$	4,250	$	16,389	$	61,712
Gross Margin %		69%		75%		82%		88%
Sales & Marketing	$	1,250	$	2,758	$	7,973	$	27,201
% of Sales		189%		48%		40%		39%
Research & Development		1,150		1,600		2,816		7,248
% of Sales		174%		28%		14%		10%
General & Administrative		750		1,750		2,750		7,040
% of Sales		113%		31%		14%		10%
Operating Income (Loss)	$	(2,696)	$	(1,858)	$	2,850	$	20,223
% of Sales		-408%		-33%		14%		29%

also founded Greatflowers.com, a grower direct online florist providing the highest quality at grower direct prices, and JewelrySpotlight.com, an online jeweler partnered with one of the largest suppliers in the world. Mr. Schiavone has extensive experience in design and development of online services and security analysis. Prior to 4Anything Network, from 1997 through 1999, Mr. Schiavone served as a principal and key strategist of InfoSec Labs, Inc., a premier security consulting company serving the financial services, manufacturing, health care, and distribution industries. In 1999, InfoSec Labs was acquired by Rainbow Technologies, Inc., a NASDAQ listed company (RNBO). From 1995-1996, Mr. Schiavone served as Senior Analyst for the National Computer Security Association, Inc. (NCSA), where he was a key contributor to the concept, design, marketing, and development of IS/Recon,™ NCSA's state-of-the-art Internet security service for intelligence collection and analysis. In 1995, Mr. Schiavone founded and served as President of Avanti Associates, Inc., a developer of voice-operated medical record-keeping software and systems designed to enable quick and reliable identification of private information and electronic media labeling for government classified and corporate proprietary environments. In 1996, the voice-operated medical record-keeping software and system business was acquired by Professional Dental Technologies. Mr. Schiavone attended Temple University.

James H. Koenig, Executive Vice President; Chief Development & Legal Officer; Director

Mr. Koenig has extensive experience in a variety of areas important to the company, including privacy laws and practice, e-commerce, direct marketing, computer technology, and law. From 1997 to 2000, Mr. Koenig served as Director of Business Development at QVC, the television and online retailer, and was responsible for general strategy and implementation of many of QVC's new and "off-air" ventures, alliances, and investments. Prior to joining the company, Mr. Koenig served as Senior Vice President, Corporate Development and Marketing, for Vicus.com, a 100-person company creating alternative health information, products, and services, where he oversaw business development, media alliances, e-commerce initiatives, advertising, and database-marketing (including e-mail permission, online, telemarketing, and direct mail campaigns). In 1996, he was part of the founding team and oversaw business and legal affairs as Executive Vice President of MaMaMedia.com, an MIT Media Lab spin-off company developing online-related entertainment and educational products for children. Mr. Koenig began his career in 1990 practicing law for 6 years at Weil, Gotshal & Manges LLP as a member of the Venture Capital/High Technology Practice Group. Mr. Koenig received a Bachelor of Science degree from the Massachusetts Institute of Technology followed by a law degree from the University of Miami School of Law where he was Business Editor of the University of Miami Law Review. Mr. Koenig has written on topics relating to business development, strategic alliances, and venture capital financing in the Internet, e-commerce, entertainment, and telecommunications industries. His most recent article E-Mail Privacy: What You Need to Know About What You Need To Know, appeared in The Council for Responsible E-Mail's compendium Permission E-Mail Marketing: Insights from Industry Leaders (v.2, February 2001).

Additionally, Mr. Koenig is a member of the Association of Interactive Media's Council for Responsible E-Mail and a member of the Privacy Committee and a general member of the Association of Interactive Media's Coalition for Addressable Media, and a voting member of the Direct Marketing Association, the Association of Interactive Media, and the Internet Alliance.

Simson L. Garfinkel, Chief Privacy Officer

Mr. Garfinkel, a journalist, entrepreneur, and international authority on privacy and computer security, has spent his career testing new technologies and warning about their implications. He has been the Chief Technology Officer of Sandstorm, founder of a community ISP, and developer of productivity software for UNIX and Windows. He is the author of nine books on privacy and information security, including *Database Nation, Stopping Spam, Architects of the Information Society, PGP: Pretty Good Privacy, Web Security & Commerce,* and *Practical Unix & Internet Security.* Garfinkel has also been a well-known columnist with features in *The Boston Globe, WIRED, Salon, Technology Review, ComputerWorld, Forbes,* and *The New York Times.* In 1997, Mr. Garfinkel's coverage of the U.S. Social Security Administration's Web site showed how lax security and poor privacy protections were endangering the financial privacy of all tax-paying Americans. The coverage sparked a Congressional inquiry; the site was shut down and redesigned as a result. Mr. Garfinkel founded VinyardNet, an Internet service provider, which he sold to Broadband2Wireless in 2000. Mr. Garfinkel is a fellow at the Berkman Center for Internet and Society at Harvard University. Mr. Garfinkel holds a Masters Degree from Columbia University and three undergraduate degrees from MIT. Mr. Garfinkel is serving part-time for the company.

Michael Miles, Chief Financial Officer

Michael Miles has over 20 years of broad financial and operational management experience with both public and private technology companies. In his role as Senior Vice President and Chief Financial Officer at Safeguard Scientifics, Inc. (January 1997-March 2000), Mr. Miles oversaw all financial, investor relations, and human resource functions during a period of accelerated growth. Mr. Miles was directly responsible for all Treasury and Capital Markets functions for Safeguard and its 40 partner companies. He was directly involved with all 14 Safeguard sponsored IPO's including Novell (1985), Cambridge Technology Partners (1992), and Internet Capital Group (1999). He has extensive experience assisting early- and middle-stage companies. Prior to serving as Safeguard's Chief Financial Officer, he served as its Vice President and Corporate Controller (January 1992-December 1996) and Manager of Financial Reporting (1984-1992). From 1980-1984, Mr. Miles served with Coopers and Lybrand. In 1980, Mr. Miles received a Bachelors of Science in Accounting from University of Scranton.

Michael Miora, CISSP, Vice President, E-Privacy Group

Mr. Miora has designed and assessed secure and private systems for Industry and Government over the past 24 years, including some of the nation's most sensitive public and private systems. As founder of InfoSec Labs (1997-1999), a recognized leader in security consulting and related services, Mr. Miora managed consulting services and programs for major companies and has performed Information/Internet Security and Privacy Assessments for companies across a variety of industry sectors. InfoSec Labs was acquired by Rainbow Technologies in 1999, for which Mr. Miora served as Vice President from 1999 to 2001. Mr. Miora has gained an international reputation for his consulting to the U.S. National Computer Security Center, the National Reconnaissance Office (NRO), and the NSA, and for providing training and seminars in Computer Security & Privacy and in Disaster Recovery. For this work, Mr. Miora held the highest government clearances. From 1995-1997, he served as the Director of Consulting Services for the National (International) Computer Security Association (NCSA/ICSA), helping ICSA achieve its status as a major force in the Information Security arena. Mr. Miora was educated at UCLA and UC Berkeley, earning Bachelors and Masters degrees in mathematics. Currently contributing chapters to the forthcoming *Computer Security Handbook* to be published by Wiley & Sons, Mr. Miora has been a speaker at conferences and has written for a variety of journals and magazines.

Cederampattu (Ram) Mohan, Chief Technical Officer

Mr. Mohan Ram Mohan has served as Vice President and Chief Technical Officer of Infonautics (a NASDAQ listed company). Mr. Mohan is the founder of the Sleuth network of products, including the award-winning Company Sleuth. Mr. Mohan was responsible for enterprise-wide operations in 2000, helping drive top-line growth and run the marketing, product management, and revenue growth functions for the company's Sleuth products. In 2000, the Sleuth products tripled from about 600,000 registered users to about 1.8 million. Since joining Infonautics in 1995, Mr. Mohan has helped architect the Internet strategy and technology for Infonautics' products, including the subscription service Electric Library. Prior to joining Infonautics, Mr. Mohan led one of the first network computing finance teams at First Data Corporation. He was also with Unisys Corporation and KPMG Peat Marwick in a variety of leadership, engineering, and technology positions. Mr. Mohan earned a Bachelor of Science degree in electrical engineering, cum laude, from the University of Mangalore, a Master of Business Administration in entrepreneurial management from Bharathidasan University, and is putting the final touches on a Master of Science in computer science from Drexel University in Philadelphia. Mr. Mohan is serving part-time for the company.

Geoff Mulligan, Senior Architect

Geoff was Senior Staff Engineer at Sun Microsystems and was principal architect for Sun's premier firewall product—SunScreen—and a founding member of the Internet Commerce Group. While at Sun, Geoff took a sabbatical to help start USA. Net and create the Net Address product, a permanent e-mail address. Before Joining Sun, Geoff worked at Digital's Network Systems Laboratory developing the DEC SEAL firewall and networking courseware, and researching e-mail systems and technology. He spent 11 years in the U.S. Air Force working at the Pentagon (he brought the first system in the Pentagon up on the ARPANET) on computer and network security, building local and wide area networks and teaching computer science at the air force academy. Mr. Mulligan holds patents in network security and electronic mail technologies. Additionally, Mr. Mulligan is a noted speaker and author and his writings include the book *Removing the Spam: E-Mail Processing & Filtering*. Mr. Mulligan received his Masters Degree in computer science in 1988 from the University of Denver and his Bachelors of Science degree in 1979 from the United States Air Force Academy. Mr. Mulligan is serving as a part-time consultant to the company.

DUE DILIGENCE

The following applies the due diligence criteria to the Postiva technology investment. The due diligence lens adopted here is that of a venture capitalist.

The "Right" Technology

As suggested in Chapter I, the "right" technology assumes that the technology investment target is part of a larger trend. Clearly, in this case, the criterion is satisfied. There is perhaps no more important technology than e-mail and all of the applications and infrastructure that support it. In addition, the use of e-mail and related communications is also driving business-to-business (B2B) transaction processing and especially business-to-consumer (B2C) transaction processing. As if that was not enough to define "right technology," we have the overlay of security (and security technology) on to e-mail and B2B and B2C transaction processing. It is arguably true that e-mail is the "killer app" of the early 21st century. But there are major problems with e-mail, e-mail security, and the way we use the communications medium to solve personal and professional problems. In other words, e-mail and e-mail security is about as ubiquitous a problem as there is. Any and all technologies that optimize e-mail are in the "right" space.

Few or No Infrastructure Requirements

Technology solutions that require large investments in existing communications and computing infrastructures are more difficult to sell and deploy than those that ride on existing infrastructures.

E-mail already comes with a hefty infrastructure price tag. The irony is that the current inefficiencies surrounding the administration of e-mail—not to mention the increasing risks associated with its use as the primary cyberspace communications medium—generate many of the hefty infrastructure bills. While the costs connected with the Postiva Platform will increase the cost of e-mail administration, the same deployment of the technology will result in a significant reduction in the overall cost of e-mail administration.

The Postiva Platform itself represents an alternative e-mail architecture that rides on existing communications infrastructures. This is good news for especially large enterprises that rely so heavily upon e-mail and are pressured to optimize existing technology investments rather than make additional ones. There is a "sell" involved here: Postiva must convince e-mail administrators and business process managers who rely on e-mail that the efficiencies and protection that the Postiva Platform provides is both cost-effective and productivity enhancing.

Budget Cycle Alignment

As Figure 3 suggests, the primary focus of the technology is initially strategic. But over a short period of time it will emerge as a tactical technology necessary to process all sorts of transactions. It is the nature of new technology to initially be seen as strategic, especially if it is unproven. But as soon as it is validated in a pilot or two it can be sold—and used—as a tactical technology.

Quantitative Impact

Impact is hypothetical, save for some data from some early pilots and some simulation-based tests. The early adopters of the Postiva Platform would have to validate the impact of the technology in terms of efficiency and security, among some other meaningful metrics.

In general, it is impossible to know precisely what the quantitative impact will be prior to a technology investment. We make estimated judgments about impact. But there are areas where impact had better occur and better be sufficient such as in the area of privacy compliance and the compliance to other regulatory requirements.

Changes to Processes and Culture

If a product or service requires organizations to dramatically change the way they solve problems or the corporate cultures in which they work, then the product or service will be relatively difficult to sell. The broad administration of e-mail occurs mostly behind the scenes in the vast majority of companies. Much of the Postiva technology will also perform behind the scenes within corporate messaging architectures and infrastructures. Compliance, for example, would occur automatically,

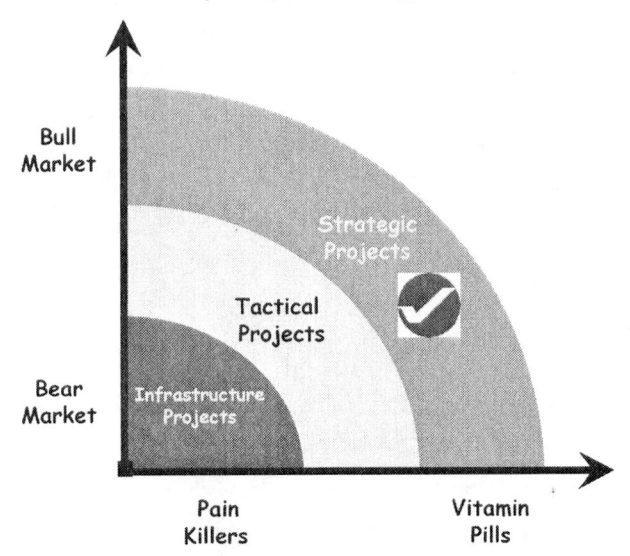

Figure 3. Investment drivers of RFID technology

while some other options—such as those around messaging prioritization—would be user-defined. By and large, the impact to process and culture is expected to be minimal.

Solutions

Companies are always on the lookout for "solutions" that solve as many problems as possible. Vendors love to produce them and VCs love to invest in them. In as much as e-mail itself is not solutions-oriented, the Postiva Platform is not a conventional end-to-end solution. It is, however, an essential element of the messaging architecture and provides much of the missing functionality and compliance that current messaging platforms fail to provide.

Multiple Exits

The filing of patents creates a potential reservoir of intellectual property that can be monetized even if sales of the Postiva Platform fail to materialize. There are also alternative products and services that have "break up" value if the primary business model fails. For example, even if the company fails to get enough traction to sustain itself over time, pieces of the company's technology could very well be sold to one of its competitors.

Horizontal and Vertical Strength

The best products and services are those that have compelling horizontal and vertical stories, since customers want to hear about industry-specific solutions or solutions that worked under similar circumstances (like for a competitor). The Postiva Platform is horizontal in every positive sense of the term. Since e-mail is both strategic and tactical, it is one of the most horizontal tools in the industry. It is also the tool that most companies rely upon to communicate and, increasingly, conduct business. There are also custom vertical solutions that can be developed and optimized for specific industries.

Industry Awareness

There's no problem here: e-mail—and all of the problems and opportunities is creates—is well known to every industry analyst on the planet. Problems with spam, privacy, security, and efficiency are front and center with all CIOs and business technology managers. Postiva, on the other hand, is not all that well known (as few start-ups are).

Partners and Allies

The key initial partner is on the trusted sender side of the house. This "Good Housekeeping"-like seal of approval is a critical component of the value of the Postiva Platform. Beyond that, the company will need to develop a suite of relationships with major (and some relatively minor) messaging, compliance, and infrastructure vendors.

"Politically Correct" Products and Services

The "e-mail problem" is a growing one. Products and services that address this problem are valuable. As concerns over privacy, security, compliance, complexity, and optimization grow, products and services that address these issues will be, almost must be, accepted in the market place.

Recruitment and Retention

Postiva is a start-up, but the list of senior managers and principals connected with the company is impressive. All of them have wide and deep experience in the e-mail/privacy/security space and the CEO has multiple successful start-up experience.

Differentiation

Differentiation is critical to success and while not every differentiation argument is fully formed when a technology is emerging, technology buyers need to understand why the product or service they are about to buy justifies their investment. Postiva is in a terrific space but there is a lot of competition—and some of it is, and will be—from large established vendors. The products and services that Postiva planned to offer actually compete with a whole host of small, medium-sized, and large vendors. Differentiation is in the TrustedSender area as well as in the technology platform itself. But there is a lot to prove in a short period of time—as is often the case with technology start-ups.

Experienced Management

Management was experienced; the overall team is world class. In fact, the team was much more experienced than most start-up teams.

"Packaging" and Communications

The management team was well aware of the importance of collateral materials, industry analyses, white papers, and the like, as part of its go-to-market strategy.

CONCLUSION

The due diligence process yielded mostly green lights. So what happened?

Postiva was unable to penetrate the marketplace in a sustained way. Why not? One important aspect of the lack of traction was the state of the financial and technology markets. The dot.com crash of 2000 caught Postiva (and hundreds of other companies) in the rubble of start-up funding problems. While the training and education consulting business did relatively well, the company found it difficult to raise additional money to invest in its platform.

But over time, the (experienced) management re-assessed how they might monetize their assets for their investors. This assessment led to the identification of its antispam technology as one of its most valuable assets. Prospective acquirers of the technology were contacted as the company simultaneously continued to try to raise money from VCs and angel investors to focus on antispam as the primary technology the company would sell. In an effort to optimize a possible sale (as part of the multiple exit strategy) or to raise money to continue investments in the antispam technology, management spun out a new company—Turntide—and focused all of its efforts to either a sale or a re-investment in the antispam technology. Postiva retained 70% ownership of Turntide.

The dual strategy worked well: Turntide was acquired by Symantec for $28 million in cash. Symantec was most interested in the antispam technology but also purchased all of the (filed) patents for the Postiva Platform. The management team—led by Vincent Schiavone—was able to leverage the potential of venture investments in Turntide against Symantec's desire for the antispam technology, which had the desired effect of increasing Symantec's interest in the new company and its antispam technology. While there were stories about how a new company—Turntide—had gone from start-up to sale in 6 months, the reality was much more complex and sophisticated. After nearly 4 years, the Postiva team had developed a business model, raised a limited amount of money, retrenched when the market went south, spun-out a new company (which it controlled), and then managed to sell the new company to a major vendor for cash. The due diligence performed on the initial investment turned out to be "right" for the wrong reasons—or should we say some of the right minor reasons. In retrospect it was the experience of the Postiva management team that enabled the company to succeed by pursuing a contingency plan to monetize its assets when the primary business model failed to get enough traction.

The lessons learned in this example are all about the weighting one might place on the management team, the space, and the intellectual property (IP) that the team was smart enough to protect—not to mention the team's skill in balancing what appeared to be a two-pronged exit (sale or investment) toward a substantial payback.

ENDNOTE

[1] This case was prepared by Vincent Schiavone with Stephen J. Andriole

Chapter X
Investing in Knowledge-Based User-Computer Interaction:
The TechVestCo Case

INTRODUCTION TO THE CASE

This is an example of investing in a product and service by a company—TechVestCo—that planned to increase its offerings to the larger business technology community. In this example, TechVestCo developed an "intelligent" aid for those who design and field easy-to-use computer-based applications. The idea is simple enough: support the design and development of software applications that are easy to use, which do not frustrate their users and lead to increased human-computer productivity. TechVestCo undertook the design and development of an interactive "workbench" that was intended to help software engineers with the design and development of "friendly" user-computer interfaces.[1]

THE NEED FOR EASIER TO USE SOFTWARE APPLICATIONS

A series of questions drove the development of the knowledge-based user interface design, prototyping, and evaluation workbench—the decision to invest in the development of a methodology and then a software tool that could be sold to software engineers responsible for designing easy-to-use software applications. Here are three:

What if there was an application, a "workbench," that understood the best practices around user interface design, prototyping, and evaluation?

What if those best practices were converted into interactive knowledge bases that assisted designers as they gathered requirements, developed prototypes, and evaluated prototypes?

What if this intelligent system ran on multiple workstations and could be networked so designers could share designs regardless of their locations?

The project resulted in a computer-based workbench that integrates knowledge-based design and development with software engineering best practices into a quasi-automated system that accelerates the design and development of "user-friendly" software applications.

THE USER INTERFACE WORKBENCH

TechVestCo designed and developed a workbench that demonstrated how knowledge-based design can support the design, development, and testing of human-computer interfaces. This workbench demonstrated how a key software engineering activity can be supported by knowledge-based design; at the same time, it also demonstrated how multiple activities can be supported by an interactive knowledge-based software engineering workbench. The platform is extensible to other software engineering activities, such as requirements management, software design, prototyping, and testing.

It was assumed that the workbench had broad commercialization possibilities: there are very few applications that integrate knowledge-based problem-solving with software engineering. Competitors like IBM/Rational have made good progress in the "automation" of some software design processes, but have not fully exploited advances in knowledge-based problem-solving for the design of human-computer interfaces. The workbench project bridged this gap using an "object-attributes-value" knowledge representation methodology (a methodology that is highly extensible to all software engineering processes).

This project assumed that:

- It is possible to leverage software engineering best practices into knowledge bases that can accelerate software design, development, applications, and deployment.
- It is possible to represent the best practices knowledge in objects-attributes-values knowledge base.
- It is possible to demonstrate the feasibility of the approach in a knowledge-based human computer interface (HCI) design, prototyping, and evaluation application.
- It is possible to extend the human-computer interface design process into a knowledge-based interactive application that will support additional software design and development processes.
- It is possible to commercialize the workbench; there are lots of software designers and developers that should be interested in acquiring the workbench to enhance their software design and development efforts.

The application supports user computer interaction designers as they identify user requirements (defined as "tasks"), build interactive prototypes, via the implementation of embedded commercial-off-the-shelf (COTS) software, and evaluate prototypes to determine whether their features should be implemented in production code.

COMPONENTS OF THE INTERACTIVE KNOWLEDGE-BASED WORKBENCH

The essence of the software design process was—and remains—requirements modeling. Figure 1 presents the elements that together yield a requirements model, which, in turn, becomes the input to the (overall and HCI) design and prototyping processes. The HCI workbench supports the identification of the HCI design features that are most likely to satisfy user requirements (given the constraints identified during the requirements modeling phase).

As Figure 2 suggests, there are a number of elements that lead to the recommended displays and HCI routines. Prototyping is an absolute prerequisite to writing software requirements specifications. We may think we know the requirements (even for a system enhancement), but the very best we can expect to do is build and test evolutionary prototypes. This is how software is developed; the TechVestCo workbench supported this process.

The process advocated (and embedded in the HCI workbench) is: prioritized requirements (given constraints) → throwaway prototyping → initial software

Figure 1. The requirements modeling process

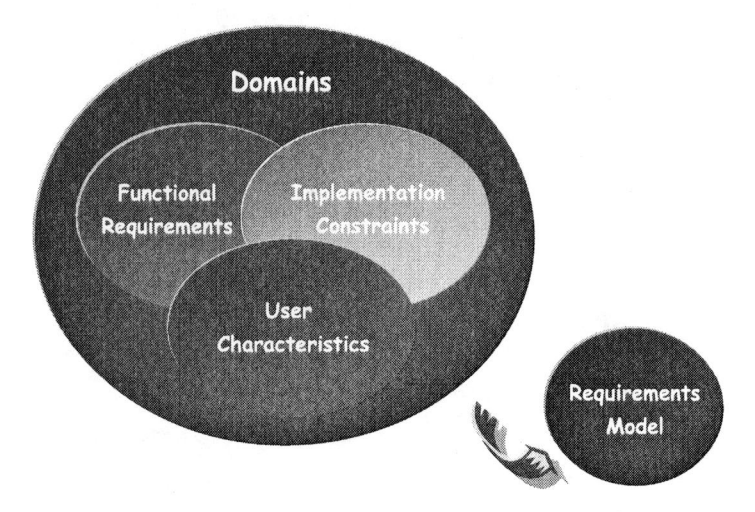

Figure 2. The display and HCI routine identification process

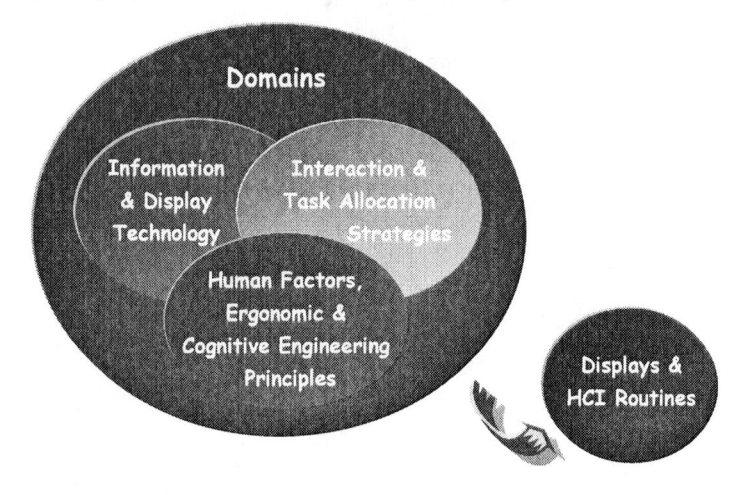

requirements → evolutionary prototyping → detailed software requirements specifications.

The key lies in "templating" the requirements modeling and prototyping processes and in getting experienced professionals to implement and manage them. The key also lies in self-documenting commercial-off-the-shelf (COTS) software that permits group design and communication. The HCI workbench supports the iterative prototyping process via the application of COTS software (which is embedded in the system). The architecture of the system permits the addition (or deletion) of COTS software as it becomes available or as new tools are identified

Figure 3. The prototyping process

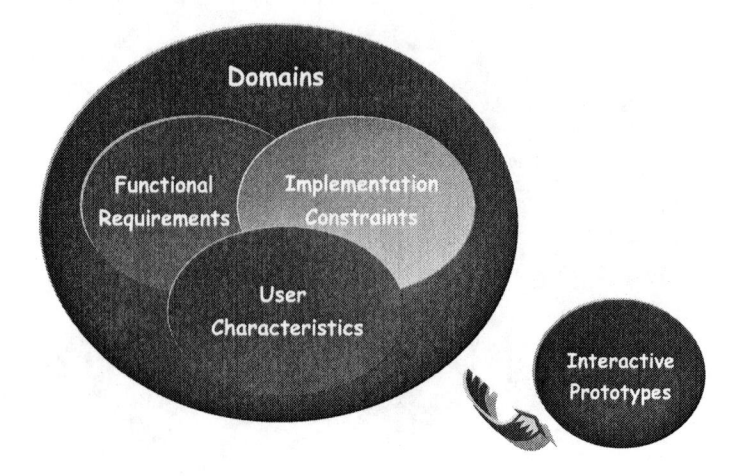

Figure 4. The evaluation process

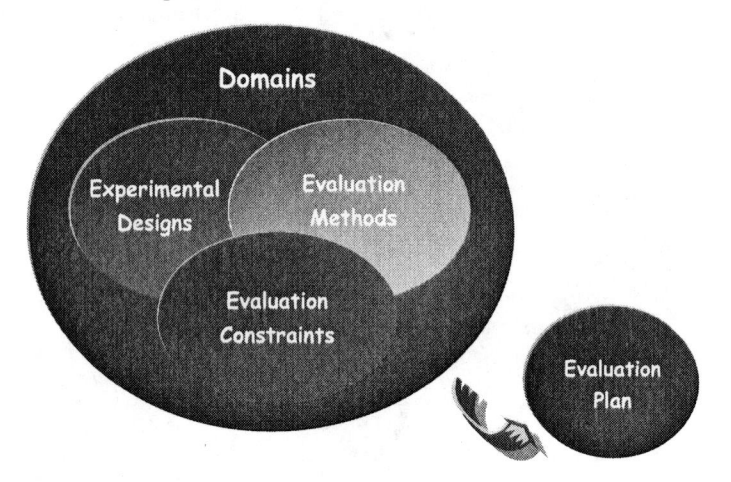

as "preferred." However, it must be remembered that not all COTS tools are equal, and that some are better suited to do some features prototyping and some less so. Figure 3 presents the prototyping "template."

As Figure 4 suggests, the evaluation process is also straightforward. The workbench has knowledge about alternative experimental designs, methods, and the constraints of the specific evaluation task at hand. From that knowledge, as well as the previous data and knowledge about requirements, displays, and so forth, it recommends the best evaluation approach.

A knowledge base lies at the heart of the system. The approach taken to the identification of design processes is anchored in the generic "objects-attributes-values" approach to knowledge representation. In effect, a set of rules were developed to aid the HCI designer as he or she develops interfaces to software applications.

Figure 5 presents the approach graphically.

Figure 6 presents the master menu for the workbench. It identifies three primary activities areas—requirements modeling, prototyping, and evaluation—and a secondary area, HCI sampling, which house video snippets of selected HCI features and COTS prototyping software. Each icon represents a functional area available to the designer. The icons represent a top-to-bottom sequential process, though the designer is not bound to proceed sequentially. The "HCI Sampling" activity area is accessible at any time.

This workbench demonstrates what is possible with knowledge-based software engineering through our HCI application. Theoretically, it would be possible to extend this approach to other aspects of the software design and development process, such as data base design or testing.

While the screen shot was created on a Macintosh, the HCI workbench runs in a Windows environment as well using a variety of off-the-shelf tools (such as rules engines).

DUE DILIGENCE

The following applies the due diligence criteria to this technology investment. The due diligence lens adopted here is that of a company assessing alternative projects

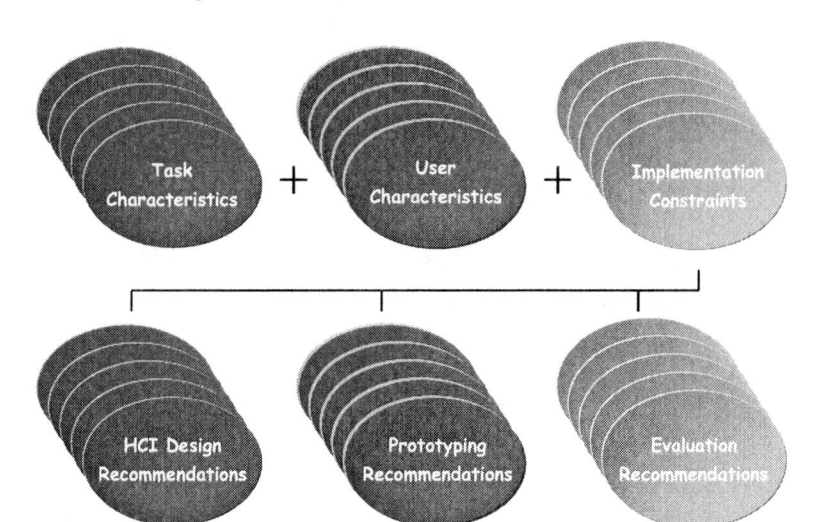

Figure 5. The knowledge base structure

Figure 6. The workbench's master menu

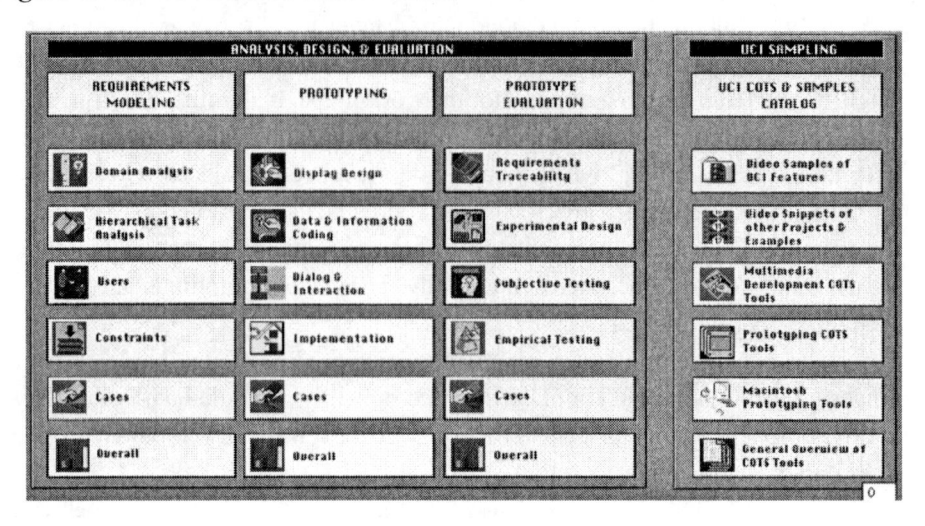

to increase its market share and increase revenues. If the workbench could be sold to the software design community, significant revenue could be generated for the company—at least that was the plan.

The "Right" Technology

As suggested in Chapter I, the "right" technology assumes that the technology product or service is productive today—and likely to remain so. It assumes that the technology "works," and is capable of "scaling" (supporting growing numbers of users). It assumes that the technology is secure. It assumes that the technology is part of a larger trend, such as the development of wider and deeper enterprise applications, like enterprise resource planning (ERP) applications.

But there is another dimension to "right." Technology does not develop in a vacuum. Those who create, buy, and invest in technology need to understand the relationship that specific technologies have with related technologies. For example, what is remote access (Citrix) technology? Is it a technology concept, a real technology, or a whole technology cluster?

Technologies can be mapped on to an impact chart which reveals that many of the technologies about which we are so optimistic have not yet crossed the technology/technology cluster chasm—indicated by the thick blue line that separates the two in Figure 7. Technologies in the red zone are without current major impact; those in the yellow zone have potential, while those in the green zone are bona fide.

The chasm is what separates the yellow and green zones. Note the location of the knowledge-based design technology on which the TechVestCo, Inc., investment was based.

Knowledge-based design—the use of production rules, neural networks, and other knowledge representation and processing techniques—has a long, if not somewhat troubled, history. There have been many attempts to leverage knowledge-based processing onto a variety of processes over the years. Suffice it to say that the track record for knowledge-based anything has not been terrific. The field of artificial intelligence (AI) has suffered greatly over the years for failing to deliver on the many performance promises made by countless academics and technology gurus.

There is clearly risk here: relatively few high profile knowledge-based applications have been successful. The light was yellow—definitely not green and perhaps even a little red.

Few or No Infrastructure Requirements

The workbench runs on Apple Macintosh and Windows compatible computers. No additional hardware or software is necessary. It is a "standalone" application requiring no additional hardware or software to operate.

Budget Cycle Alignment

The decision to invest in the design and development of a knowledge-based user-computer interface workbench was ultimately driven by the design to generate revenue for TechVestCo. The plan was to sell the workbench to companies that wanted to design and develop user-computer interfaces. The workbench is positioned as a painkiller since it reduces the time, effort, and cost of developing effective interfaces. From TechVestCo's perspective, the workbench represents a strategic investment especially since it sought to demonstrate the applicability of knowledge-based design; buyers of the workbench would likely see the workbench as tactical.

Quantitative Impact

Ideally, impact reduces some form of "pain," though obviously the impact of "vitamin pills" can be appealing. Quantitative impact also helps differentiate products and services. The UCI workbench yielded 30% to 50% impact defined in terms of time, effort, and cost. But the real impact—the development and deployment of user-computer interfaces that provide improved navigation of software features—was not fully determined at the time of the product's launch.

Figure 7. Technologies, impact and the chasm—and Citrix

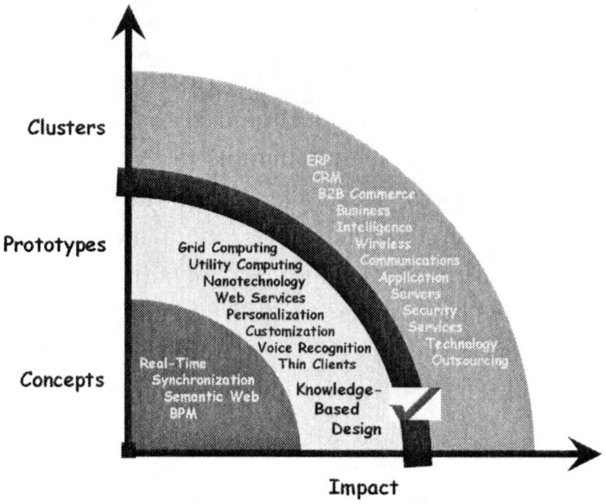

Changes to Processes and Culture

If a product or service requires organizations to dramatically change the way they solve problems or the corporate cultures in which they work, then the product or service will be relatively difficult to sell. Fortunately, the workbench requires no changes to processes or culture, except that software designers would be able to have their UCI design efforts supported by an application designed to do just that. While this sounds like there would be no impact on process or culture, there was a risk that some software designers would resist the help even if it were useful.

Solutions

The workbench is standalone, but definitely not a "solution" in the broad software design and development process. It represents a piece of the process that ideally can be accelerated by the embedded knowledge base and user-interface design process—but it is not in any sense a solution to the software design, development, or testing process.

Multiple Exits

An investment in the workbench is in a sense nonretrievable. Once the dollars are invested, the expectation is that the product can be sold for a profit to software designers and developers. If the product fails to excite the market place, there is not much that can be done to recoup the investment. While some valuable knowledge would be gained from the development of the workbench, this knowledge acquisition could not be described as an acceptable exit.

Horizontal and Vertical Strength

The best products and services are those that have compelling horizontal *and* vertical stories, since customers want to hear about industry-specific solutions or solutions that worked under similar circumstances (like for a competitor). Without a good vertical story, it will become more and more difficult to make horizontal sales. CIOs expect their vendors to understand their business. Smart vendors organize themselves horizontally and vertically to appeal to their clients.

The UCI workbench is a horizontal and vertical application. Once used in a specific vertical industry—for example in the development of user-computer interfaces for applications in the insurance industry—the workbench will store the interface for future re-use. The more UCI applications developed, the deeper the industry relevance will become. (In fact, the workbench has already been optimized for the defense command and control domain.) At the same time, the workbench is extremely horizontal and capable of supporting the design, prototyping, and evaluation of interfaces in many vertical areas.

Industry Awareness

There is considerable industry awareness of the potential of knowledge-based applications, but there is less than adequate understanding or recognition of successful knowledge-based applications. The industry is in fact skeptical of knowledge-based applications specifically and artificial intelligence generally.

Partners and Allies

The workbench would be considered a niche product by a relatively unknown vendor. There are no real partners or allies in the channeling of the product. Support comes from the developer of the workbench.

"Politically Correct" Products and Services

Most technology managers will not risk their careers on what they perceive as risky ventures—even if the "risky" product or service might solve some tough problems. Was the decision to develop the workbench "risky"? Was it part of a conservative culture? The decision was clearly risky, especially when considering the track record of the company with the development, packaging, marketing, sales, and support of shrink-wrapped software.

Recruitment and Retention

As suggested, the TechVestCo, Inc., team was qualified in the design and development of software applications but less so in the sales and marketing of commercial software. The company could well attract and retain professionals in the former category but had an unproven record in attracting and retaining professionals in the latter.

Differentiation

At the time that the product was developed there were no competitors.

Experienced Management

The developers of the workbench were experienced software application developers. They also understood the UCI design, prototyping, and evaluation process extremely well. But did they have experience in the design, development, packaging, marketing, and sales of shrink-wrap software? The record indicated that experience here was thin.

"Packaging" and Communications

The collateral materials for the workbench were targeted more toward software engineers than to managers of software development teams. While the technology "wow factor" might have been adequately described, the business value of the workbench was less convincing.

CONCLUSION

The due diligence process yielded some serious concerns. Not surprisingly, the investment failed to pay dividends. Given the due diligence results, the workbench should never have been developed. There were way too many yellow and red lights and relatively few green ones. It is not surprising that the workbench failed to excite the software design and development community.

ENDNOTE

[1] The initial prototype of the workbench was funded by Rome Laboratories of the U.S. Air Force. Work was performed by Stephen J. Andriole, Lee Ehrhart, and Charlton Monsanto. Subsequent investments in the application were made by TechVestCo, Inc.

Chapter XI
Enterprise Investing in Wireless Technology:
The Villanova University Case

INTRODUCTION TO THE CASE[1]

This is an example of enterprise investing in a specific technology—wireless communications technology—in an effort to provide access to network and core applications to Villanova University administrators, faculty, and students. Like many universities, Villanova is investing in wireless technology to provide mobility to its community of technology users.

BACKGROUND

Villanova University, 12 miles west of Philadelphia, Pennsylvania, was founded in 1842 by priests and brothers of the Order of St. Augustine. It is comprised of five colleges: Liberal Arts and Sciences, Engineering, Business, Nursing, and the School of Law. Villanova's national and international reputation, as well as its students' academic experience, has been enhanced by its distinguished faculty, which has steadily garnered national recognition, including Fulbright Fellowships, Guggenheim Fellowships, and a host of teaching awards. Villanova was recently recognized in the renowned *Princeton Review* for the depth and breadth of technology on campus today, eventually cited as number one of the top 25 "Most Connected Campuses."

Copyright © 2009, IGI Global, distributing in print or electronic forms without written permission of IGI Global is prohibited.

CHALLENGE

Villanova University has embraced the enhancement of the educational experience for faculty and students as one of its key strategic goals. University administrators hope to achieve this goal through advanced, convenient, flexible technology, according to Stephen Fugale, chief information officer at Villanova. "One of the most logical means for achieving this is through wireless," he explains. "There is a clear desire to be online anytime and anywhere as part of the learning experience on campus."

SOLUTION

To test the viability of a wireless network, the university decided to pilot a program late in 2002, using technology from Cisco Systems.® Several Cisco Aironet® 1200 Series wireless access points were installed in the College of Commerce and Finance. One was installed in the College of Law, two in the student union, and several others in computer services offices.

The 802.11a standard, with a data rate of up to 54 Mbps, offers greatly enhanced performance and eight distinct channels for enhanced scalability. Although this standard is not compatible with 802.11b devices, it is immune to interference from devices that operate in the 2.4-GHz band, such as cordless phones, Bluetooth devices, microwave ovens, and hand-held barcode scanners. The IEEE 802.11g standard provides backward compatibility with IEEE 802.11b equipment, preserving users' investment in their existing WLAN infrastructure. However, because 802.11g is limited to the same three channels as 802.11b, scalability may become a factor as wireless LAN (WLAN) user density increases.

The Cisco Aironet 1200 Series Access Point supports the IEEE 802.11b standard, has an 11-Mbps data rate, and provides a migration path to the new IEEE 802.11a and IEEE 802.11g standards—a feature that adds scalability and investment protection.

In addition, the Aironet 1200 Series can be upgraded in the field; customers can order it with the 802.11b radio, for instance, and then add or swap out radios to the new standards as their application and bandwidth requirements evolve.

One of the most interesting applications of Villanova's pilot wireless program involved the Business School's Executive MBA (EMBA) program. The program lasts 24 months, and sessions are held in the Villanova Conference Center, located in a large hotel and conference center about a mile and a half from the main campus. Four Cisco Aironet 1200 Series access points were installed in the center, which is connected to the campus through the Internet.

The pilot program was not Villanova's first experience with wireless networking. A Cisco competitor had previously donated ten wireless access points to the business school, which the university's Network Services group set up as a WLAN for business students. Some of the other colleges had set up their own independent networks as well.

"Not only were there several unconnected WLANs on campus when I arrived (in 2002), but several different network vendors were involved. We wanted to trim this down," Fugale says.

That meant unifying the wireless environment and establishing some preferred strategic partnerships that would help us manage the total cost of ownership, centralize lines of support, and above all, ensure we received top-of-the-line equipment in terms of quality, serviceability and expandability. (Fugale, personal communication)

The decision to standardize on Cisco Aironet wireless technology was based on three key precepts, Fugale (personal communication) adds:

First, Cisco is clearly the market leader. Nobody questions the company's broad array of products and capabilities. Second, I have had experience with Cisco in the past and can attest to its solid commitment to follow-up customer support. Finally, Cisco has taken the time to understand Villanova and our needs.

The first phase of a multistage rollout, which extended the WLAN to all parts of the campus, began in 2003. Nearly 100 additional access points have been installed on campus, in addition to the eight Aironet access points which were deployed at the business school during the pilot phase. The Cisco Aironet access points connect to Cisco Catalyst® Intelligent Ethernet 3550 switches and through them to a Cisco Catalyst 6506 router, which provides routing for the wireless subnet. "We wanted the Catalyst 3550s for a couple of very important reasons," Center explains.

One was their ability to provide inline power over Ethernet cables. This saves money that we would otherwise have had to spend on additional wiring. We also wanted to tie the access points together at single locations, such as putting in 24 per floor at the business school and eight at another location and so forth. In this way, the switches let us aggregate the access points together to provide power. (Center, personal communication)

"In addition, this lets us take advantage of some of the more advanced features of the 3550 switches and the 1200 access points," he continues.

For example, when somebody connects to the wireless network, his or her user identifier will determine which VLAN [virtual LAN] the person gets put on. A technician can log on and have full access to whatever he wants, but a student or faculty member gets whichever VLAN he or she is authorized to connect to. (Center, personal communication).

Finally, all of Villanova's firewalls are Cisco PIX® firewalls.

RESULTS

The initial phase of Villanova's wireless rollout also included 25 access points within the university's Falvey Memorial Library, as well as the installation of access points to create a number of "hot spots" in the center of the campus. Among these "hot spots" is the Belle Air Terrace cafeteria at the Connelly Center, the student union. Another student dining area called the Italian Kitchen, is also covered, as is the faculty dining room at Dougherty Hall.

The next phase will be to expand wireless access to the College of Liberal Arts and Sciences. Coverage will also be expanded to the College of Nursing and the School of Law.

The university's ultimate goal is to provide wireless coverage to the entire campus, beginning with the academic environment, then adding the remaining public areas where students congregate and study, and finally, in the dormitories. Administrators are even considering eventually extending the university's wireless coverage to support students living off campus.

DUE DILIGENCE

The following applies the due diligence criteria to the Villanova wireless technology investment. The due diligence lens adopted here is that of an organization acquiring and deploying a technology that will enhance its operations. The process was led by the university's CIO, who was looking for the most cost-effective solution possible to satisfying the university's mobility requirements.

The "Right" Technology

As suggested in Chapter I, the "right" technology assumes that the technology investment target is part of a larger trend. What is wireless access technology? Is it a technology concept, a prototype technology, or a whole technology cluster?

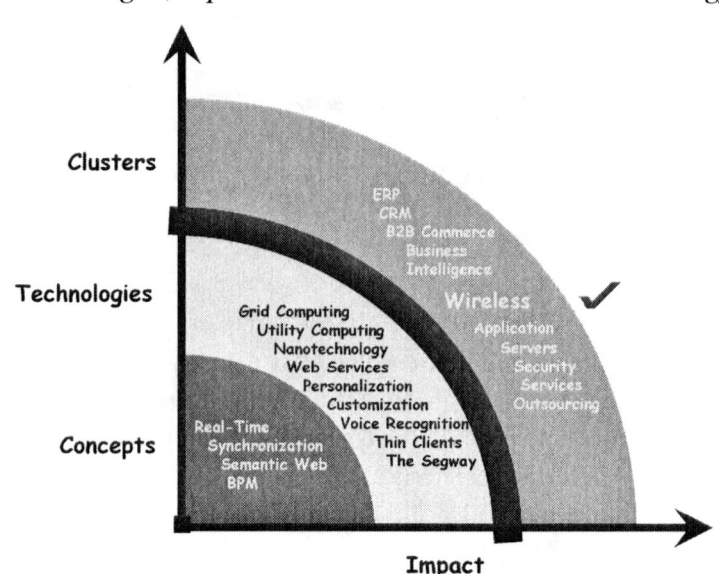

Figure 1. Technologies, impact and the chasm—and wireless technology

Technologies can be mapped on to an impact chart which reveals that many of the technologies about which we are so optimistic have not yet crossed the technology/technology cluster chasm—indicated by the thick blue line that separates the two in Figure 1. Technologies in the red zone are without current major impact; those in the yellow zone have potential, while those in the green zone are bona fide. The chasm is what separates the yellow and green zones. Note the location of wireless technology. Wireless technology has crossed the chasm even though there is still some uncertainty about standards, the technology itself, cost, security, and support.

The light was green here, even though wireless technology is still evolving. There is also lots of precedent for the deployment of wireless networks in many colleges and universities. In fact, a solid argument can be made that as wireless bandwidth increases there will be virtually no land-lines left on university campuses. The timing for all this? Hard to tell, but certainly within 5 to 10 years we can expect most—if not all—tethered university services to disappear.

Few or No Infrastructure Requirements

Technology solutions that require large investments in existing communications and computing infrastructures are more difficult to sell and deploy than those that ride on existing infrastructures. If technology managers have to spend lots of money to apply a company's product or service, they are less likely to do so—if the choice is another similar product or service that requires little or no additional investments. Unfortunately, wireless access is initially expensive. The existing infrastructure at Villanova was a hodge-podge of wired access points around the campus. While it was not all standard and transmission speeds varied, the infrastructure was in place and "worked." Moving to wireless access for students and faculty required a sizable infrastructure investment. In this case, an infrastructure investment was necessary to adopt the new technology and the capabilities it enabled. Worse, given the nature of the rollout, the university is still maintaining two access environments—one wired and one wireless. In fact, it plans to do so for at least several years. But does this make the investment "bad"? No, because the long-term investment—given longer-term technology trends—justifies the investment. Eventually, the world will go wireless and Villanova, like just about all university campuses, will benefit from a relatively early investment in wireless technology.

In addition to the new wireless architecture is the support necessary to keep the wireless environment up and running (not to mention secure). Support requirements (and costs) will decrease over time for two reasons: first, the overall access environment will eventually tilt toward wireless technology as the tethered environment winds down its useful life. Secondly, the cost of supporting wireless environments will decrease over time—especially if the university decides to outsource support for what will become a commodity.

Budget Cycle Alignment

As Figure 2 suggests, the initial focus of the Villanova wireless investment was strategic. The wireless initiative was initially intended to capitalize in the advancement of wireless technology. The key to the investment, however, was its consistency with the overall strategic technology plan. Was the investment a "vitamin pill" or a "pain killer"? The university is constantly looking for ways to improve the way it interacts with faculty, students, administrators, and alumni. Wireless technology is a key component of the of the interaction strategy. It is also consistent with technology trends, the university's strategic technology plan, and the budgeting cycles that permit the allocation on discretionary investment dollars.

Quantitative Impact

The expected impact of the wireless technology deployment includes reduced cost, improved remote and mobile communications, and reduced support infrastructure requirements. The Cisco technology that Villanova deployed was certainly tested in other environments prior to the Villanova implementation. Impact was thus

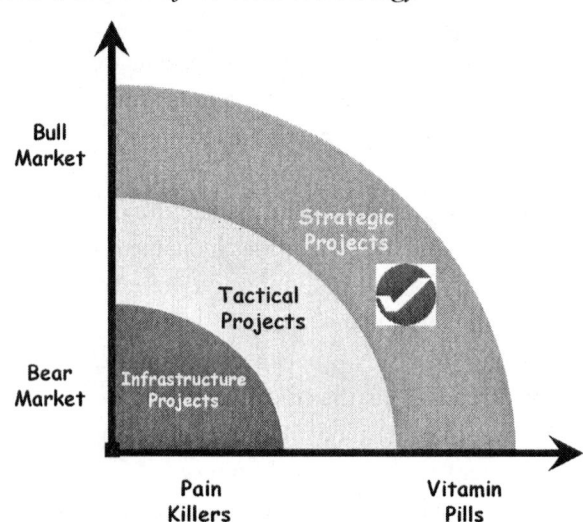

Figure 2. Investment drivers of wireless technology

predictable—to a degree. Metrics will be developed to determine exactly what the impact has been over time. Impact data from similar deployments were available from Cisco. The Villanova CIO also contacted other universities to assess the impact their wireless implementation projects have generated.

Changes to Processes and Culture

The availability of wireless access to networks, applications, and e-mail changes the interaction process and to a limited extent the culture of the communications and computing environment. Mobile computing on a campus—or within any organization—changes the way members of an organization access data, communicate and, in particular to a university environment, learn. Some of these changes are good but some were problematic. For example, wireless communications enables constant communications (unconstrained by the need for access to wired networks). While this of course is desirable, wireless communications also enables communications among students during class (or other meetings) where instant messaging and other online activities are already rampant. Enabling students to go from class to class constantly communicating distracts them from learning (obviously). The university is actually evaluating some in-class jamming capabilities to make sure that students participate in the learning process instead of e-mailing each other or ordering jeans at www.jcrew.com.

Other changes are positive. Wireless access to networks, data bases, and applications makes the normal transaction processing on campuses easier, faster, and agile.

This change to the existing processes and culture is beneficial to the organization making it easier to conduct university business.

Solutions

Wireless technology is not in and of itself an "end-to-end" solution to the university's communications requirements. It is an enabling communications technology that contributes to the efficiency of university operations of all kinds.

Wireless technology-based products and services are not end-to-end solutions for the technology communications infrastructure or architecture, but it does represent—as bundled by Cisco—a segmented "solution." This means that the integration and interoperability requirements of wireless technology are largely solved by Cisco. While the nature of wireless technology is not "stand alone" or "end to end," Cisco has bundled the elements of the technology enough for it to qualify as an integrate-able partial solution to a host of larger requirements. There was a strong sense that Cisco would remain as the primary wireless communications vendor over time and therefore be capable of assisting with the integration and interoperability of its wireless technology with other communications and applications technology.

Multiple Exits

The deployment of Cisco's wireless technology was essentially a win/lose approach to wireless communications at the university. In other words, while there were alternative vendors and technologies the university could have turned to if the Cisco deployment failed, it would have been extremely costly to do so. The other "exit" was to default back to a completely wired campus. Neither of these exits was considered acceptable or—in the case of a total Cisco failure—likely. Cisco was selected because of its reputation, experience, products, and service.

Horizontal and Vertical Strength

The best products and services are those that have compelling horizontal *and* vertical stories, since customers want to hear about industry-specific solutions or solutions that worked under similar circumstances (like for a competitor). Without a good vertical story, it will become more and more difficult to make horizontal sales. CIOs expect their vendors to understand their business.

Cisco is primarily a horizontal vendor: its products and services are used in multiple vertical industries. At the same time, like most large technology vendors, it organizes and customizes its products and services for specific clients in specific vertical industries including academia. Villanova was not the first campus that Cisco made wireless—not by a long sight. In fact, Cisco had a great deal of academic experience that was leveraged on to the Villanova deployment.

Industry Awareness

Everyone is aware of wireless technology and of Cisco.

Partners and Allies

CIOs expect a broad network of support for the products and services they purchase. Cisco has a long history of partnering with lots of third party vendors and consultants. The Cisco channel of partners is wide and deep. There is no danger of becoming orphaned with a Cisco product or service. All lights were green here.

"Politically Correct" Products and Services

Technology investors will not risk their careers on what they perceive as risky ventures—even if the "risky" product or service might solve some tough problems. Wireless communications is generally perceived as a hot technology—the "right" technology for a lot of applications. Villanova University can be characterized as a conservative, cautious investment culture. Leadership is also appropriately cautious but also willing to move in new directions when evidence suggests that the return on the investment will be strong.

Recruitment and Retention

Cisco is a technology blue chip company that has little trouble attracting the best and the brightest. Buyers of Cisco's technology products and services can expect lots of available talent directly from the company and also from its substantial network of partners. The Cisco team that supported Villanova was excellent.

Differentiation

Differentiation is critical to success and while not every differentiation argument is fully formed when a technology is emerging, technology buyers need to understand why the product or service they are about to buy justifies their investment. Cisco's products and services were determined to be solid and competitive in the context of wireless communications. Performance data was widespread and industry analysts helped assess the strengths and weaknesses of Cisco and competing wireless technology vendors.

Experienced Management

The key here is to see the right mix of technological prowess and management experience available to develop and deliver a successful product or service. The Cisco team was experienced and smart. The company is solid and has an enviable reputation in the industry.

"Packaging" and Communications

As a company, Cisco is mindful of the importance of packaging and communications. Cisco sponsors wireless conferences and meetings and already has a substantial number of white papers that describe its approach to the application and support of the new technology. Their Web site (http://www.cisco.com/en/US/products/hw/wireless/products_promotion0900aecd801a118c.html) demonstrates the importance the company attaches to wireless form and content.

CONCLUSION

The due diligence process yielded nearly all green lights. The deployment went smoothly; impact was excellent, and the number of problems was minimal. The due diligence process here was generally successful and demonstrates how a fundamentally conservative approach to technology investing can yield positive results. The process was "conservative" primarily because the vendor and the technology were well established and proven. Villanova was by no means the first campus to go wireless. Risks were minimal and costs were manageable.

ENDNOTE

[1] This example was prepared by Stephen Fugale with Stephen J. Andriole

Section III

Due Diligence Tools and Techniques

Chapter XII
Tools of the Trade

There are several approaches, methods, and tools that will enhance the technology due diligence process. Three will be discussed here:

- Technology trends analysis methodology
- Off-the-shelf tools for due diligence analysis
- Due diligence project management
- A due diligence template

TECHNOLOGY TRENDS ANALYSIS METHODOLOGY

The methodology used to develop our trends analyses involves three steps.

First, it is loosely based on content analysis, a formal technique based on document scanning, and the cataloging of found categories. For example, if the subject of "Linux" appears in 100% of the scanned material, then its content score is high, and if "Linux support" appears in 75% of the material as part of the general Linux coverage, then a pattern can be identified. These patterns become the source of potential trends. The primary materials scanned to do the high level content analysis include the following:

Content Sources

- Various Investment Bank Analyses and Reports
- Various Company Reports
- Forrester Reports
- Aberdeen Reports
- Selected Gartner Reports
- Selected Books
- *Application Development Trends*
- *Bank Systems and Technology*
- *Baseline*
- *BizEd*
- *BioIT World*
- *Business 2.0*
- *Business Finance*
- *Business Week*
- *C@ll Center CRM Solutions*
- *Communications Convergence*
- *Communications of the ACM*
- *Communications News*
- *Communications Solutions*
- *Communication Systems Design*
- *Communications Technology*
- *Computer Reseller News (CRN)*
- *Computer Technology Review*
- *Computer Telephony*
- *Computerworld*
- *CT Media*
- *CSO*
- *CRM*
- *CRN*
- *Customer Interaction Solutions*
- *Customer Interface*
- *Customer Support Management*
- *CFO*
- *Direct*
- *Document Processing Technology*
- *DocuWorld*

- *DM Management*
- *DM Review*
- *eBusiness Advisor*
- *EC Technology News*
- *e-learning*
- *Electronic News*
- *Electronic Commerce World*
- *Entrepreneur*
- *Enterprise Development*
- *Enterprise Systems Journal*
- *Enterprise Systems*
- *eWeek*
- *Fast Company*
- *Federal Computer Week*
- *Forbes*
- *Fortune*
- *FSB - Fortune Small Business*
- *Frontline Solutions*
- *Government Computer News*
- *Government Executive*
- *GPS World*
- *HP World*
- *IBM Think Research*
- *Inc.*
- *Inc. Technology*
- *Information Week*
- *Infoworld*
- *Insurance & Technology*
- *Integrated Communications Design*
- *Integrated System Design*
- *Intelligent Enterprise*
- *Internet Week*
- *Inter@ctive Week*
- *Internet World*
- *Internetworld.news*
- *IEEE Computer*
- *IEEE Pervasive Computing*
- *IEEE Software*
- *IEEE Spectrum*
- *IEEE Transactions on Systems, Man & Cybernetics*

- *Knowledge Management*
- *KM World*
- *Microsoft Executive Circle*
- *Military & Aerospace Electronics*
- *Mobile Computing*
- *NASA Tech Briefs*
- *.Net Magazine*
- *Net Economy*
- *Network Computing*
- *Network Magazine*
- *Network World*
- *New Architect*
- *Optimize*
- *PC World*
- *Profit: The Oracle Applications Magazine*
- *Presentations*
- *Red Herring*
- *Sales & Marketing Automation*
- *Software Development Times*
- *Sm@rt Partner*
- *Sm@art Reseller*
- *Software Development*
- *Storage, Inc.*
- *Syllabus*
- *Tech Briefs*
- *TechDecisions for Insurance*
- *Technology Horizons*
- *Technology Review*
- *Technological Horizons in Education Journal*
- *Telecom Business*
- *Telephony*
- *Teleprofessional*
- *Teradatareview*
- *Transform Magazine*
- *The Industry Standard*
- *VAR Business*
- *XML & Web Services Magazine*
- *Washington Technology*
- *Wired*

Second, the patterns are assessed against an existing data base of trends developed over a long period of time.

Third, for validation purposes the patterns and trends are discussed with "real" analysts and practitioners including investment banking analysts, CIOs, CTOs, and technology vendors.

All of this is synthesized into the trends that appear to be valid at a point in time. Obviously, trends change—though it is the persistent leading edge ones that investors are most interested in.

CIOs, vendors, and VCs must all conduct technology trends analyses, though for somewhat different reasons. In the technology investment business, trends analysis should be a core competency.

Perhaps the best way to illustrate how technology trends can be tracked and documented is to illustrate a couple of trends analyses. The first example is Pervasive Computing and the second is Intelligent Systems Technology. These analyses appear in Appendices A and B at the end of the book.

OFF-THE-SHELF TOOLS FOR DUE DILIGENCE ANALYSIS

The due diligence process can be supported by any number of off-the-shelf software tools.[1] Many of these tools can be used to organize, define, weight, and score investment options with reference to the 15 due diligence criteria. Templates can be set up that can be used over and over again. Many practitioners of multicriteria decision-making have favorite software tools that they use to address all kinds of option selection problems. While these tolls can be helpful, they are by no means required. If your organization is comfortable with multicriteria decision-making and has used some of the available tools, then it might make sense to apply them to due diligence processes, but if the tools are unfamiliar then it makes little sense to invest in the training and mindshare necessary to get the most out of them.

DUE DILIGENCE PROJECT MANAGEMENT

Discipline works: it makes (and saves) money. Project management should be a core discipline in every due diligence process that you launch. The real key is consistency and persistence. Initiatives fail because investors roll out programs, processes, and policies that vary from group to group or organization to organization, or because they lose interest over time, something that professionals can smell long before the plug is officially pulled. I remember well sitting in audiences listening to senior executives talking about a company's major new initiative only to hear the lifers muttering "this too shall pass." So what should your due diligence project management expertise look like? What skills do you need—really? Here they are:

- The ability to assess a due diligence project's likely success or failure. Can the project be successful? What are the risks? Who is the best person to lead the project? Is a good team available? What are the immediate problems we have to solve?
- The ability to execute project fundamentals, such as milestones, deliverables, schedules, cost management, reviews, and so forth. These skills are not necessarily resident in your organization or on your due diligence project management team. You might consider getting a number of your good project managers certified in the latest thinking, processes, and tools. The Project Management Institute (www.pmi.org) is a good place to start.
- The ability to kill bad projects; the ability to determine if a due diligence project is hopeless or salvageable. How do you kill projects? This is like: "should I sell that stock that is down 50% from when I bought it"? Or, can this project come back from the dead (the stock is only hibernating ... it will be back)?

There are at least two reasons to kill a project:

- Project execution can be quantitatively measured. Here all you have to do is measure estimated vs. actual project performance. If the schedule, milestones, costs, deliverables, and risks are 33% or more out of sync with your estimates, the project is out of control and unlikely to recover. If two or three of the indicators are 20% to 25%, the project should be flashing yellow and tracked closely to see if it goes red.
- If a project is drifting from its strategic or tactical objectives, or if execution is poor, a judgment must be made about the likelihood of turning the project around. Like the stock you bought that tanked, can the project come back? The hardest decision to make is one that turns the lights off for good. But if project execution is poor—and getting worse—it may be time to pull the plug.

How do you "see" all of your projects? Some companies and venture organizations have weekly project meetings, some have monthly ones, and politically anal companies only do it informally, privately. What everyone needs is a *dashboard* that immediately shows which due diligence projects are on track and which are under-performing—and which are candidates for capital punishment.

Dashboards are not hard to build; they are even easier to buy. Microsoft Project can be used to feed off-the-shelf reporting applications or you can customize one to show projects as red, yellow, or green, as well as the trends. You should standardize on both the project management and dashboard application. You should also make sure that accurate information gets into the dashboard, which should run on your desktop, laptop, and PDA. In other words, it should be possible to check on major projects anytime you want. Some of these tools—like The Project Control Panel (developed by the Software Program Managers Network; www.spmn.com/pcpanel.html])—extract data from Microsoft Project and inject it into a Microsoft Excel tool that displays project status. Other tools, like Portfolio Edge from Pacific Edge Software, enable you to track multiple projects at the same time.

You need to make the rules. For example, you might have a 10% to 15% estimated/actual variance rule that triggers weekly project meetings. You might have a rule that says that variation on project deliverables is more important than schedule variation; and you might have one that triggers some project survival rules. You might have a rule that states that if more than seven due diligence criteria yield bad results, the investment option should be immediately abandoned. The key is to field a set of rules that work for your organization, culture, and due diligence project experience.

A DUE DILIGENCE TEMPLATE

Technology due diligence is a process that involves qualitative and quantitative assessments around 15 specific investment criteria. Due diligence is the formal term for the process by which we screen and select alternative investment options. The approach described here is part quantitative, part qualitative, part analytical, and part intuitive—because due diligence itself is part art, science, and luck.

Due diligence is organized around a set of constant criteria that can be applied to technology investment decisions of all kinds. Depending on where you sit, the criteria can serve you well—especially if you "customize" them to the decision at hand.

The focus of this book is on technology due diligence that results in a technology investment of one kind or another. The investment targets include everything from software applications, personal computers (PCs), laptop computers, cell phones, personal digital assistants (PDAs), communications hardware and software, data, security, and technology services. The lenses used to vet investment opportunities and challenges are organized around the specific requirements that all technology investors—including, especially, CIOs, VCs, and technology vendors—need to satisfy to achieve their objectives.

Is there is a cheat sheet we can use? Are there Cliff Notes? Is there an "idiot's guide" to due diligence? No, but there is a template that we all can use to make sure that we organize due diligence projects most likely to succeed.

Here is a checklist of items that together comprise a template. Use it to organize your next due diligence effort.

- **What are you buying?**
    - Software applications
    - Data, information, content, and knowledge
    - Communications
    - Security
    - Infrastructure
    - Technology services
    - Advanced technology
- **What impact are you seeking?**
    - Internal efficiencies
    - Operational impact
    - Strategic impact
    - Market share
    - Competitive positioning
    - Revenue
    - Profit
    - Acquisition
    - Public offering
- **How are you organized?**
    - Balanced team
    - Technology expertise
    - Organization expertise
    - Management expertise
    - Sales & marketing expertise
    - Optimal consultants
    - Schedule
    - Access
    - Timing
- **Business case organization**
    - Who
    - When
    - Format
    - Execution

- **Define due diligence criteria**
  - Right technology trend?
  - Low infrastructure requirements/low change?
  - Aligned budget cycle?
  - Quantitative impact?
  - Small changes to process & culture?
  - End-to-end "Solution"?
  - Multiple defaults?
  - Horizontal/vertical stories?
  - High industry awareness?
  - Right partnerships & alliances?
  - "Politically Correct"?
  - Recruitment/retention strategies?
  - Differentiation?
  - Good management?
  - Packaging and communications?
- **Weight, score, and total due diligence criteria (Weight 0 . 1.0; Score 1 – 10; Total Weight X Score)**

	Weight	Score	Total
Right Technology Trend?			
Low Infrastructure Requirements/Low Change?			
Aligned Budget Cycle?			
Quantitative Impact?			
Small Changes to Process & Culture?			
End-to-End "Solution"?			
Multiple Defaults?			
Horizontal/Vertical Stories?			
High Industry Awareness?			
Right Partnerships & Alliances?			
"Politically Correct"?			
Recruitment/Retention Strategies?			
Differentiation?			
Good Management?			
Packaging & Communications?			

- **Business case development**
    - Investment
    - No investment
    - Seek more information

This template can be summarized in a diagram (see Figure 1).

There are also some relative weights that we should respect—as Figure 1 suggests. As Figure 2 indicates, CIOs, vendors, and VCs see the world differently.

Note that the "differences" are not huge; instead, note that there are shades of differences among the major investor groups. Some of these are negligible; others are meaningful—like the importance CIOs attach to keeping their infrastructures unperturbed.

The key point is the need to define, weight, and score criteria according to a dominant investment perspective.

Based on the previous XI chapters—and the template in Figure 1—you should be ready to proceed with just about any technology due diligence project that presents itself, regardless of your investment perspective. Hopefully, the journey has been worth your investment of time and the technology investments you make will yield meaningful returns.

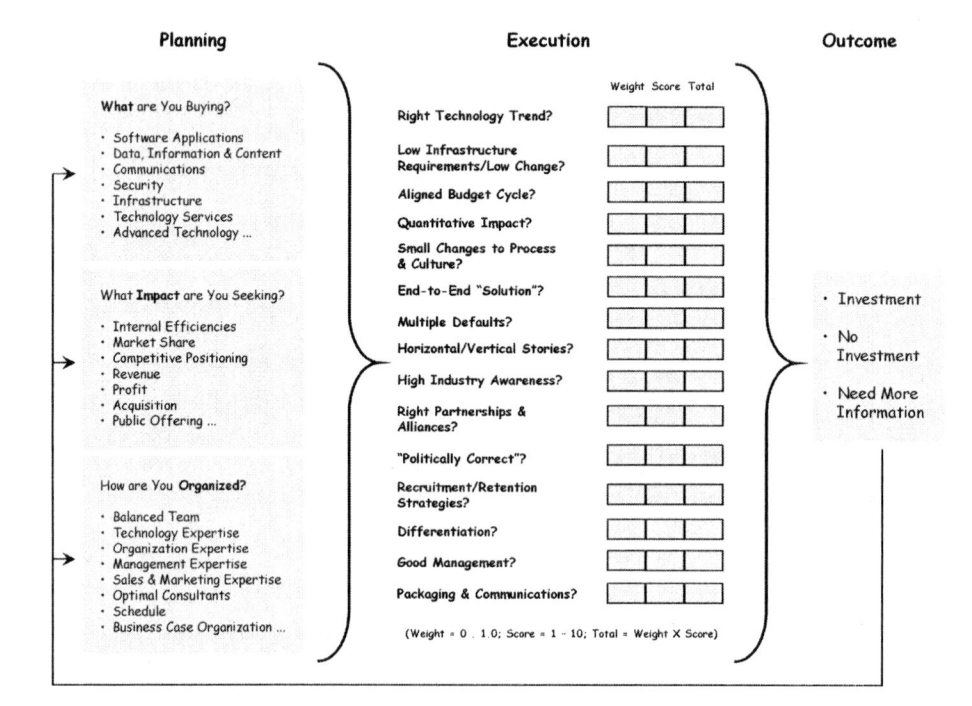

Figure 1. A due diligence template

Figure 2. Relative criteria weightings

	CIOs	Vendors	VCs
Right Technology Trend?	H	H	H
Low Infrastructure Requirements/Low Change?	H	M	M
Aligned Budget Cycle?	H	M	L
Quantitative Impact?	H	H	H
Small Changes to Process & Culture?	H	M	M
End-to-End "Solution"?	H	M	M
Multiple Defaults?	L	H	H
Horizontal/Vertical Stories?	M	H	H
High Industry Awareness?	H	H	H
Right Partnerships & Alliances?	H	H	M
"Politically Correct"?	H	M	M
Recruitment/Retention Strategies?	H	M	M
Differentiation?	M	H	H
Good Management?	M	M	H
Packaging & Communications?	M	H	H

REFERENCE

Maxwell, D.T. (2006, October). Decision analysis: Aiding insight VIII. *OR/MS Today*, 31(6).

ENDNOTE

[1] See Maxwell (2006) for a review of the numerous multicriteria decision-making tools out there.

Section IV

Appendices

Appendix A
Technology Trends Analysis:
Trends in Pervasive Computing

As discussions about the commoditization of technology pervade corporate America, a monumental trend is gathering steam. That trend is pervasive computing. One way to think about pervasive computing is to position the Internet and the World Wide Web (WWW) as a prototype, which is really what it was in the 1990s. The speculation around the new digital economy was valid, on the one hand, because it was based on a vision that was essentially accurate. But on the other hand, speculation was based not on an evolutionary march toward ubiquitous seamless connectivity but rather a revolutionary prerequisite that the prototype Internet simply could not provide. At the same time, "Internet" opportunities are still ahead of, rather than behind, us, and these opportunities are likely to be wrapped in pervasive computing.

The vision persists and will evolve into a suite of architectures and applications that will revolutionize business-to-business (B2B), business-to-consumer (B2C), business-to-employee (B2E), and business-to-government (B2G). In fact, it will be impossible to perform the following tasks without relying on pervasive computing technology:

- Buy and sell
- Negotiate and partner
- Advertise and market
- Manufacture

- Distribute
- Communicate
- Entertain
- Heal
- Govern
- Learn

In the midst of all of the hype, all of the technology and all of new ways we are expected to communicate and compute, there is precious little insight into how it all ties together. If the world becomes "virtual" through Internet ubiquity, how will it impact us? What can we expect?

This analysis identifies and describes the technology trends that will determine how we will work with personal digital agents who will do our bidding over the Web, how we will learn new things through virtual reality-based simulations distributed around the world, how social referenda will occur in real-time, how we will buy the dress of an academy award nominee while watching the awards ceremony, how we will walk down the 16th hole at Augusta National Golf Club chatting with a virtual Tiger Woods about club selection, how we will manufacture from our computer consoles, and how we will treat all varieties of illnesses remotely and compassionately.

This analysis identifies and describes technology trends that when taken together define the macro trend, pervasive computing, a trend that will enable all sorts of activities that today are discrete, disconnected transactions. Pervasive computing will make transactions continuous and seamless.

The analysis describes the macro trend through the multiple trends lens of software, services, and communications, the major drivers of computing and communications applications, architectures, and infrastructure. What we have done here is identify and describe the trends most likely to impact the transaction processing future wrapped under the pervasive computing mantra.

But just as importantly, the analysis outlines an action plan that will help companies prepare themselves for the inevitable connectivity that will change the way we all do business.

It is important to stand back and assess what we have done with technology over time. If you are a large insurer, retailer, or manufacturer you have probably still got a lot of legacy applications that service major parts of your back-office business.

You also probably have some first generation client/server applications that you deployed in the early to mid-1990s that you are supporting—and no doubt you have added some Internet applications, rounding out your back/front/virtual office applications suite. You have also probably "kluged" them together with some integration tools and tricks.

The challenge now is to reassess your computing and communications environment once again, this time with reference to the pervasive computing "wave." Figure 1 describes the various computing eras in some detail (with some additional details about pervasive computing in the bullets that follow Figure 1).

There are vertical industry implications to all of this: some industries have been slower to move along the continuum than others while others have embraced the newer technologies enthusiastically. When we overlay vertical industries on to company age and size, a picture emerges that describes where most companies live. For example, if a company is over 50 years old, is in the insurance or financial services industry and has revenues over $5 billion, it is probably stuck between Eras 2 and 3.

All of this leads to a vision of the future that is measurably unlike anything we have experienced before. In other words, the future, defined in terms of a 3 to 5 year window, will not represent the steady extrapolation from well understood events but rather a revolutionary change that will impact every aspect of our personal and professional lives. Unlike other "revolutions," however, this one is based upon extrapolative infrastructure trends: the revolutionary explosion will come when automated, always-on applications get deployed.

This is the vision with which we all must become familiar. What will it mean when our applications are automated? Continuous? Integrated? How will your busi-

Figure 1. Major computing eras and pervasive computing

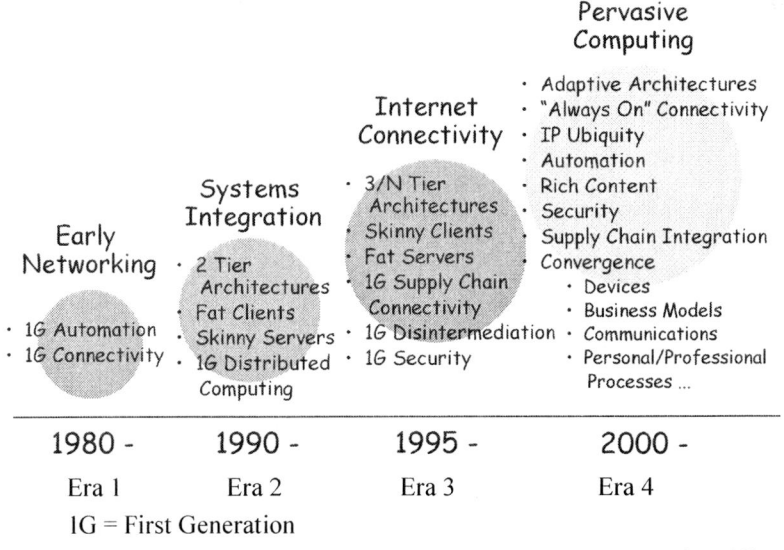

Early Networking	Systems Integration	Internet Connectivity	Pervasive Computing
• 1G Automation • 1G Connectivity	• 2 Tier Architectures • Fat Clients • Skinny Servers • 1G Distributed Computing	• 3/N Tier Architectures • Skinny Clients • Fat Servers • 1G Supply Chain Connectivity • 1G Disintermediation • 1G Security	• Adaptive Architectures • "Always On" Connectivity • IP Ubiquity • Automation • Rich Content • Security • Supply Chain Integration • Convergence • Devices • Business Models • Communications • Personal/Professional Processes ...
1980 - Era 1	1990 - Era 2	1995 - Era 3	2000 - Era 4

1G = First Generation

continued on following page

Figure 1. continued

1. **Adaptive Architectures**

    Computing & communications architectures & infrastructures capable of adapting to heterogeneous environments; architectures that integrate & interoperate across platforms ...

2. **Always On Connectivity**

    Via combinations of narrow and broadband connectivity, continuous personal & professional transactions; ability to continuously access and transact ...

3. **IP Ubiquity**

    IP addresses proliferate across the physical & virtual worlds: cars, homes, buildings, clothing, appliances, etc.; real-time ability to locate all entities ...

4. **Automation**

    Continuous transaction processing via (all) client, server & network-based cooperating intelligent agents; full customization & personalization; the rise of "exception processing" ...

5. **Rich Content**

    Voice/text/video interactive multimedia content; immersable content ...

6. **Security**

    Total security solutions (authentication, authorization, administration ...) ...

7. **Supply Chain Integration**

    Full supply chain integration; real-time inspection of supply chain processes; emergence of personal & professional supply chains ...

8. **Convergence**

    Convergence occurring simultaneously in multiple sectors, including:

    - Devices

        Convergence of access & transaction processing devices, including especially telephones, PDAs, pagers, PCs, embedded processors ...

    - Business Models

        Convergence of personal, professional & hybrid transactions & workflow/collaboration processes ...

    - Communications

        Convergence of all forms & content of communications; integration of off-line & on-line communications ...

    - Personal & Professional Processes ...

        Integration of non-stop, seamless transaction processing across all personal & professional domains ...

ness models and processes adapt to ubiquity? Will your internal processes support continuous transaction processing? These are just a few of the questions we will all need to answer.

The trends that we will examine here include:

- **Software trends**
    - Enterprise/Internet application integration (EAI/IAI) integration
    - Transaction platform development
    - Supply chain connectivity
    - Personalization and customization/business intelligence
    - Automation
    - Rich content aggregation/management
    - Personal and professional portals
    - Architectures: Embedded applications and peer-to-peer computing
    - Voice recognition/natural interfaces
    - Web services/service oriented architectures
- **Services trends**
    - Outsourced service providers
    - Application integration service providers
    - Rich content management service providers
    - Development services
    - Infrastructure engineering services → solutions
- **Communications trends**
    - Wireless applications
    - Network security solutions
    - Bandwidth management and optimization
    - Telecom
    - Broadband
    - Network applications and services
    - Optical networking
    - Touch technologies

These are the computing and communications trends we believe will exert the most impact on pervasive computing. They are also the trends we believe investors in technology—whether they be CIOs, vendors, or VCs—should track.

PERVASIVE COMPUTING TECHNOLOGY TRENDS

There are three broad trends that will enable and extend pervasive computing:

- Software trends
- Services trends
- Communications trends

Software Trends

The software trends we will explore include:

- EAI/IAI/exchange integration
- Transaction platform development
- Supply chain connectivity
- Personalization and customization/business intelligence
- Automation
- Rich content aggregation/management
- Personal and professional portals
- Architectures: embedded applications and peer-to-peer computing
- Voice recognition/natural interfaces
- Web services/service oriented architectures

These trends define the first leg of pervasive computing enabling technology, trends companies should track as indicators of the breadth and depth of pervasive computing's impact on business. The above list is significant because of its simultaneous arrival and because of the combinatorial effect of the technologies on the list: individually they are all important but together they enable a whole new future.

EAI/IAI/Exchange Integration

There are at least three pieces here. But according to industry analysts, no single vendor has all of the integration pieces in place—yet:

- 1. "Federated Database" EAI (Applications Integration Through Access to Multiple Data Bases ... eg, Cohera & Information Builders, Inc.)
- 2. "Process Automation" EAI (Applications Integration by Building Workflow/Process Layers on Top of Existing ERPs & Enterprise Processes ... eg, Vitria)
- 3. "Brokered" EAI (Products that Integrate Applications by Brokering Information ... eg, New Era of Networks & Active Software)

The trends here suggest that there will be software to create new applications as well as platforms that integrate the capabilities of many of the leading vendors. BEA Systems, IBM, Oracle, TIBCO, Vitria, and WebMethods have good capabilities in scalability, universal connectivity, business process tools, and standards compliance. What if a vendor platformed the best of the best? The integration of disparate exchanges and exchange processes (alternative auction forms, for example) is beginning to emerge. Once companies conquer their internal/external integration problems they will move to address their real-time transaction processing integration problems.

What all this means to technologists and technology and business managers is that integration will be a major driver of pervasive computing and that progress here is faster and better than expected. It is important to track this trend because applications integration—when it is truly seamless—will catapult our ability to integrate processes and transactions among customers, suppliers, and employees. A pilot or two here makes sense, especially since everyone has a smattering of disparate applications that need to get integrated. EAI tools and techniques will extend business models and processes, making them ubiquitous, always on, and intelligent.

The assumption here is that while we will certainly get to the next level of connectivity, it is just as likely that we will not throw too much away. Instead, EAI tools and techniques will be used to link applications that will ride on the pervasive computing infrastructure.

The trends analyses indicate that the collapsing of supply chains continues at an incredible pace aided by a variety of software tools and "platforms." Industry analysts predict that the major transaction engines will survive, but that other vendors will not survive the inevitable Darwinian shake-out in the space. Our trends analysis indicates no less than twenty transaction processing engines are working today—and that it will be impossible for them all to own meaningful market share. Track the five market share leaders; ignore the others.

Market makers-in-a-box are also emerging, like from IBM (Websphere Commerce Suite). The ability to integrate different market maker software will continue to be highly valued since these engines are not natively interoperable. It is this space that companies like WebMethods have targeted. It is also an integral component of the pervasive computing revolution.

EAI tools and techniques will morph into Internet application integration (IAI) tools and techniques, and both will facilitate the integration of transaction exchange engines. First generation Internet applications are already disconnected from each other and the myriad legacy systems on which they often depend.

So what are the key application technologies and standards you need to track? With apologies right up front for the following techno-speak, you need to track the major application development and integration technologies offered by the major vendors and their henchmen, which include: XML (extensible mark-up language) and its extensions, Java (the generic programming language) and its extensions, and Microsoft's .net technology (designed to integrate data and applications). Tracking these macro trends will pay dividends downstream.

Let us talk about Web services, which exhibits all of the characteristics of a trend that may or may not have long-term legs. The idea is simple: get the industry to adopt a set of common technology standards to make applications (and data) integrate and interoperate. Wouldn't that be nice? There are at least three XML-based standards that define Web services: Simple object access protocol (SOAP), Web services description language (WSDL), and universal description, discovery, and integration (UDDI). Sorry, here we go again. SOAP permits applications to talk to each other; WSDL is a kind of self-description of a process that allows other applications to use it; and UDDI is like the Yellow Pages where services can be listed. The simplest understanding of Web services is a collection of capabilities that allow primarily newer applications to work with each other over the Internet. Because of the relative agreement about the standards that define Web services there is potential efficiency in their adoption. Conventional glue, for example, may consist of middleware, EAI technology and portals, where Web services—because it is standards-based—can reduce the number of data and transaction hops by reducing the number of necessary protocols and interfaces. Eventually, the plan is to extend Web services to your entire collaborative world, your suppliers, partners, customers, and employees. As you may have already inferred, Web services—theoretically at least—reduces the need for conventional integration technology.

So what should you do about Web Ssrvices, what some call the industry's newest silver bullet?

Web services is fueling the development of service-oriented architectures (SOAs), the organization of software functions and activities that cooperate within an application or even over the Internet. SOAs will change the way we think about software dramatically over the next 5 years.

Platform development, another key software trend, is closely related to the applications integration trend. As platforms become more "standardized," integration problems will decrease. One of the prerequisites to pervasive computing is applications that work together. But note that perfect seamless integration is not necessary for pervasive computing. What is necessary is predictable, reliable integration and evolving proprietary and nonproprietary platforms will also help a lot.

Transaction Platform Development

In spite of the mindshare penetration of Ariba, CommerceOne and other trading/exchange platforms, the "killer app" in e-procurement is an adaptive, real-time platform that dynamically supports "perfect" exchanges, that is, exchanges that involve multiple buyers and sellers in a dynamic pricing environment. Real-time is the killer characteristic of perfect market-making platforms. Trading platforms perform additional tasks, such as organizing buyers and sellers into cooperative trading parties, but the field is still lacking the killer app in the space. At the same time, the major players are evolving their business models in this direction. Part of the problem for these vendors is the complexity of the technology necessary to move from their respective models to a true exchange, since perfect exchanges require fundamentally different software architectures than exchanges based on supplier-only, seller-only, or multiple buyer or seller aggregation. But clearly their strategic plans will move in this direction—and fast (arguably, had the dot.com's not crashed, we would be there already). Pay close attention to this trend: it is a key component to supply chain connectivity (which is enabled by pervasive connectivity).

Online payment options are growing in number and complexity. While consumer preference remains with credit card purchasing (over e-cash or debiting), trends indicate that online billing will be integrated with other billing methods, tools, and techniques. Vendors are improving tools to permit buyers to register and then pay without credit cards. While credit card payment for B2C transactions rules today, it will not tomorrow. Estimates are that within 3 to 5 years, e-cash and related alternatives to credit cards will account for more than 70% of all payments. The engines to support these payment processes represent key enablers of the digital economy and pervasive computing. Exchange and payment platforms will permit seamless, automated, and "frictionless" B2B transactions.

Storage area management (and it is cousin, content management) are huge pervasive computing requirements trends. This area will continue to explode over the next couple of years. In a sense, Akamai represents an architectural response to frequently-requested content management. Akamai preaches the most frequently requested content in servers located strategically along the Internet. The Akamai effect will repeat itself on sites all over and storage and content management will be necessary to keep Web sites humming. Products and services in this area will be in great demand; a killer app here would generate enormous productivity and efficiency—and would have to distinguish itself from piecemeal solutions.

If you look at the discussions about e-CRM, supply chain management, sales force automation, automated marketing, and the evolution of the modern e-call center (which requires an IP backbone), there is an opportunity to build an integrated platform that would combine all of this functionality in a single application. The underlying technology here would have to involve some EAI/IAI/Web services/SOA tools, some telecommunications services, middleware, and lots of communications technology; the integrated application would ultimately evolve to a "whole customer management" platform. This does not exist today and the trends analyses suggest there would be a huge market for it almost immediately, especially if the rollout of broadband communications continues to keep pace with recent commercial deployment. Watch for the integration of what we today call customer relationship management (CRM), customization, personalization, and content management platforms, among others.

Supply Chain Connectivity

As we move toward continuous, connected commerce, supply chain software (and services) will rise dramatically in importance. Unlike the EAI/IAI space, there is not an endless list of players; rather a few dominant ones (i2, Manugistics, and the supply chain modules offered by the major ERP vendors).

The functionality of supply chain management and planning (SCM/SCP) software is rising dramatically. Next generation tools will integrate supply chains across vertical industries and SCM/SCP software platforms.

Supply chain connectivity will occur because of the efficiencies that integrated supply chains yield. When suppliers know that wholesalers and retailers are selling—and for what price—they are in a position to adaptively organize their production processes and schedules. Collaborative forecasting and planning, when it is fully integrated, will fundamentally change the way businesses produce and distribute goods and services.

Supply chain applications will continue to evolve and converge, as ERP vendors, SCM/SCP specific vendors and systems integrators all continue to offer tools to connect producers, distributors, wholesalers, retailers, and customers.

Personalization and Customization/Business Intelligence

Pervasive computing will breathe even more life into mass customization and personalization (and the data mining it requires):

- **Customer, Supplier & Employee "G2" & "Manipulation"**
  - New Models for Identifying, Profiling & Interacting with eCommerce Customers, Suppliers & Employees
  - Distributed Data Mining for Customer/Marketing Personalization
  - Vertical Cross-Selling Inference Models
  - Models of the "Extended Customer"
  - Infomediation Tools ...

Personalization and customization will converge at customer/supplier/ employee touch points which will require the integration of CRM, eCRM, SFA, customer self-service, telesales, campaign management, interactive marketing, and other applications.

There are also opportunities to integrate CRM platforms into networking and call center service models yielding a people-facing solution. Applications, platforms and service models that recognize this trend are bell weathers of pervasive computing progress. The general data mining world will also be resurrected by customization and personalization trends—and the pull for pervasive computing (built upon the major CRM platforms).

There is lots of noise about business and competitor intelligence. This assumes access to data/information/knowledge/content in ways that support all kinds of analyses. Cognos and MicroStrategy represent two successful software vendors in this space. In some respects, this trend represents the intersection of data integration, knowledge management, and enterprise information portal technology. But there is a deeper implication here: all data/information/knowledge/content can be integrated into a platform that would—essentially (and pervasively)—be a flexible distributed data warehouse that supports whatever data users want.

The trends here reveal that all applications that interact with people—customer relationship management, supply chain, sales force automation, customization, personalization, and business intelligence software—will collapse into integrated suites. There are tons of applications that do essentially the same thing as well as market leaders with large market shares (like Siebel Systems). The hottest action is in the broadly defined personalization area, which combines CRM, customization, one-to-one marketing, permission marketing, relationship marketing, and multiple touch point software. But the personalization space is crowded. As is usually the case, about 10% of the companies own 90% of the market share. But remember that a long list of vendors in a space also validates opportunities: if the trend were not valid there would not be so many vendors trying to occupy it.

When we turn to CRM and e-CRM applications, we see a similarly large number of vendors. There is still room for the development of behavioral models that correlate demographic and (online, real-time, and off-line legacy) data into informed CRM, SFA, up-selling, cross-selling, promotional and other inference models (though privacy concerns will continue to rise). All of these applications require integration. Front office CRM and e-CRM applications must be linked with operational customer information data bases, back office core processing applications, business intelligence (BI) software, and customer data warehouse(s)—and then delivered via the Web through wired and wireless connections.

There is room for a technology that would link disparate CRM/customization applications. This is a technology that will catapult pervasive computing applications integration and extensibility.

Automation

There is perhaps no more important technology for pervasive computing than intelligent systems technology (we will talk much more about this in the next section of this chapter). The trends here continue to indicate that the Web will become increasingly automated and that intelligent agents will drive this trend.

Intelligent agents will play a major role in pervasive computing-based transaction processing. There are several levels of functionality here. One is the development of horizontal agent technologies and architectures applicable to multiple computing and communications infrastructures. Another is the deployment of vertical agents with deep domain knowledge about specific industries.

The general capabilities appear below:

- **Intelligent Agents**
  - Software Agents Capable of Autonomous, Self-Initiated, "Social" Able to Communicate with Users (& Other Agents), Reactive, Dynamic, Asynchronous (Actions Independent of Linear/Linked Events), Event-Driven (Can Proact & React to Events), Self-Executing (Can Run Themselves) & Self-Contained (Have What They Need to Run Themselves)
  - Focused on Specific B-to-B & B-to-C Problems

Agents are already powering applications. In the network monitoring area, Computer Associates (CA) and IBM are applications using predictive analysis tools. There are tons of automation opportunities. Some major capabilities will be spun around automation, intelligent systems technology and cooperating intelligent agents. Pervasive computing will be about continuous computing and continuous computing will be about automation. Agents can be applied to security, shopping, manufacturing, e-mail, content management, and countless other application areas.

There are a number of opportunities in the automation area. Trends analysis reveals an increase in attention paid to the resurrection of artificial intelligence (AI). While CA continues to tout their "neugents"—applied principally to network and systems management—other vendors are exploring how to leverage intelligent systems technology in a variety of applications. CA has announced its intention to extend neugents into their entire suite of applications. RightNow Technologies integrates intelligence into customer service applications through e-mail and related applications.

Many companies would benefit from the application of some form of intelligence to their applications and services. The market would reward a killer app here, since there is currently no single company (or even group of companies) that can claim market leadership or market share in the space.

Artificial intelligence (AI) will return with a vengeance. Why? Because Web-based automation is inevitable—it is cheaper, faster, and more efficient than human-computer interaction. The initial trends here have been task-based. "Bots"—like MySimon—assist shoppers and collaborative filtering agents actively assisting Internet searches. This is just the beginning. Watch for the vertical spread of intelligent processing followed by the widespread horizontal application of intelligent technology.

Trends analyses suggest that now that raw data is available and beginning to integrate, the need for better behavior inference models is rising dramatically. Why? Because mass customization, impulse buying, and collaborative forecasting and replenishment (CFAR) all require insight into the buying patterns of individuals and organizations. Some of these models are based on complex psychological models of human behavior, while others are based on simpler extrapolative models of organizational buying patterns. Now that we are close to enterprise data integration, the need for robust intelligent models will rise.

Rich Content Aggregation/Management

- **Converged, Rich Content & Content Management**
  - Industry-Specific (Vertical) & Generic (Horizontal) Content Creation, Integration & Aggregation
  - Distributed Content Synchronization & Management
  - Enabled Content
  - Content Repositories & Repositories Management Services
  - Universal Content Access

This area continues to expand—especially as bandwidth increases and the richness of applications rises:

Companies like RealNetworks are well positioned in the streaming media area with tools that synchronize and stream rich media. Validation of the space comes from Cisco's (and other vendors') commitment to develop rich content management solutions.

Tools and services that span content creation (tools + repositories), content management (development, QA + deployment), and content delivery (Web servers, application servers + eCommerce analysis/metrics tools) will be required to support pervasive computing.

Content management and distribution is also becoming complex. In fact, it is impossible to separate technologies like load balancing and caching from content management and distribution. Ultimately, this takes us to performance enhancement and quality of service (QoS) capabilities as distributed business models get deployed.

Significantly, some of the largest technology players—like Cisco—are moving aggressively into the content management and distribution space. Last Year, Cisco announced a suite of content networking products. IBM, EMC, and other large vendors have also announced hardware, software, and services capabilities in the space.

Some of the niche vendors in the space include:

For document content management:
- EMC/Documentum
- Hummingbird communications
- Open text

For electronic publishing:
- Interleaf
- Inso
- Arbortext

For software configuration management:
- IBM/Rational software
- Continuous software

For Web content management:
- Allaire
- Broadvision
- Eprise
- Interwoven
- Vignette
- IntraNet Solutions

The ability to exploit and optimize these and other vendors' technologies and tools will become increasingly important as the vendors themselves add services to their repertoires.

The new content delivery business model assumes multiple touch and access points, including wireless, Web site-based, interactive TV, and data interchange through different communications mechanisms, including e-mail, call centers, and data warehouses.

The space is morphing into integrated tools, technologies, services and solutions—all necessary to support pervasively distributed data, information, content, and knowledge.

Personal and Professional Portals

Always on, continuous computing will be facilitated by personal and professional portals. Yahoo! Will—like all of the other general purpose portals—continue to evolve into professional and personal portals. There will also be lots of additional portals that support a variety of transactions.

The portal landscape is evolving quickly but like many other software areas lacks integration. The major corporate portals (like Plumtree, Hummingbird, SAP AG, and Oracle), the commercial portals (like My Yahoo! and Netscape), the publishing portals and the personal portals are all point solutions to disembodied processes. As pervasive computing takes hold, there will be a need for cross-portal integration. (There will also be a need for portals with greater functionality and integration capabilities than the current products on the market today have.)

Enterprise information portals (EIP) are growing rapidly as a delivery solution to disparate data bases, applications, remote access, and other problems that demand the integration of function, access, and processing. In fact, EIPs are emerging as a meta-application that umbrella all applications within and beyond a company's firewall.

Portal's come in various shapes and sizes, including corporate, consumer, vertical, and commerce portals. The emphasis here is not on the creation of business models that are portals but on the processes by which portals are designed, developed, hosted, and supported. Opportunities for the application of off-the-shelf portal software are growing rapidly. When well conceived, they represent the ultimate in applications integration, or the interface to all varieties of inside and outside the firewall applications and data bases.

Portal software is offered by:

For ERP portals:
- SAP
- Oracle/PeopleSoft

For collaboration:
- IBM/Lotus
- Microsoft
- Novell

For business intelligence:
- Brio/Scribe
- Hummingbird
- Plumtree

Industry analysts predict that the implementation of EIPs would occur within 95% of all corporations by 2008—up from 60% in 2003. This rapid growth suggests a huge opportunity for design, development, hosting, and on-going support—even if the adoption predictions are a bit exaggerated. Subsumed in the EIP trend is applications integration tools and technology.

Architectures: Embedded Applications and Peer-to-Peer Computing

Programmable digital signal processors (DSPs) and microcontrollers have made all sorts of embedded application development possible for decades. If we assume that pervasive computing is continuous computing, that commerce will become increasingly automated and that applications will have to adapt to dynamic transaction processing, then the market for embedded applications should grow considerably. Real-time scheduling, interfacing, and hardware-software optimization are but a few of the applications that can enhance performance.

Peer-to-peer architectures are also taking hold. While everyone's familiar with Napster-like applications, there are already a handful of start-ups exploiting the capabilities of peer-to-peer (such as Microsoft's Groove Networks). Pure peer-to-peer—where peers have equal capabilities—and hybrid peer-to-peer models, where servers participate in the task allocation and sharing process, represent terrific potential especially when targeted at specific horizontal tasks (like virus protection and encryption) and vertical tasks (like financial transactions). Peer-to-peer architectures will enable all sorts of pervasive computing applications.

Voice Recognition/Natural Interfaces

It is been almost three decades since the first promises about voice recognition were made by researchers at the Defense Advanced Research Projects Agency (DARPA, the same Defense Department agency that created the underlying technology for the Internet). The idea was simple enough: develop hardware and software that could understand what people said—as well as what they meant by the words they used—and then, through "natural language understanding" respond coherently in the same, or even another, language, as appropriate. Full voice input and output was, and remains, the goal, which includes "semantic understanding," or the ability to understand the meaning and context of language, not just the structure of sentences ("syntactic understanding"). Impressive progress was made early on but real-time knowledge-based language understanding has proved difficult—so hard, in fact, that the most impressive progress was made on syntactic side (with limited vocabularies with less-then-perfect speech input recognition devices).

So where are we now, and why should we care? Pervasive computing will require lots of "natural" interaction. Just assume for a moment that it is possible to converse in continuous sentences with machine that understands exactly what you say and what you mean—regardless of the language you are using. In order for this to work several things have to be true. First, for privacy and personalization purposes, the system must distinguish among speakers. Next, it must understand the discrete commands they give as well as the continuous sentences they speak. Numerous knowledge bases help the system understand structure, inferences, and purpose.

Let us look at the pieces of this capability and where we are today and the applications that might benefit the most from improving voice and speech recognition technology.

Discrete speech input applications—those that recognize specific commands (like "Open Powerpoint")—are improving. Some have vocabularies for thousands of words, and some permit users to customize vocabularies to meet the requirements of specific vertical industries. Discrete speech recognition applications work well in specific contexts especially when lots of people will be using the commands in so-called "speaker independent" applications. (Other "continuous speech" applications work best when they are well-trained by their users, which occurs when users speak extensively to the application so the application can "learn" the sounds the user is likely to make when specific words are spoken.)

Efficient continuous speech recognition is harder to achieve. While the goal is to support the kinds of communication that people naturally experience, sometimes the delay between spoken words and their appearance on the screen (or the delay between a spoken word and some reaction) becomes annoying, primarily because delays are unnatural. Research suggests that the most productive users of continuous speech recognition applications are professionals who are also experienced with the use of dictation systems, where pauses, punctuation, and other language structures are part of the process.

We are still some years away from true natural language understanding where software "reasons" what the speaker means when a request like: "show all of the flights from New York to Miami that leave at midnight and cost less than $500" is made. While there is specificity in the query, the speaker is actually interested in getting to Miami relatively cheaply late at night, and would be delighted to receive a response from a natural language understanding system like: "there are no flights from New York to Miami at midnight that cost less than $500, but there are five flights that leave between 11PM and 1AM that cost in the range of $400 - $600

… would you like to see them?" In order to provide this kind of response the application needs to understand what the speaker really wants and in order to do that it must understand travel, priorities and goals. Imagine how sophisticated natural language understanding applications would have to be to understand all possible queries in every conceivable context. Or put another way, while humans take for granted our ability to understand queries like: "how many aces were served," and the requisite (tennis, not poker) contextual interpretation of the query, deep natural language understanding systems will have to understand thousands of inter-related contexts before they can interpret and infer meaning and purpose. Today we have several applications that can convert sounds into parts of speech that can be recognized, displayed and stored—and trigger some predetermined action. While these applications are far from intelligent they are extremely powerful and for selected tasks very productive.

How can this technology be exploited? There are any number of uses that might benefit business including:

- Data base, e-mail, knowledge access
- Call center customer service
- Account balance checking
- Order processing
- Manufacturing production control
- Personal tasking
- Speech-to-text/text-to-speech

Access to information or transactions can be through devices mounted in desktop and laptop computers, hands-free embedded devices (such as in automobiles), personal digital assistants (especially the voice/phone-enabled ones), assembly line devices, and even voice portals to enterprise resource planning (ERP) applications.

Part of the adoption challenge lies in the integration of voice into existing access and communications capabilities. As always, integration and interoperability become important opportunities and constraints in the deployment and support of speech recognition applications. Sophisticated users of information technology appreciate that enabling technologies like speech recognition are best exploited as extensions of existing business processes already supported by technology. Voice-enabled customer service, for example, represents a great way to extend a company's care and handling of its employees, customers, and suppliers so long as it works well with the other care and handling tools in use at the company.

Who are the players here? The major speech recognition vendors include IBM and Philips. These vendors have been in the market for a long time. A number of additional vendors have come and gone focusing on speech-enabling the Web.

They include: BeVocal, Foodline.com, General Magic, Onebox.com, SpeechWorks, Tellme and Vocal Point, among others.

Track these companies and their products. All of the progress in CRM, customization, personalization, and automation will require sophisticated natural interfaces. Voice is the most natural interface, so track the technologies that enable natural communications among data, applications, and infrastructures.

Web Services and Service-Oriented Architectures

In spite of the word "services," Web services represents quintessential integration. In a nutshell, Web services refers to a suite of standards-based tools and techniques that permit continuous e-business. In a sense, the current trend in Web services is toward a relatively imprecise coupling of applications and functionality through some flexible technologies, like XML. In some respects, it is the ultimate IAI methodology, though in some others it is an evolutionary response to the relatively clumsy integration efforts that occurred in the late 1990s. Web services also represents an attempt to integrate disparate platforms and architectures with the 21st century's version of wrapper technology, XML.

As suggested above in the context of integration and interoperability, Web services exhibits all of the characteristics of a trend that has long-term legs. The idea is simple: get the industry to adopt a set of common technology standards to make applications (and data) integrate and interoperate. Wouldn't that be nice? There are at least three XML-based standards that define Web services: Simple object access protocol (SOAP), Web services description language (WSDL), and universal description, discovery, and integration (UDDI). SOAP permits applications to talk to each other; WSDL is a kind of self-description of a process that allows other applications to use it; and UDDI is like the Yellow Pages where services can be listed. The simplest understanding of Web services is a collection of capabilities that allow primarily newer applications to work with each other over the Internet. Because of the relative agreement about the standards that define Web services there is potential efficiency in their adoption. Conventional glue, for example, may consist of middleware, EAI technology and portals, where Web services—because it is standards-based—can reduce the number of data and transaction hops by reducing the number of necessary protocols and interfaces. Eventually, the plan is to extend Web services to our entire collaborative world, suppliers, partners, customers, and employees. Web services reduces the need for conventional integration technology.

Track what the big vendors are doing in the space. Already IBM, Oracle, Microsoft, and other vendors have announced their commitment to Web services standards, though their actual commitments remain to be precisely defined.

Web services is a fascinating technology development with enormous potential. But the reason why you need to track progress here so closely is because of the relationship between Web services—standards-based integration—and collaborative business models. Web services has real cornerstone potential. It is likely to become a major enabling technology.

Service-oriented architectures (SOAs) represent the next step toward integration and interoperability. Building on Web services standards, SOAs make it possible for software components to be shared and assembled on the Web to solve specific transaction processing problems. In its their ambitious form, SOAs will permit the virtual and temporary assembly of software components necessary to achieve some specific functionality—like checking remote inventory—and then immediately disassemble themselves, after they have determined how much the fee should be for the execution of the transaction. Libraries of components will live on the Web and be available for work at a moment's notice.

There are all sorts of opportunities in the SOA world for CIOs, vendors, and VCs. This is open territory for users, creators, and investors.

Services Trends

The services trends consistent with the theme of pervasive computing fall into the following broad categories:

- Outsourced service providers
- Application integration service providers
- Rich content management service providers
- Development services
- Infrastructure engineering services → solutions

Outsourced Service Providers

Pervasive computing will require service expertise that may extend well beyond the capabilities of many companies. Outsourcing in one form or another will continue to increase. The big story here continues to be applications services and hosting. Just about every vendor is moving into this space. The packaged software vendors, the telecommunications vendors, and the systems integrators are all offering applications-for-rent. Pressures on traditional software licensing models is growing and the nature of the offerings themselves is changing dramatically.

As small and medium-sized businesses accept the application service provider (ASP) model, which they will, more services will be required as these companies outsource more and more of their technology. While the large enterprises will lag in their adoption of this model—for a variety of capital and cultural reasons—they will eventually succumb to the logic of renting vs. buying, building, or maintaining.

Commoditization remains a real threat to the profitability of the current ASP model. Without higher margin services, the "basic" ASP model—like the basic cable TV service—is relatively unprofitable. So the ASP services model will expand to include a variety of capabilities that range from applications support to security (just like cable companies try to sell their existing customers more and more high margin services, like video-on-demand, and premium channels).

The most obvious software support trend continues to be this move toward ASP—and commoditization. AppCity, Inc., announced "no-charge" apps and expects to launch proprietary applications (in business, information, lifestyle, and shopping) that will be free to users (who will suffer through banner ads—the principal way AppCiy expects to generate revenue). While there will be "free" competition, especially on the B2C side, the trends indicate that this model will be short-lived on the B2B side though price compression will continue and many these companies will probably go under.

Another trend is the segmentation of the ASP market which includes single-app ASPs, choose-your-own app ASPs, flexible pricing-based ASPs, and vertical ASPs, among other variations on the ASP theme. Surebridge, for example, offered customers the option of renting-to-own apps licenses or a strict rental model. Single application ASPs, which are often proprietary, offer customers bare-bones support and can offer them inexpensively because the apps were architected for the Web from the outset and therefore cheaper to build, modify, and support.

Many of the ASP models, as well as the purer hosting models will merge into full service total solutions providers (TSPs) that will initially emerge as horizontal TSPs and then morph into divisions that will be vertically targeted. Successful vertical TSPs will—if they can achieve critical mass—be acquired by the leading TSPs. Competition here is likely to be fierce, because the largest technology service providers, through partnerships with leading independent software vendors (ISVs), are able to move into the space relatively easily. Their movement into the space will, however, occur in exactly the opposite way the current ASPs/TSPs are trying to penetrate the market, that is, first into the small and medium-sized businesses and then into the large enterprises. The larger technology providers are today primarily in the large enterprises. Within 3 years, however, we can expect all resistance to fall. The race is now on to see which TSPs win and how the markets will segment. Companies that cannot morph fast enough (like Exodus Communications, a leading pioneering ASP, failed in the late 1990s) will become M&A targets. Watch also for the storage area network companies to first become "independent" ASPs and then become part of a larger TSP offering.

Full service TSPs will see their margins grow over time, though there will clearly be "division envy" within these organizations, with some divisions commanding high margins and others relatively small ones. The lower margin TSP services will become loss leaders for many TSPs, while the higher margin services will pay all of the bills—and then some. There is no great danger of overall revenue/profit compression in the space: the introduction of new technologies and services will keep the margins healthy for some years to come, although there may be an overall reduction in profitability—and even regulatory pressure on the industry—as we move toward a pure technology "utility."

This concept of "utility" is important for pervasive computing. As commerce becomes continuous, we are all going to need reliable, scalable, secure operational support—which itself will be defined broadly to include the introduction of new applications. Stated somewhat differently, pervasive computing will require "utilities." The total solutions providers (TSPs) will evolve into these utilities.

The Web professional services companies (also known as interactive agencies) will also have to continue morph toward additional services. The first additional service to be added will be strategic services, where Web consultancies will offer full-strategic business consulting services, encroaching on what companies like Accenture, PWC, and IBM Global Services now do. This is a natural initial service addition for the e-business and Web professional services companies. After that, they will either develop full implementation and support service offerings or partner to achieve an end-to-end capability.

Another major trend: once ASPs achieve a critical mass of capabilities why would they not expand into related service areas, like e-call centers? Trends analysis suggest that the morphing will not stop among ASPs, hosters, and services companies—all of which are primarily horizontal—but will morph into "vorizontal" areas like call centers, customer relationship management (CRM), sales force automation (SFA), and the like.

While there are lots of Web applications performance metrics point products, there is a need for a metrics framework that would track internal and external performance metrics, such as how well a Web site's applications are performing as well as how much business they are generating. The necessary technology to provide this service is general "sniffer" technology as well as data warehousing and mining technology, or so-called "web-intelligence" technology. In short, while Media Metrix does a great job, there is a ton of additional analysis that will be necessary to keep sites up and productive, especially as transactions become automated.

The outsourced service provider market is clearly changing. The models have evolved so far that companies, like Jamcracker and Agiliti, actually integrate the disparate services of multiple vendors and multiple service providers. This kind of services integration, by the way, is completely consistent with pervasive computing requirements: this discussion is about the economics of making it all work. The facts are simple: as more and more services are outsourced, more and more companies will require integrated, managed solutions.

Traction is easier to achieve in vertical markets, where applications and computing and communications infrastructures are distinct. The earliest shakeout in the ASP market took place among the pure plays, the horizontal ASPs. The more vertically inclined ASPs have fared relatively better.

Vertical xSPs (VSPs) can focus on specific sectors more easily than pure plays, since small vertical players will often evolve into larger ones (on the same scalable infrastructure). VSPs like Casecentral (legal), DocumentForum (legal), EchoPass (high-tech), HotSamba (manufacturing), InfoCure (health care), Trizetto (health care), TalkingNets (telecommunications), Virtual Financial Services (financial services), and General Growth Properties (retail) focus on the unique (though large market) requirements of specific vertical industries. It is also important to note that focusing on small and medium-sized business will by definition require a more complete solution than focusing on large enterprises, which already have substantial infrastructures they usually need to amortize.

The recommendation here is to track companies that:

- Focus on specific vertical industries
- Offer integrated solutions
- Have extensive partnerships with software and telecommunications vendors and so forth

Application Integration Service Providers

The enterprise application integration (EAI) and Internet application integration (IAI) service market continues to grow. It is likely to do so for several years to come - especially as early Internet applications achieve "legacy" status—and especially as enterprise information portals (EIPs) continue to grow in popularity.

The EAI/IAI services market will expand to include the enterprise resource planning and extended resource planning (ERP/XRP) markets, the front/virtual office market, the supply chain market, and the horizontal middleware market. Specializing in one or all of these markets without a corresponding focus on applications and related solutions services will, however, increase pressure on the pure EAI/IAI services vendors. The question for leading EAI/middleware vendors—like BEA Systems—is whether to dominate the software market, the software + services market or the solutions market.

Advances in Web services will level much of the EAI/IAI playing field. As more and more software vendors build Web services compatibility into their offerings, the proprietary EAI/IAI tools will lose much of their appeal— another reason to track developments in Web services and service-oriented architectures closely.

Development Services

The development market is changing quickly. As suggested above, the e-business professional services market has included implementation + strategy and in so doing is promising enterprise and Internet applications development as part of the package.

Development platforms and architectures are also evolving. Web services will emerge as the killer integration standard. It is just too powerful to ignore and even the proprietary incarnations of the technology—like Microsoft's .Net and Sun Micro Systems J2EEE—are more "standard" than not.

The big news is the penetration of Linux in selected markets. The adoption curve for Linux is very similar to the early adoption curve for Windows: it is adopted first as a file and print server and then as an applications server. Windows (and of course UNIX) are way ahead in the apps server market; but Linux is earning its stripes in lots of companies. Part of entry into the club is cost-effectiveness: UNIX has for years been seen as "expensive," while Windows relatively "inexpensive." Support costs for Linux are impressively low.

The Linux service market is also expanding. As its penetration continues, demand for services around the OS are also increasing. Large technology service providers, like IBM, CSC, and Perot, are in relatively strong positions to offer Linux support easier than smaller ones focused more on e-business services or Linux-only service vendors like RedHat (which will have to include support for heterogeneous environments to survive).

Linux is playing a role in the pervasive computing era. It has adaptive capabilities, is relatively inexpensive to install and maintain, and as an open architecture provides integration with other open applications and even some proprietary ones.

A key trend is the development of Linux applications and the development and deployment of integrated development environments (IDEs) to accelerate development of Linux applications. Inprise supports Linux apps in its Java IDE—an important step in the legitimization of Linux and in the development of open applications—that will accelerate pervasive computing progress.

Another important trend is the arrival of systems management tools and services that support Linux. Computer Associates has evolved its Unicenter TNG framework to support Linux. Veritas and Mission Critical Linux are developing systems management tools for Linux as well. Similarly, another key to Linux adoption, clustering, is beginning to appear through offerings from companies like TurboLinux, SGI, and Veritas. As Linux knocks down the development/ROI metrics, systems management and clustering barriers to adoption, its penetration will continue to rise so long as other events, like UNIX and Windows effectiveness, evolve as expected. There is still a raging debate about the long term viability of open vs. proprietary software and the proprietary vendors are systematically opening their architectures to respond to the Linux and other open systems threats to their markets. Likely long-term outlook: a combined open/proprietary standard will emerge that will render the debate less important than it is today. The trends toward TSP utilities and Web services will also make many of these architecture religious wars irrelevant.

Infrastructure Engineering Services → Solutions

One of the areas that is receiving a great deal of attention is Internet infrastructure engineering services, especially all of the services that revolve around the Internet protocol (IP). A typical IP-centric service portfolio would include applications services, networking services, and integrated access services.

As more and more business models become increasingly distributed and IP-based, we can expect this class of service provider to prosper. In terms of the above discussion of ASPs/TSPs and the hot horizontal areas, infrastructure architects would fall in the communications/connectivity solutions space with capabilities everyone will need.

Communications Trends

Pervasive computing communications trends include:

- Wireless applications
- Network security solutions
- Bandwidth management and optimization
- Telecom
- Broadband
- Network applications and services
- Optical networking
- Touch technologies

Wireless Applications

Wireless technology and applications continue to explode. The wireless applications space is also still poised to grow dramatically. Applications that integrate with the existing large market share of tethered applications, that extend these applications, and that represent whole new wireless Internet-based functionality will enable large segments of pervasive computing. If we assume the continued development of 2-way packet data networks, wireless local area networks, and 2-way paging networks, smart phones, wireless modems, and PDAs, we can expect opportunities for second generation and 2.5 generation wireless applications—from Web clippers and microbrowsers to intelligent agent-based applications that execute serious B2B and B2C transactions. (Investments in 3G or third generation networks with enormous wireless bandwidth will only slightly lag 2.5 generation investments, which may be adequate for serious pervasive computing—or at least the initial phases of the revolution—but 3G → 4G applications will continue to evolve and eventually dominate network architectures.)

Next generation networks will redefine next generation applications. The interrelationships among latency, security, and bandwidth changes the nature of voice, e-mail, video conferencing, large file transfer, and Intranet applications. The previous discussion about voice input and output assumes robust IP-networks to support wireless voice interaction.

Other applications include mobile inventory management applications, product location applications, service management applications, mobile auction and reverse auction applications, mobile entertainment applications, m-distance education applications and, of course, m-music applications.

As always, there are significant integration and interoperability opportunities, especially as they involve connections among mobile devices, communications towers, mobile switching centers, wireless protocol-to-IP gateways, Web servers, the Internet itself, application/middleware servers, single sign-on servers, and all of the front and back office applications and data stores inside small, medium, and large businesses (including ERP apps, CRM apps, e-mail/messaging/groupware applications, data base applications, and the like). Applications that glue all of these parts together wrapped in reliable, scalable functionality represent real opportunities in the pervasive computing era. In other words, wireless middleware and wireless middleware-based applications will play huge roles enabling pervasive computing. Web services standards will help here.

There are also wireless integration requirements that must be satisfied. As the pure play e-business professional services companies move into the space, so too are the device manufacturers and carriers targeting opportunities in what is certainly more than an incremental opportunity. Those that make ubiquitous devices have a vested interest in the quality, security, and reach of IP-networks—just as pervasive computing requirements evolve.

Wireless technology and applications are exploding. This is, arguably, the hottest sector of the new economy and Internet infrastructure (still followed closely by optical networking).

The wireless applications space is poised to grow dramatically—and in some counter-intuitive ways. For example, wireless voice portals that exploit speech recognition and text-to-speech (TTS) technology that will support B2B and B2C commerce, have begun to penetrate the market.

Today the wireless network space is the largest growth space, while the applications space is the smallest, but by 2008 – 2010 the ratio will reverse, with applications dominating the space.

Capabilities will soon exist to develop applications that integrate with the existing large market share tethered applications, that extend these applications, and that represent whole new wireless Internet-based functionality.

The Internet content adaptation protocol (ICAP) will make content viewable from cell phones, pagers, PDAs, and other mobile devices. Other standards—like the wireless application protocol (WAP), Bluetooth, global systems for mobile communication (GSM), 802.X (WiFi), and WiMax—are all defining the direction of wireless communications, content management and display, and support.

Mobile e-commerce (or so-called "m-commerce" banking, payment, shopping, entertainment, and so forth) will drive wireless applications and services; virtual carriers will emerge, carriers who will outsource the wireless pipe operation to commodity carriers and focus instead on owning and servicing a large customer base and the multiple revenue streams it will generate.

Network Security Solutions

The security area continues to search for an integrated solution that addresses authentication, authorization, administration, and recovery for applications inside and outside of the firewall. Security is its own outcome but increasingly privacy is becoming a major driver of security applications and services. We are also seeing the arrival of "pure play" security solutions companies. The tragic events of September 11, 2001, have raised the digital security stakes considerably.

Biometric authentication will finally come into its own. As advances in public key infrastructure (PKI) and digital certificate technology continue, the need for a biometric overlay appears to be growing as a hybrid or partial solution to a variety of security problems. While it is true that biometric authentication offers the only reliable authentication technology, it is also true that the appeal of biometric authentication has historically been associated not with routine access but special purpose, highly sensitive access. The events of September 11, 2001, will increase the appeal of biometric and other authentication technologies that up to this point have enjoyed limited application.

The real payoff lies in integrated security solutions. This is still a relatively uncrowded market, though companies like Verisign are rapidly redefining themselves as comprehensive security solutions companies.

Bandwidth Management and Optimization

This area segments into several areas we all need to track:

- Broadband provisioning software
- Quality of service (QoS)
- IP (Vertical) virtual private networks (VPNs)
- TSB/BLECs/BPL

The provisioning software market is growing. Regardless of the connection technology or the bandwidth provided, there is a need for broadband provisioning software. Continuous commerce will demand it.

Solid provisioning software solutions would provision all flavors of broadband (optical, cable, etc.) on all forms of backbone network protocols (frame relay, IP, ATM, and so forth). While the need is great, the space is relatively uncrowded, though Cisco has claimed the space (along with a few other vendors).

QoS opportunities are still impressive, though some analysts are beginning to distinguish between QoS and QoE (quality of experience), the latter characterized by an external view of network and communications performance. Other analysts see QoS evolving to QoE. The ideal applications here are intrusive and non-intrusive, real-time, introspective and predictive. They also integrate among carrier operation support services (OSSs) through selected middleware applications.

IP VPNs (virtual private networks) continue to be high on the list of fast, cheap remotely accessible voice/data WANs (wide area networks). There are opportunities to verticalize VPNs (building on the horizontal VPN technology products and services now on the market. The combination of managed vertical VPNs, remote access, and dedicated Internet connectivity in a single service represents an integrated solution. Opportunities also exist to create services around vertical VPNs that provide private data and voice traffic. There is also a natural marriage among QoS/QoE, security solutions and the deployment of vertical VPNs.

The TSP/BLEC (tall, shiny buildings/building local exchange carrier) market offers opportunities to provide in-building managed IP services. The DSL (digital subscriber line) and other broadband carriers and service providers are into the space (sometimes pulling fiber or coaxial media directly into buildings). But there are "pure plays" in the space as well. These service providers are somewhat unique in their partnerships with real estate investment trusts, building owners, property managers and major players in the industrial real estate market. The unique (real estate-based) point of entry into the business represents a legitimate barrier to entry to the traditional carriers and communications service providers. It is likely that hybrids will emerge with BLEC partners providing access to value-added services.

The trend is interesting because it represents an alternative path to customers. Broadband over power lines (BPL) is yet another broadband delivery technology that bears watching. Recent government regulatory decisions have made BPL delivery systems viable.

Telecom

Traditional telecoms continue to migrate their PSTN platforms to next generation network architectures based on IP. The primary lag effect? The huge costs sunk in fiber cable, SONET, SDH transmission equipment, undersea investments, and analog copper loops, among other indirect sunk costs. There is also the belief, that trends analysis suggest is misdirected, that the demand for higher priced broadband services will lag for a significant percentage of the population. Is this a conspiracy to amortize these investments? Unlikely. If there is money to be made in next generation telecommunications, especially broadband, those telecommunications solutions will emerge.

As data traffic rises, the need to revamp existing infrastructures and architectures will also rise; telecoms are trying to have it both ways: they want to optimize their existing infrastructures while migrating to the next generation (which is happening faster outside the local loops than inside). They all know they need to migrate to a consolidated packet backbone infrastructure. Many companies are already well into the long march, a march whose outcome is absolutely essential to pervasive computing.

Broadband

Digital subscriber line (DSL) technology still looks strong—especially given the infrastructure realities described above. Given that there is nearly 200 million installed telephone copper wire lines in the U.S., and that DSL rides on existing copper, there is every reason to believe that Internet access will be largely accomplished through DSL.

The need for broadband will also be driven by the demand for new classes of applications, such as video teleconferencing, telemedicine, interactive TV (broadcast quality video), 3D, and virtual reality-based applications. These applications cross the B2C and B2B markets where entertainment will meet learning and synchronous and asynchronous communications.

DSL will compete with cable modems and wireless access to the Internet. Wireless looks strong—stronger than cable modem-based access to the Internet—but is lagging in deployment because of its newness and because of persistent line-of-sight requirements. Nevertheless, as wireless technologies mature, we can expect it to take its place alongside DSL and cable within 2 to 3 years. If there is one industry that the cable industry fears it is the satellite access/content industry.

The demand for DSL services will grow as DSL is deployed, but the services will extend well beyond the core capabilities of DSL to include applications monitoring, hosting and even customization. In effect, just as ASPs are morphing into "solutions providers" (that must offer always-on, reliable connectivity) so too will smart DSL providers morph into a full service providers. Cable companies will move into the communications space through voice-over-IP offerings.

There is no question that telecommunications and broadband capabilities will drive significant aspects of pervasive computing. Cost-effective broadband connectivity will power the digital economy.

Network Applications and Services

By 2008–2010, we will see the full integration of voice and data with IP as the adhesive. Voice will be carried in every which way, and voice-over-IP will emerge as a mass market alternative to the PSTN over the same timeframe.

Unified messaging (UM; integrated phone, fax, pager, e-mail) is still a killer communications application. All next generation networks will provide UM. Convergence is already driving this (as evidenced) most clearly in the newest phones/Web access/pager/PDA devices. Stand-alone—or integrated—opportunities here should be monitored.

Optical networking (see below) will continue to gain steam. Once gear is deployed that can send signals over long distances without regeneration, and once the fiber-optic backbone is fully deployed, then applications and services opportunities will explode.

Network operating systems (NOSs) and network operating centers (NOCs) are changing dramatically, and quickly. As the Internet becomes the new public data network, the demand for services supporting this network is rising faster than the industry can define or provide them. Our trends analyses suggest that intranets will converge with extranets to create "virtual enterprise networks"—which will require complex support services. A key capability will be heterogeneous platform integration and support services.

Network and systems management is receiving more and more attention as more and more applications are integrated and extended outside of corporate firewalls. Performance tools that monitor and manage networks are numerous. In fact, there are over 50 network and systems management point solutions and over 10 "frameworks" that offer integrated network and systems management capabilities. Vertical frameworks will also emerge as network and systems management solutions.

Service providers in this space will have to offer multiple mobile and wireless network service capabilities, including wireless LANs, wireless loops, celluer/PCS, mobile IP, wireless ATM, and satellite access and support. Coverage issues, bandwidth, integrated applications, among other criteria, will determine which of these network topologies actually "wins." Networks will be heterogeneous and service providers will have to integrate and guarantee the quality of multiple network combinations.

The wireless service industry is attracting lots of attention as wireless penetrates the market. Some of these companies are applications companies, while others are more wireless network service providers.

Optical Networking

Every indication is that the optical networking space will remain hot. Like the wireless space, the optical network space—and all of the derivative spaces (like photonics)—is developing capabilities in equipment, technology, and services.

There are a number of specific capabilities in the space. They are rank-ordered below.

- Optical service provisioning
- Passive optical network services
- Gigabit Ethernet over fiber switches
- Optical switches and transmission equipment
- Optical switching components
- Optical integrated circuits

In addition, the deployment of optical technology should be endowed with intelligence. Sycamore's tagline—"intelligent optical networks"—makes good technology and business sense, and provides a key pervasive computing enabling technology.

The end game is an optical mesh network with an IP overlay that supports intelligent optical transport, flexible (multi-gigabit) service delivery, wavelength traffic engineering, optimization and management, end-to-end provisioning, and restoration—all of which is essential to the communications infrastructure necessary to support pervasive computing and continuous commerce.

Touch Technologies

Technologies that touch employees, customers, suppliers, and partners are exploding through applications that support new call center models, new customer relationship management processes, and new transaction processing capabilities. The proliferation of Internet access, broadband, and wireless access devices is fueling the requirement for enabling touch technologies.

On the list are voice over IP (and DSL), speech recognition, and client-side intelligent agents (discussed above). These technologies can power a variety of applications that dovetail with access and transaction processing trends. These applications support:

- Integrated messaging and chat (through instant messaging and related technologies)
- "Assisted browsing"
- "Web callback"
- Integrated eCRM voice/video/data
- Voice-enabled search
- Interactive training

These kinds of applications are enabled by touch technologies we should track.

Touch technologies close the loop between quality of service (QoS) applications and quality of experience (QoE) applications, or quality behind and in front of the firewall.

A PERVASIVE COMPUTING ACTION PLAN

The trends analyses we have conducted indicate some profound changes in the structure of the technology industry, changes that will have major effects on the way we do business. Taken together, they define a whole new suite of capabilities, capabilities that will define what computing Era 4—pervasive computing—will actually look like.

There is a shift occurring in the nature of the technology "platforms" being provided to small, medium-sized, and large customers. In the software area, we are seeing a proliferation of "me-too" packages: in fact, it appears as though the software industry is evolving faster than anyone predicted to packaged applications as the preferred software deployment strategy for all-sized organizations. This means that the software industry will be run by "professional" software developers and deployers (ASPs/TSPs), not by systems integrators (who will, nevertheless, still have plenty to do). The shift is from ownership of source code to the deployment and support of the code. ASPs/TSPs—which includes just about every service provider out there—will become the primary software delivery mechanism for the industry. They will not develop primary software but will implement the software from other vendors, at least initially. Some ASPs/TSPs will implement their own software (such as Oracle, Siebel and SAP) competing, for a time, with the multiple channels that they have encouraged.

In effect, software will become "servitized": buyers will not buy software, software implementation, or software support services from multiple vendors. Nor will they build complex applications in-house. If this trend—supported by adoption rates and surveys—continues, then the future will belong to those who combine software (plus all varieties of clients), support, integration and interoperability services. Enterprise applications integration will occur within these mega software service organizations (who will become expert at the use of EAI and IAI tools, Web services, service-oriented architectures, and all related middleware).

In addition to software-related services, we can expect the same buyers to force the integration of telecommunications (integrated voice, video and data) hardware, software, and services. Since much of today's business is e-business, and since so much of e-business will be supported by TSPs, connectivity is inseparable from the development, delivery, and support of business models. This means that first generation ASPs will have to continue to include communications support to their customers and second generation ASPS (TSPs) will unquestionably provide "tethered" and "un-tethered" connectivity to their clients' employees, customers, and suppliers.

We are also seeing the rejuvenation of so-called "old economy" companies who are rethinking their use of technology. Just as "neutral" B2B exchanges appeared at an incredible rate, just as quickly selected vertical industries pre-empted the neutral penetration of their supply chains by establishing their own exchanges. In a sense this represents the revenge of the verticals—who are using off-the-shelf software packages to implement their own controlled exchanges.

Finally, we are seeing a growing role for business and e-business strategists. But here too the prediction is that disembodied services—regardless of where they are along the margin continuum—will always lose to integrated services and, ultimately, integration solutions. Creative insight into supply chains and the creation of new business processes will drive the application of new technologies, but, again, it is the integrated services and solutions that will be the most appealing.

When taken together all of this spells pervasive computing. Without question the Internet protocol (IP) has revolutionized the way we store, process and communicate data, information, content, and knowledge.

The movement toward pervasive computing is well afoot: access devices will increasingly not be traditional laptops or desktops. PDAs and converged communications devices will dominate e-business and mobile commerce within 2 to 3 years. Software is changing and architectures are becoming more and more distributed:

just 10 years ago (when first generation client/server architectures were deployed) applications were still centralized, if not actually, then certainly conceptually. First generation client/server applications were viewed as extensions of mainframe-based applications, not as revolutionary architectures destined to replace their data center-based parents. Today new applications are conceived as distributed, deployed as distributed, and supported in the same way, often by a third party geographically distributed from the companies that built or bought them. The real impact is just this, a predisposition toward distributed transaction processing—and pervasive computing provides the enabling infrastructure to make the whole centralized/distributed architecture moot. But much more importantly, personal and professional transaction processing requirements can only be fulfilled by ubiquitous connectivity, always-on connectivity, and automated applications, among other capabilities.

Appendix B
Technology Trends Analysis:
Trends in Intelligent Systems Technology

This trends analysis is about the resurgence of artificial intelligence or "AI." It is also about the next generation of "decision support" applications that are already appearing in horizontal technology architectures—like network and systems management frameworks—and in vertical applications, like those intended to make customer service representatives more effective.

Some of us may remember AI as the darling technology of the 1970s and 1980s. A ton of money was thrown at the field and countless PhD dissertations were written by budding knowledge engineers, expert systems developers, and natural language parsers. The U.S. Department of Defense (DOD) led the pack with money, hype and a variety of applications that turned out to be more prototypical than real. Nevertheless, a couple of decades later we are revisiting all this and companies like Computer Associates, Microsoft, and IBM, among many others, are embedding "intelligence" into their applications. This is a technology that we must understand.

We may also remember the early decision support systems (DSSs) that populated vertical industries in the 1980s and early 1990s. Some of these applications were fueled by complex analytical methodologies like Bayesian statistical models and statistical optimization algorithms. The users of these systems had to be technologically sophisticated just to understand what the applications did, let alone how they worked. Well, browsers are now decision support systems that permit search, transactions processing and data information management—all by pointing and clicking.

The take-away from this analysis should be a plan to track intelligent systems technology and intelligent decision support very, very closely. Why? Because the two will impact nearly all of our business models, our technology investments, and our overall business-technology convergence strategy. In short, if we had to pick one area that might very well have the greatest impact on business over the next 5 years, the application of intelligent systems technology might well be it.

But what is "AI"? What is intelligence systems technology? What is intelligent decision support? What are the key technologies necessary to make it all work together? Let us us step back for a moment in time and look at what DOD thought the application potential might look like. Here is a mock message from a very smart intelligent "command and control" system whose job it is to monitor international environments.

DEPARTMENT OF DEFENSE/JCS MESSAGE CENTER
UNCLASSIFIED
ACTION
INTELLIGENT SYSTEMS MANAGEMENT COMMAND
DISTR
SECDEF SECDEF FILE
1315124Z JUL 2011

FM USCINCEUR SURROGATE VAIHINGEN GERMANY
TO CINCUSAREUR SURROGATE HEIDELBERG GERMANY//AEADC/ AEAYF//
CINCUSNAVEUR SURROGATE LONDON ENGLAND
HQ USAFE INTELLIGENT SYSTEMS RAMSTEIN AB GERMANY//RC//DO&I
INTELLIGENT SYSTEMS MANAGEMENT COMMAND WASH DC
UNCLAS E F T O
ECJ3 11292
SUBJECT: I&W THRESHOLDS & TACTICAL PLANNING

1. TO CONFIRM UNUSUAL MILITARY MOVEMENTS IN PERSIAN GULF; SENSORS SUGGEST MOBILIZATION ON GRAND SCALE

2. RECOMMEND IMMEDIATE EXPANSION OF SURVEILLANCE IN AREA (CONSULT THE SYSTEMS DEFENSE PLAN FOR THE LOCATION OF SMART COUSINS ALONG THE SURVEILLANCE NETWORK AND ACTIVATE INTELLIGENT RPVs).

3. TACTICAL DEFENSIVE PLANS ALREADY COMPLETE. ENEMY COURSES OF ACTION ESTIMATED. FRIENDLY COAs DETERMINED AND UNDER WAY.

4. SET FOR MULTI-SOURCE, CONTINUAL, AND INTERPRETIVE INPUT AND FEEDBACK ACROSS THE NETWORK. IMPLEMENT ADDITIONAL PROCESSORS IF NECESSARY; NOTIFY HUMAN DECISION-MAKERS ONLY WHEN LIKELIHOODS OF ATTACK CROSS THE .70 THRESHOLD.

MINIMIZE NETWORK CHATTER; MAINTAIN CONTACT THROUGH MULTI-LEVEL SECURE DATA BASES ONLY FROM THIS POINT ON. REFER ALL HUMAN QUESTIONS SIMULTANEOUSLY TO ALL PROCESSORS ON THE NET.

The above represents some intercepted conversation between several intelligent computers in a hypothetical alert posture in 2007. Can computers make the kind of decisions suggested by the above simulation? Can they make intelligence estimates? Can they wage war?

There are today what appear to be exciting applications of AI and decision support. Some of them sound almost as far-fetched as the one above—but are they really that far-fetched? Can intelligent systems manage networks? Can they serve customers? Can they grow business?

Following is another message—but this time it is from an intelligent network.

NETWORK OPERATIONS CENTER
CONFIDENTIAL/ENCRYPTED
INTELLIGENT SYSTEMS NETWORK COMMAND
JUL 6, 2011
0900 EST

SUBJECT: NETWORK OVERLOAD

TO CONFIRM EXTREME USAGE OF THE VIRTUAL PRIVATE NETWORKS & UNPRECEDENTED NUMBER OF SECURITY ATTACKS ON THE FIREWALLS

RECOMMEND IMMEDIATE RECONFIGURATION OF NETWORK TOPOLOGY: IF TRAFFIC AND/OR ASSAULTS CONTINUE FOR 90 MINUTES OR MORE, RECONFIGURATION WILL OCCUR
OBSERVATIONS

NETWORK AND SYSTEMS MANAGEMENT FRAMEWORK IS UNDERPERFORMING TO SPECIFICATIONS AND THE NEGOTIATED SERVICE LEVEL AGREEMENT (SLA)

DESKTOP AND LAPTOP ACCESS DEVICES ARE NON-STANDARD AND DEGRADING NETWORK PERFORMANCE

DESKTOP APPLICATIONS ARE ILL-CONFIGURED FOR OPTIMAL PERFORMANCE
MANAGEMENT DECISION-MAKING IS LAGGING PRICE-PERFORMANCE CALCULATIONS: ESTIMATED LOSSES DUE TO INACTIVITY = $8M TO DATE; PROJECTED LOSSES = $2M PER MONTH

5. PERSONNEL RECORD ANALYSIS INDICATES GROWING SKILLS GAP AMONG ALL PROFESSIONAL LEVELS

INTELLIGENT DECISION-MAKING AND TRANSACTION SUPPORT

All of these possibilities sit squarely in the middle of larger questions about decision support and intelligent transaction processing. Over the past 20 years, we have seen decision support systems, executive information systems, data mining applications, online analytical processing (OLAP) applications, and now Web-based search engines all purporting to be the "digital executive officers" to human decision-makers.

We know, for example, that the design, development, and use of decision support systems will change dramatically in the early 21st century. Our expectations about what DSSs should do are rising as rapidly as the requisite technology is evolving. By 2010 problem-solvers will routinely use DSSs to deal with all sorts of simple and complex problems. They will also benefit from systems capable of providing much more than database support and low-level inference-making. Just as importantly, the distribution of DSS power will be expanded beyond specialized professionals; decision support systems will be available to us all on and off the job. This "distributed knowledge management" will forever change the way we think about personal and professional problem-solving and decision-making.

It is safe to say that most DSSs today support decision-making only indirectly. There are DSSs that manage projects, provide easy access to operational data, and otherwise deal with relatively structured problems. This distinction between "structured" and "unstructured" targets of opportunity is important to understanding the range of today's DSSs and the promise of tomorrow's. Early proponents of

DSS's technology hoped that their systems would help decision-makers generate, compare, and implement transaction options, but most DSSs indirectly support options analysis. Real-time option generation and evaluation has evaded designers—except in some rare instances where a single purpose system was developed to address a very well-bounded problem, like buying a car.

Early DSSs were data-oriented. The field migrated slowly toward model-oriented applications, and over the next three to five years should move toward systems that synthesize data, models, knowledge, and interfaces into some creative capabilities. Next generation systems will also be capable of addressing structured and unstructured problems.

The action will be in intelligent systems technology, user-computer interface (UCI) techniques, and display technology. The new applications perspective on decision support will be extremely broad, reflecting the capabilities of new systems that will be embedded (in larger information systems) and functional on many levels. Future decision support systems will permit decision makers and information managers, resource allocators, administrators, strategic planners, network administrators, and inventory controllers to improve their efficiency.

What will be driving what? Will new corporate and governmental requirements suggest new DSS requirements, or will new DSS technology suggest new requirements? Would next generation DSSs look the way we expect them to look if they were conceived in an applications vacuum, or is the interpretation and anticipation of applications driving the "form" that future DSSs will take?

The questions are relevant to all parts of the information and manufacturing economy. In the automobile industry, for example, robots fulfill specific requirements because robotic technology has evolved into a cost-effective alternative to human labor. Word processing is now cheap, efficient and distributed. But has word processing changed, eliminated, or reduced requirements? It is interesting that many social theorists have argued that the development of photocopiers has dramatically changed the way information is produced and distributed. Some argue that copiers have triggered a paper explosion that has over-satisfied the need for information. In other words, copiers, just because they exist, now satisfy more requirements than can be identified. Will DSSs suffer similar "success"?

While definitions of decision support continue to grow, so too will our understanding of computer-based problem-solving. Decision support, while very broad in concept and application, will be subsumed under the general rubric of computer-based problem-solving which, over time, will also experience radical change. Expectations about what computers can do for users will continue to rise. Consistent with the evolution in expectations about the power of DSSs, computer-based problem-solving systems of all kinds must satisfy analytical requirements. For example, a network administrator needs to know when a network goes down and the tools he or she uses to monitor networks, pinpoint problems and take fast corrective action have become increasingly automated and in some cases quite intelligent. Applications monitors now "sniff" networks for problems and report them back to the human administrators that must decide what to do next. But is this a good allocation of intelligence and responsibility? Intelligent decision support systems would automatically find and correct problems and simply report what happened to their human masters in a weekly report.

Over the past few years the DSS community has seen the preeminence of knowledge-based tools and techniques, though the range of problems to which heuristic solutions apply is much narrower than first assumed. It is now generally recognized that artificial intelligence (AI) can provide knowledge-based support to well-bounded problems where deductive inference is required. We know that AI performs less impressively in situations with characteristics (expressed in software as stimuli) that are unpredictable. Unpredictable stimuli prevent designers from identifying sets of responses, and therefore limit the applicability of "if – then"solutions. We know, for example, that so-called expert systems can solve low-level diagnostic problems, but cannot predict the technology industry's structure in 2020. While there were many who felt from the outset that such problems were beyond the capabilities of AI, there were just as many sanguine about the possibility of complex inductive problem-solving.

The latest methodology to attract attention is neural network-based models of inference-making and problem-solving. Neural networks are applicable to problems with characteristics that are quite different from those best suited to conventional expert-systems-based AI. Neural nets are non-sequential, non-deterministic processing systems with no separate memory arrays. Neural networks comprise many simple processors that take a weighted sum of all inputs. Neural nets do not execute a series of instructions, but rather respond to sensed inputs. "Knowledge" is stored in connections of processing elements and in the importance (or weight) of each input to the processing elements. Neural networks are allegedly nondeterministic, nonalgorithmic, adaptive, self-organizing, naturally parallel, and naturally fault tolerant. They are expected to be powerful additions to the DSS methodology arsenal, especially for data-rich, computationally intensive problems.

The "intelligence" in conventional expert systems is preprogrammed from human expertise, while neural networks receive their "intelligence" via training. Expert systems can respond to finite sets of event stimuli (with finite sets of responses), while neural networks are expected to adapt to infinite sets of stimuli (with infinite sets of responses). It is alleged that conventional expert systems can never learn, while neural networks "learn" via processing. Proponents of neural network research and development have identified the kinds of problems to which their technology is best suited: computationally intensive; nondeterministic; nonlinear; abductive; intuitive; real-time; unstructured/imprecise; and nonnumeric.

It remains to be seen if neural networks constitute the problem-solving panacea that many believe they represent. The jury is still out on several aspects of the technology. But like AI, it is likely that neural nets will make a measured contribution to our inventory of analytical models and methods.

Intelligent systems technology holds great promise for the design and development of smart DSSs. Natural language processing systems—systems that permit free-form English interaction—will enhance decision support efficiency and contribute to the wide distribution of DSSs. When users are able to type or ask questions of their DSSs in much the same way they converse with human colleagues, then the way DSSs will be used will be changed forever.

Expert systems will also routinize many decision-making processes. Rules about investment, management, resource allocation, and office administration will be embedded in expert DSSs. It is sure unlikely that individuals will go onto the Web and execute trivial transactions. Smart support systems will automatically execute hundreds of pre-defined "authorized" transactions.

DSSs will be capable of adapting from their interaction with specific users. They will be able to anticipate problem-solving "style," and the problem-solving process most preferred by the user. They will adapt in real-time, and be capable of responding to changes in the environment, like a shortage of time.

DSS designers will also benefit from a growing understanding of how humans make inferences and decisions. The cognitive sciences are amassing evidence about perception, biasing, option generation, and a variety of additional phenomena directly related to DSSs modeling and problem-solving. Information technology will be informed by new findings; resultant DSSs will be "cognitively compatible" with their users. Because so much behavior is moving to the Web, it will be possible to build giant demographic correlators that will understand who we are and what we want from what we do. Inferences about our "styles," preferences, needs, and wants will be easily inferred from adaptive trails of our behavior on- and off-line.

Next generation DSSs will also respond to the situational and psycho physiological environment. They will alter their behavior if their user is making a lot of mistakes or taking too long to respond to queries. They will slow down or accelerate the pace, depending on this input and behavior. The field of cognitive engineering—which will inform situational and psycho-physiological system design strategies—will become increasingly credible in the early 21st century. The traditional engineering developmental paradigm will give way to a broader perspective that will define the decision-making process more from the vantage point of requirements and users than chips and algorithms. Principles of cognitive engineering will also inform the design and human computer interfaces (see below).

It is extremely important to note the appearance of DSS development tools. Already there are packages that permit the development of rule-based expert DSSs. There are now fifth generation tools that are surprisingly powerful and affordable. These so-called "end-user" systems permit on-site design and development of DSSs that may only be used for a while by a few people (one is described below). As the cost of developing such systems falls, more and more throw-away DSSs will be developed. This will change the way we now view the role of decision support, not unlike the way the notion of rapid application prototyping has changed the way application programs should be developed.

Hybrid models and methods drawn from many disciplines and fields will emerge as preferable to single model-based solutions largely because developers will finally accept diverse requirements specifications. Methods and tools drawn from the social, behavioral, mathematical, managerial, engineering, and computer sciences will be combined into solutions driven by requirements and not by methodological preferences or biases.

Twenty years ago no one paid much attention to user interface technology. This is understandable given the history of computing, but no longer excusable. Since the revolution in micro computing, software designers have had to devote more attention to the process by which data, information, and knowledge are exchanged between the systems and their operators. There are now millions of users who have absolutely no sense of how a computer actually works, but rely upon its capabilities for their very professional survival. Living on the Web, these users search, compare, contrast, order, file, and report in ways they could not have even imagined just 5 years ago.

Software vendors are sensitive to both the size of this market and its relatively new need for unambiguous, self-paced, flexible computing. Where are the ideas coming from? The field of cognitive science and now "cognitive engineering" is now, justifiably, taking credit for the progress in user computer interface (UCI) technology, since its proponents were the (only) ones asking why the user-computer interaction process could not be modeled after some validated cognitive information processing processes. It is no accident that much UCI progress can be traced to findings in behavioral psychology and cognitive science; it is amazing that the cross-fertilization took so long.

UCI progress has had a profound impact upon the design, development, and use of DSSs. Because many of the newer tools and techniques are now affordable (because computing costs have dramatically declined generally), it is now possible to satisfy complex UCI requirements even on personal computer-based DSSs. Early data-oriented DSSs displayed rows and rows (and columns and columns) of numbers to users; modern systems now project graphic relationships among data in 3D high resolution color. DSS designers are now capable of satisfying many more substantive and interface requirements because of what we have learned about cognitive information processing and the affordability of modern computing technology.

The most recent progress in UCI technology is streaming multimedia, or the ability to store, display, manipulate, and integrate sound, graphics, video, and good-old-fashioned alphanumeric data. It is now possible to display photographic, textual, numerical, and video data on the same screen. It is possible to permit users to select (and de-select) different displays of the same data. It is possible to animate and simulate in real-time, and cost-effectively. Many of these capabilities were just too expensive 5 years ago and way too computationally intensive for the hardware architectures of the 1980s and early 1990s. There is no question that multimedia technology will affect the way future DSSs are designed and used. The ability to see and hear data, information, and knowledge, as well as the ability to see and hear descriptions of that content from friends, colleagues and service representatives "in person" over the Web, will dramatically change the way we think about information sharing, collaboration, and problem-solving. The gap between the way humans "see" and structure problems will narrow considerably through the application of multimedia technology and the collaborative communications applications—like NetMeeting—that exploit these new capabilities.

The use of multimedia graphical displays of all kinds will dominate future UCI applications. Growing evidence in visual cognition research suggests how powerful the visual mind is. Complicated concepts are often easily communicated graphically, and it is possible to convert complex problems from alphanumeric to graphic form. There is no question that DSSs will exploit hypermedia, multimedia, and interactive graphics of all kinds.

Speech input and output should also emerge over the next few years as a viable UCI technology. While predictions about the arrival of "voice activated text processors" have been optimistic to date, progress toward continuous speech input, and output has been steady. Once the technology is perfected, there are hundreds of applications that will benefit from keyboard-/mouse-/touchpad-less interaction.

UCI technology will also permit the use of more methods and models, especially those driven by complex—yet often inexplicable—analytical procedures. For example, the concept of optimization as manifest in a simplex program is difficult to communicate to the typical user of a browser-based application. Advanced UCI technology can be used to illustrate the optimization calculus graphically and permit users to understand the relationships among variables in an optimization equation. Similarly, probabilistic forecasting methods and models anchored in Bayes' Theorem of conditional probabilities while computationally quite simple are conceptually convoluted to the average user. Log odds and other graphic charts can be used to illustrate how new evidence impacts prior probabilities. In fact, a creative cognitive engineer might use any number of impact metaphors (like interactive thermometers and graphical weights) to present the impact of new evidence on the likelihood of events.

Finally, advanced UCI technology will also permit the range of decision support to expand. As anytime communications bandwidth between applications and users is increased, the range of applied opportunities grows. UCI technology permits designers to attempt more complex system designs due to the natural transparency of complexity that good UCI design fosters.

Some argue that the interface may actually become "the system" for DSS users. The innards of the system—like the innards of the internal combustion engine—will become irrelevant to the operator. The UCI will orchestrate process, organize DSS contents and capabilities, and otherwise shield users from unfriendly interaction with complex data, knowledge, and algorithmic structures. In some sense, we have already achieved aspects of this prediction through the ubiquitous browser.

Next generation DSSs will be smaller and cheaper, and therefore more widely distributed. They will be networked, and capable of up-loading and down-loading to larger and smaller systems. Input devices will vary from application to application as well as the preferences of the user. As suggested above, voice input will dramatically change the way DSSs are used in the future; voice activated text processing will expand the capabilities of DSSs by linking decision support to a whole host of applications in a "natural" unobtrusive way, though it is likely that truly robust voice activated systems will not appear until the around 2011.

DSSs will have embedded communications links to databases and knowledge bases, other systems on decision support networks, and the outside world via conventional communication systems. The Web will be the ubiquitous playing field for smarter and smarter DSSs.

DSSs will be used very differently in the future than they are today. They will function as clearinghouses for our professional problems. They will prioritize problems for us, and they will automatically go ahead and solve many of them. They will become problem-solving partners, helping us in much the same way colleagues now do.

They will also be deployed at all levels of the virtual organization. The distribution of DSSs will permit decision support networking, the sharing of decision support data, and the propagation of decision support problem-solving experience (through the development of a computer-based institutional memory of useful decision support "cases" that will be called upon to help structure especially recalcitrant decision problems). Efficient organizations will continue to develop inventories of problem/solution combinations that will be plugged into their decision support networks.

Finally, next generation intelligent DSSs will bridge the gap between our professional and personal lives (already blurring over the Web). They will have the capability to manage our personal and professional affairs. The blurring of the traditional lines between our professional and personal worlds may not be desirable, but it is sure likely.

INTELLIGENT SYSTEMS TECHNOLOGY AND THE RANGE OF COMPUTABLE PROBLEMS

This trends analysis suggests that artificial intelligence can evolve impressively if it assumes a role within the larger family of qualitative and quantitative methods we now use to find data, prioritize actions and execute transactions. But first, let us assume that not all analytical and computational problems are amenable to computer-based solutions. Second, it assumes that AI, like operations research, decision analysis, and "conventional" computer science, all have their strengths and weaknesses; and that the misapplication of one or more of these methods classes undermines our problem-solving performance and credibility. Third, let us assume that it is possible to "match" the "right" method to explicit problem definitions.

A close inspection of the so-called "problem domains" of artificial intelligence, however, reveals that a great many of them are "well-bounded" and characterized by deductive inferential processes. It seems that expert systems are best used to solve problems about which we already know a great deal and whose problem-solving processes are explicit, unambiguous and, therefore, deductive. Is there a problem with this role for AI? Of course not; in fact, the matching of AI to such problem types represents a step in the right direction.

Unfortunately, we often mismatch analytical methods and problems. We use heuristic solutions for mathematical problems; we use neural networks for well-bounded bank loan application evaluations, and we insist upon human information processors for air traffic control. We have few heuristics for problems/methods matching. Methods are selected often (and unfortunately) on the basis of preference, inertia, or because the color of the money (what individuals, vendors, VCs, or foundations want to fund) is optimization, AI, or Bayesian.

The decision to use AI, operations research, decision analysis, or whatever, is predicated upon the nature of the problem at hand. It should first be stated that not all problems are "computable" via the application of any individual (or sets of) method(s). Some problems are so recalcitrant that they defy modeling (and certainly quantification). Some problems can be effectively addressed via methods, tools and techniques resident in human cognitive processes and embedded in software; others are so inductive or abductive that they fall well beyond the scope of a "computable problem."

AI is maturing and our understanding of where and how it can be applied is growing. We now know that knowledge-based solutions are not always appropriate, and that sometimes the worst method is artificially intelligent-based. We also know that there are many kinds of problems uniquely suited to the tools and techniques of AI. But most importantly, AI should be regarded as a player on a larger team, a member of the methods family.

The above suggests that AI has some unique capabilities and enormous potential in some specific areas. It does not suggest that those problems that have remained beyond the reach of analytical methodology are all of a sudden amenable to AI.

AI TOOLS AND TECHNIQUES

AI is an interdisciplinary subfield of systems engineering, psychology, electrical engineering, and computer science that seeks to recreate in computer software the processes by which humans solve many simple and some complex problems. AI "knowledge engineers" extract expertise from professionals like doctors, geologists, and signal processors, and then structure it in a way which permits relatively flexible problem-solving in a specific area, such as medical diagnosis, geological drilling, or data analysis.

AI systems differ from conventional ones in a number of important ways. First, conventional systems store and manipulate data within some very specific processing boundaries. AI systems store and apply knowledge to a variety of unspecified problems within selected problem domains. Conventional systems are passive, where AI systems actively interact with, and adapt to, their users. Conventional systems cannot infer beyond certain preprogrammed limits, but AI systems can make inferences, implement rules of thumb, and solve problems in much the same way we routinely decide whether or not to buy a Ford or a Chevy, or accept a new professional challenge.

The representation of knowledge is the mainstay of AI, expert systems, and intelligent systems technology research and development. If knowledge and expertise can be captured in computer software and applied at a moment's notice, then major breakthroughs may be possible in the production and distribution of knowledge—and in the execution of transactions. If it is possible to capture the best medical diagnosticians, the best managers, the best intelligence analysts, and the best customer service representatives in a set of flexible and friendly computer-based systems, then productivity and efficiency will explode (as rapidly as costs decline).

AI systems designers use a set of unique tools to represent knowledge and build intelligent problem-solving systems. Imagine for a moment the detailed subject outlines that appear in every volume of the Encyclopedia Britannica. Then imagine a computer program—not at all unlike the ones resident in human brains—capable of searching through the outline for information in order to solve a specific problem. AI search routines permit information to be structured as knowledge and permit system users to apply the knowledge to a variety of problem-solving tasks. Today's search engines use AI to find the right information for users and to serve as Web navigators.

Conventional and special purpose software languages permit intelligent systems designers to represent knowledge in several ways. The most widely utilized knowledge representation technique involves the development of cognitive rules of thumb usually expressed in "if ... then" form. In the world of high finance, for example, it is very easy to imagine a rule-based investment advisory system containing a rule like, "if gold falls below $700 per ounce, then invest 25% of available resources in gold." A system with a thousand such rules might prove very efficient since it could continually monitor investment conditions around the world and, via its rules, allocate resources accordingly.

As you have no doubt already surmised, the key to the power of all rule-based systems lies in the accuracy and depth of their rules. Bad rules produce bad conclusions, just as bad human probability estimates frequently result in bad outcomes. It is the job of the knowledge engineer to make sure that the rules represent substantive expertise to the fullest extent possible. This requirement, in turn, means that rule-based systems can never stop developing. In order for them to keep pace with the field they are trying to represent electronically, they must routinely be fed new rules.

One of the earliest AI research goals was to develop computer-based systems that could understand free-form language. The "natural language processing" branch of AI represents knowledge by endowing software with the capability to understand the meaning of words, phrases, parts of speech, and concepts that are expressed textually in English, French, German, or whatever language is "natural" to the intended system user. It is now possible to converse directly with a computer in much the same way we converse with human colleagues. Natural language systems are today in use in DOD to track ships at sea, organize and manipulate huge data bases, and bridge the gap between smart but otherwise rude expert systems. In cyberspace, search engines support varying levels of natural language interaction.

It is important to distinguish between the tools and techniques of AI and the substantive areas targeted by the AI R&D community. Tools and techniques consist of special purpose software languages, rules, semantic and inference networks, natural language processing, and even unique hardware systems. But not every area is vulnerable to these tools and techniques.

The intelligence and network operations center messages at the beginning of this trends analysis suggests the potential of AI. Implicit in the messages is the ability to automatically monitor a large environment, issue warnings based upon an understanding of the dynamics of threat, communicate rapidly within a network of similarly intelligent computers, make decisions about the need for further action, and formulate tactical plans based upon rules. The accomplishment of any of these tasks is impressive; the goal of the intelligent systems technology community is to automate them all.

At the very least, intelligent systems are intended to augment problem-solving; the more robust ones may very well consistently outperform human decision-makers.

Up to this point we have felt the effect of individual AI systems. But what will be the effect of numerous AI systems working in concert? We have yet to calculate the impact of networked intelligent systems. The message at the beginning of this report suggests that smart systems will—in time—be able to communicate among themselves simultaneously on many levels. They will be able to acquire and process data, dispatch collection resources, assess strategic and tactical situations, develop plans, make decisions, deploy forces, win wars, and save our networks. But they will also make mistakes. We now know that unless the knowledge engineering phase of AI systems development is taken with extreme care, the system may demonstrate flashes of brilliance in one area and incredible stupidity in others. The tools necessary for the design and development of very "deep" knowledge bases have yet to be developed.

AI systems will simultaneously serve as the ultimate force multipliers and smart surrogates. They will extend the power of existing unintelligent computer, mechanical, and weapons systems, augment and eventually even replace human problem-solving, and assume larger and larger operational responsibility. Unlike human expertise, which cannot be easily duplicated, expert systems can be copied as often as the need arises. It will thus be possible to distribute expertise where human experts might never venture. It is impossible to estimate the effect of the widespread distribution of knowledge. While it might result in improved performance across the board, it might also create an undesirable disincentive among humans to learn how to solve problems on their own. The ease of AI systems use may prove too seductive to the average professional.

As implied above, an extremely promising area is the design and development of knowledge processing architectures based upon human neural networks and other biological system concepts. The new thrust in biological emulation generally may well revolutionize the way information, data, and knowledge is stored and processed in machines.

Still another promising area is the merger between larger computer-aided software engineering (CASE) tools with evolving approaches to knowledge engineering. As the largest bottleneck in the design and development of expert and other intelligent systems, knowledge engineering might be served by the quasi- and full-automation of some knowledge elicitation procedures. A movement toward CASE might permit knowledge engineers to determine early in an intelligent systems project's life the nature of depth of necessary knowledge bases; it might also permit them to accelerate the knowledge-based systems prototyping process.

It is clear that the real payoff for AI and related technologies lies in the extent to which they can link creatively to other technologies and applications. Hybrid "wholes-greater-then-the-sum-of-their-parts" offer tremendous potential, but in order to move in this direction designers must become problem-oriented. For example, the insertion of smart monitors in network and systems management applications has given Computer Associates' marketing executives some differentiation through their "neugents," intelligent agents that improve the cost-effectiveness of it automated network and systems management framework, Unicenter.

Some things to consider:

- Complex processes that have resisted conventional systems and analysis procedures can sometimes be addressed by intelligent systems technology, especially complex real-time data integration and mining problems, problems that require immediate action-reaction, and really tough problems due primarily to the sheer volume of data or the size of decision spaces.
- Automation is now possible in selected areas and will become more possible in additional domains; we know, for example, that it is possible to automate marketing campaigns.
- AI, expert systems and intelligent systems technology may well generate huge returns on investments, and provide a technology-based response to increasing competition, the volatility of business models and the pace of technology change.
- Increasingly, instead of developing whole new systems, we will be asked to improve existing systems. There is enormous leverage in the re-engineering of existing systems via the insertion of expert systems and other intelligent systems technology.
- There are many functions, tasks, and subtasks that can be addressed through intelligent systems technology inherently more efficient than conventional computational technologies.
- As staff reductions become more commonplace (to reduce operating expenses), we will need to find ways maintain productivity; intelligent systems technology provides one approach.

Intelligent systems are morphing into "intelligent agents." To most observers, a software agent is intelligent if it is most of the following:

- **Autonomous:** Self-initiated
- **Social:** Able to communicate with users (and other agents)
- **Reactive:** Able to answer questions and initiate action
- **Dynamic:** Are time and space sensitive
- **Asynchronous:** Action independent of linear/linked events
- **Event-driven:** Can pro-act and react to events
- **"Inactive" user interaction:** Users can ignore
- **Self-executing:** Can run themselves
- **Self-contained:** Have what they need to run themselves

Intelligent systems can deal with a variety of tasks:

- What humans regard as complex
- Time criticality
- Well-bounded, previously modeled dynamic events
- Information overload
- Directed search
- Routinized, repetitive behavior

WHY YOU NEED TO UNDERSTAND IT

Hopefully, it is now clearer why we regard intelligent systems technology as a "keeper" technology and why we see intelligent decision support making a major comeback over the next several years.

Our emphasis on this technology can be explained by the potential impact it can have across a variety of horizontal technology areas as well as just about every vertical industry we can imagine.

We are already using a variety of pretty sophisticated intelligent tools and techniques—embedded in a variety of applications on our desktops, in our servers, and even in our mainframes. We are suggesting a second, much more powerful wave of intelligent applications that will revolutionize how we compute and communicate.

THE RANGE OF APPLICATIONS

The range of applications is broad. There are several classes of applications with specific ideas within each class.

Design Aids

This first class of intelligent applications covers a lot of ground. There are at least as many distinct flavors as the list below suggests.

Intelligent systems for:
- Requirements analyses
- Prototyping
- Software development
- Testing and evaluation
- Hardware and software configuration management
- Manufacturing design (all classes)
- Pharmaceutical design
- Architectural design

Embedded Systems

The embedded systems world is also growing dramatically. The range of applications here includes:

Embedded systems for:
- Search and retrieval
- Data organization and structure
- Hardware configuration
- Hardware and software adaptation

Agents

This area is perhaps the hottest. The ubiquity of the Web has triggered all sorts of ideas for the application of intelligent agents to all sorts of tasks, including, but certainly not limited to:

Intelligent agents for:
- Routine tasks
- Information overload reduction
- Personal and professional time management
- Collaboration and group problem-solving
- Network and systems management
- Asset management

- Buying and selling
- Supplying
- Planning
- Negotiating

CASE STUDY

Here is a case study of how artificial intelligence can be used to develop tactical battle plans.

Intelligent Military Planning

This case focuses on Army Corps planning. The idea was simple: provide tactical planners with an intelligent computer-based assistant. The result is described below.

The Military Planning Processes

Military planning at the Corps level, like the planning that occurs at all command levels, is mission-directed, top-down, and structured. Corps commanders and their staffs must take a number of steps before developing a "concept of operations." They must first convert mission guidance into sets of goals and subgoals that are unambiguous and realistic. They must prepare the battlefield by integrating intelligence about area characteristics, weather, and adversary objectives and capabilities. They must identify likely adversary courses of action (COAs), and then determine the best way to deploy, attack, or defend.

There are reams of "how to" planning documents available to Corps commanders. But the essence of successful planning can be traced to the quality of available intelligence, the capabilities of the opposing forces, and the judgment of the commander.

In many important respects, tactical planning is an art. Aspects of the planning process cannot always be taught unless the planner has had actual field experience. At the same time, this is not to suggest that parts of the process cannot be computerized or improved via the implementation of some "scientific" procedures.

Good Corps planning is iterative. Successful Corps commanders are simultaneously creative and pragmatic. They are also good choreographers, required to balance the priorities of command against the realities of limited resources, imperfect intelligence, and a formidable opponent.

The planning process itself could not be implemented without certain "props," such as good maps, clear acetate, tactical symbols or pins, and plenty of grease pencils. In spite of frequent criticisms about the use of such "low technology," historically it has been very difficult to improve it—even with "high technology" fixes.

In a typical planning exercise, maps, acetate overlays, notes, files, references, and messages are used to develop and assess the plan. Data, information, judgment, and experience are compressed by a small group of planners into a single "concept of operations."

Attempts to develop computerized planning aids began about two decades ago, but significant progress has only been made during the past 5 years. But even the recent aids suffer from a common problem—the force-fitting of specific analytical methods onto the planning process regardless of whether or not they are appropriate. Some planning aids, for example, use Bayes' theorem of conditional probabilities to calculate likely adversary COAs. While Bayes' theorem is powerful, it is inappropriate to COA assessment because it forces the planner to think about likelihoods in some ways that require planners to ignore what has been taught about tactical planning.

Other problems can be traced to the kind of interaction that many aids impose upon the planning process. One aid, for example, requires Corps commanders to input hundreds of numeric scores which the aid needs to calculate the "best" concept of operations.

Still others failed because they were simply too large to support anything but tactical training. Many were also too expensive to enjoy widespread distribution.

The challenge required to first determine if Corps planning should be computerized. Could computers help or would they add yet another layer of complexity upon a process already well understood and executed? Next the team faced the challenge of identifying the points in the planning process where an aid could make the best contribution. Were there enough to justify the investment? What about methods? Was the inventory large enough to perform critical tactical planning tasks? What about cost?

The ultimate challenge, of course, was simple. Could the team move significantly beyond acetate and grease through the application of intelligent systems technology?

The TACPLAN Aid

The first step toward the development of the intelligent assistant involved the conduct of a series of requirements analyses designed to determine exactly how Corps commanders formulate tactical plans. We videotaped six expert planners as they formulated a concept of operations for a defensive Western European scenario. Actual planners were split into two groups and asked to develop a solution to the same tactical planning problem.

This data was supplemented with Field Manuals, Officers' Handbooks, and the general literature on planning. After we studied all of the data, we developed a taxonomy of planning tasks and subtasks and then identified the analytical methods likely to satisfy the tasks.

The very first issue we faced required the team to profile the role that they wanted the aid to play. Should it in fact "aid" the planner or should it attempt to relieve the planner from certain tasks? Should the goal be to develop an aiding system or an automated planner?

The requirements suggested clearly that the development of an automated planner was undesirable—given what we were able to discover about tactical planning. The team determined that the best way to proceed was to conceive of TACPLAN as an intelligent aid only, and to require its behavior to be as unobtrusive as possible.

TACPLAN uses several conventional and unconventional methods to help Corps commanders develop a tactical plan. First, it assumes that tactical planning is amenable to a divide-and-conquer strategy, where planning problems are broken down into smaller and smaller problems. It also assumes that the smaller problems and subproblems should be solved in a specific order. It assumes, for example, that COA assessments should not be made until area characteristics and relative combat capabilities have been thoroughly analyzed. It also suggests that planners codify their primary and secondary goals before any analysis takes place. TACPLAN supports these and other assessments via a method known as multi-attribute utility assessment (MAUA). MAUA is a technique that supports analysis by identifying, weighting, and scoring criteria vis-a-vis COAs, terrain features, and the like.

The technique yields ranked lists of the most likely COAs, the most inhibiting terrain characteristics, and the concept of operations most likely to satisfy the commander's goals.

TACPLAN also has intelligence. It "knows" about terrain characteristics, preferred doctrinal options, force structures, and the relationships among area characteristics, doctrine, and combat capabilities. Its knowledge is stored in planning rules that are consulted whenever a planner makes a judgment, designates a likely adversary COA, or makes an assumption about how far and how fast combat units can move over specific kinds of terrain.

Remember that TACPLAN helps planners; it is in no way intended to replace them. Its intelligence is passive and unobtrusive. TACPLAN does not ask the planner what he wants to do and then do it for him; instead, the aid watches the planning process and only alerts the planner when something it "knows" (via its rules) about the process or about the problem at hand has been ignored or contradicted. It then alerts the planner to the problem. If the planner chooses to ignore TACPLAN's advice then the process proceeds (though the planner is asked to explain why he or she believes that TACPLAN's advice is wrong).

Planners manipulate data and information, test alternative hypotheses about the best way to deploy forces, and call up different (terrain, order of battle, etc.) displays which appear as overlays on the screen. They can add or subtract overlays, while a set of "rules" about the tactical planning process observe the planning process and alert them when a "violation" occurs.

TACPLAN runs on Windows machines. Its display guides planners through the plan building process, asking a series of questions about the planning problem. Planners are asked to assess area characteristics, relative combat capabilities, adversary and friendly COAs, and candidate concepts of operations. While all of this is going on the display is building the plan graphically. Planners can interact directly with the display by adding, deleting, or moving units around the map, by drawing COAs directly onto the graphic map, or by recalling and overlaying old plans onto a current problem.

The strength of the display lies in its realism and the extent to which it improves upon the use of paper maps, clear acetate, and grease pencils. TACPLAN's display presents actual maps of Corps areas of interest. Tactical symbols and COAs are stored as overlays on the graphic map.

The maps and annotations are stored digitally. Since the maps have been digitized planners have the capability to zoom in or out on the display. Since the digital overlays are computer controlled, they expand or shrink depending on the planners' field of view.

TACPLAN permits planners to develop tactical plans in much the same way that they now develop plans, but with some important differences. First, the aid structures the process. Second, it integrates elements and subelements of the process. Third, it monitors and instructs the process by checking planning judgments against its own knowledge base. Fourth, it permits planners to work with the same medium—actual maps—that they use when developing tactical plans without the aid of a computer. Fifth, it permits them to annotate onto the map image (not unlike the way they annotate on acetate with grease pencils), store their annotations, and recall them whenever they want (capabilities that acetate and grease pencils cannot match). Sixth, TACPLAN records the planning process for future study or application. In fact, it is possible to develop an inventory of Corps plans that can be recalled and displayed whenever a similar problem is faced. Seventh, TACPLAN is flexible and adaptive. Over time, the rules that govern its behavior will be modified and the knowledge base itself will be expanded. If, for example, a particular piece of advice is consistently ignored by Corps commanders, then the rule responsible for the advice can be changed. Since TACPLAN records the planning process it also records the disagreements, disagreements that can be used to make the aid more intelligent.

Finally, TACPLAN is inexpensive and transportable. Nearly all of its components can be purchased off-the-shelf. Special emphasis was placed on the design of an aid that would not consume huge resources in development or use. In fact, TACPLAN is in many respects a "rapid prototype" that can be modified easily, quickly, and inexpensively.

Appendix C
Technology Trends Analysis:
Trends in Business Technology Integration

Let us look at the relationship with technology through a different, sometimes personal lens—sometimes even angry lens—and let us see if it frames what the new business technology relationship should look like.

Let us assert that the wheels have come off the digital revolution. While there have been important productivity gains attributable to the widespread deployment of computing and communications technology, a lot of the productivity has been the result of extraordinary human effort. Worse, the digital revolution was supposed to be as much about fun and fulfillment as it was productivity and efficiency. It has not played out that way: what started as a genuine new wave phenomenon quickly morphed into a darker process distorted by professionals with no lives.

Remember when companies began selling older personal computers to their employees as fringe benefits? They quickly began to give them away so they could connect everyone to corporate networks—so they could work in the evening and on weekends from home. Remember when it was acceptable not to reply to e-mail messages immediately? Now digital tyrants want instant messaging to come to the workplace—after they have marveled at the compulsive effect it is had on their kids.

The angry lens sees this as a plot.

Digital technology is infiltrating our lives as quickly as they are deteriorating. Access to technology—and expectations about the role it should play in our lives—is driving massive change in our personal and professional behavior. Most of us are clueless about the effect the pace of technology change is exerting on how we work, learn and live. The digital revolution—like all of the others—is about money, power, productivity, and manipulation. But stop and think about who is in control. Your work week is longer, your privacy is gone, and you spend several hundred hours a year servicing your personal technology infrastructure. You check voice-mail and e-mail ten times a day—and at least once a day while on vacation. Your personal digital devices interrupt everything. You carry a pager, a cell phone and a personal digital assistant (PDA), and you cannot wait until they all converge into one really reliable, really smart monster device. We are digital now, we are netizens. But what the hell does that mean? How much of this is good and how much is a pain?

Digital technology intersects with our culture and lifestyle in unprecedented ways. We are working harder—not smarter—and managing lots more stress. Our personal lives have been under serious attack from our professional lives for at least a decade, just about the time the digital revolution kicked into high gear. We are wired. We are mobile. When we finally get home we exploit the same connectivity to relax, to be entertained, but have no demilitarized zones between work and play. The bozos who champion convergence actually believe that you and I will want our movies interrupted with phone calls received through our television sets.

Switch over to the corporate environments in which we all work. While we have gotten good at procuring PCs and other devices, we are still not clear about usage protocols. We are still not sure about how to manage all this stuff and we are forever searching for ways to measure the relationship between technology and business productivity. We have connected everyone but we are still not exactly sure why we did so.

Let us look at this convergence phenomenon—the poster child for digital schizophrenia—more closely. After the Internet bubble burst, and especially after the business-to-consumer (B2C) and business-to-business (B2B) business models were savagely rejected by investors, lots of gurus jumped onto the convergence bandwagon as the next new, new thing. The thinking here is that all those digital devices that we maintain at home and work should be connected, that convergence is a logical technological step, and a very good thing for personal and professional life management. After all, who would not want voice-mail from one's spouse to have priority over your boss'? Or a reminder about your kid's soccer game? But it does not stop there. The way convergence plays out is anything but pretty. Our cell phones will evolve into smart professional digital assistants (PDAs) that will process e-mail, voice-mail, pages, and other "opportunities" from work, retailers, wholesalers, and anyone else who can get access to our digital persona. Access will be facilitated by embedded global positioning systems (GPSs) that will track us, integrating our locations with databases that understand our preferences. What all this means is that a little voice will let us know that a sushi bar is only 15 feet away and if we are interested in the diversion our pending appointment can be rescheduled. A really smart assistant will know that it is been 2 months since the last sushi experience and will automatically reschedule the next appointment. Of course while eating we will be interrupted by any number of communiqués that have priority over food and relaxation. So we will gobble down the fish and run into the digitally summoned cab to take us to an unanticipated location where we will discuss things we will think about for the first time.

Convergence will make it possible for us to immediately know the cholesterol impact of the filet we are about to eat, when our kids are speeding, when our friends are avoiding us, when we are missing meetings, and how the hell to almost get away from it all. The net effect is total immersion into a digital prison.

Technology has so completely preoccupied us with work that there is little time for anything else. The very process of continuous connectivity saps our availability for activities outside the world of digitalia, yet we are told that continuous connectivity will be wonderful. The soccer game you were able to make because a device located and reminded you also interrupts the game, diverting attention from the simplest pleasures to much more complex and often unsolvable problems. What used to be the normal ebb and flow of demarcated personal and professional lives has been replaced by continuous problem-solving that intrudes on everything else—especially those activities so far removed from aggressive competition—like intimacy.

All of this is really about what might become the most important lost opportunity in modern history: technology can free us—but we are well on the way to technology enslavement.

While it is relatively easy to at least entertain arguments about a dubious relationship between technology that connects (annoys?) us and our personal lives, the professional argument is more difficult to make. Why? Because we have all been conditioned to believe that technology helps business, that without technology there would be no business. Both true, but it is essential that we place all assessments within the context in which they are made. For example, in the 1970s technology began to support back office processes in some significant ways; in the 1980s it began to move to the front office. The 1990s gave us distributed computing and of course the World Wide Web. What is worrisome is that technology has not merely "evolved": the changes that occurred in the 1990s are revolutionary—yet our perceptions about their value are absolutely evolutionary.

DO YOU THINK?

The 21st century ushered in lots of stuff. Thinking was not high on the list. As the pace of our lives has increased (with the aid of digital technology) we have abdicated thinking for reacting. Digital technology has created only defensive opportunities: so much information is pumped into us that it is all we can do to just react. Problem-solving is data-driven, not managed by creative insight. The whole process by which we identify, decompose, and solve problems is dependent upon the data we receive, not insights we create. How many times have problems at work been parked until FedEx delivers a package or another e-mail arrives: "we need to hear from Charlie before we do anything ... Charlie has the data!" Without realizing what is happened we have fallen into patterns that reduce the importance of natural intelligence and creativity for those that reward reactive speed. Even personal decisions—like deciding where to go to college—are data-driven. Web sites enable "if – then" processing: if you want an urban, small, artsy school, then you should go to college X – and not college Y. An expert system embedded in the Q&A process actually suggests where students should and should not go. How many open houses are skipped because Web sites say the school is a bad fit?

Thinking takes effort. Reacting is easier and faster, especially with digital gadgets. The more we connect employees the more we reward reactive agility.

The significance of this shift is enormous. Interpersonal relationships, primary, secondary, and higher education, and lifelong training are all affected by reaction- vs. pro-action-based understanding and communication. Instead of inductive learning we have compromised to deductive processes, where we react to known facts—presented to us through our digital toys. The infiltration of digital media into the classroom is sometimes good but potentially very dangerous. Discussions about science, geography, and history often take the form of asking students what they think about a demonstrated phenomenon instead of how the phenomenon came to be. We all know that discussions can be jump-started by providing examples, strawmen, demonstrations, and the like, but we also know that such techniques are often reserved for less-than-creative groups who need stimuli to get started. Digital technology and media make it possible to insert examples and demonstrations of all kinds into curriculum whenever an opportunity presents itself. Teachers familiar with multimedia tools can import pictures, video clips, and other media into their lessons that are minutes old, and then ask their students to react. Kids love this stuff, but are they learning how to compose or decompose? How about grown-ups getting trained?

ACCEPTABLE ADDICTION

You are 40 years old. You are a professional. You use technology every day to do your job and manage certain aspects of your life. You are smart and productive. You are connected. But you are harried, conflicted, and overwhelmed. What happened to you? Probably two things: you fell for the American dream believing technology was one of the springboards to the dream's realization. Ask yourself: would you rather have a restored 1954 Corvette or a 1+ gigahertz personal computer? What did you say? How about the value of private, disconnected time? More valuable than day trading? Or incessantly checking e-mail and voice-mail?

According to survey data, if your household makes over $75,000 a year and has a couple of kids, your technology infrastructure looks something like this:

- 1 Work Provided Cell Phone
- 1 Personal Digital Assistant (PDA)
- 1.5 Work Personal Computers
- 2.5 Personal Home Computers
- 1 Printer at Work (Access to Many More)
- 1.5 Home Printers
- 1 Conventional Fax Receiver at Work
- 1 Computer-Embedded Fax Receiver at Work
- 1 Home Fax Receiver
- 1 Work Pager
- 1 Family Pager
- 1 Home DVD Player
- 2 Home Video Cassette Recorders
- 3 TVs
- 1 Home MP3 Player
- 2 Home Telephone Lines
- 5 Wired Phones
- 3 Family Cell Phones
- 1 Work Phone Line
- Voice-Mail at Work
- 1.5 Voice-mails at Home
- 1 Work Internet E-Mail Account
- 3 Home Internet E-Mail Accounts
- 1 Analog Camcorder
- .5 Digital Camera
- 1 Video Game Player

This "average" inventory is sometimes supplemented with a global positioning system (GPS) in a car and a cell phone, the use of intelligent agents to find the cheapest prices on the Web, and a home theater fed by cable, satellite, DVD, and Internet signals. If your kids are even semigeeky, they have their own Web pages. Lots and lots of technology. Does it all work as advertised? Does it work at all? At work you should be asking some really tough questions about the return-on-technology-investments—not to mention the total cost of ownership of all this stuff.

If this list resembles your current technology holdings, you are out of control. A time and motion specialist would have a field day assessing your utilization efficiency. With this much junk, it is impossible to target it all at activities and processes that enrich your life. Hell, you probably spend a couple of days a year just replacing batteries. Does any of this sound familiar? Is your professional environment similarly chaotic? Do you know exactly how many computing devices are in your company?

You are also knee deep in the idea of technology. Those who manipulate us for a living use multimedia to associate technology with "new," "smart," and "successful." Until recently, the New York Stock Exchange was "out" while the Nasdaq was "in." Desktop computers are "out," PDAs are "in." The number of on-board processors in your luxury car is a status symbol. If you are technologically illiterate, you are inferior.

Even if you do not realize it, you are in deep. You have bought into fulfillment through technology at home and at work.

Our relationship with technology is sadomasochistic. It is warm enough to seem friendly, but it is a Trojan horse. It is an acceptable, politically correct, but deadly, addiction.

What if someone told you that the average cost for a corporate PDA was $250, but the annual support costs per device exceeded $2,000? Would you still buy PDAs for everyone, even if the deployment was "cool"?

HAVING FUN YET?

About 40 years ago someone said that the biggest problem we would have would be the disposal of surplus leisure time. Technology was going to create this surplus. But it has not turned out that way: while technology makes aspects of our lives fun and efficient, it also often betrays us. How many hours a week do you tinker with your personal technology infrastructure? What is the real cost of continuous connectivity? Do you even have any idea of what you spend annually at work on all forms of "hard" and "soft" technology support?

A lot of our infatuation with technology is stupid. It is also expensive. Look around your house. How many computers do you have? Are they all the same, or do have to maintain a bunch of nonstandard machines? Does your fax machine connect to your PC and personal digital assistant? Why not? How many telephone lines do you have coming into your house? How man AOL accounts do you have? Do you have a broadband connection to your home? And if you have a broadband connection, do you still have an AOL or another Internet service provider (ISP) account? Why?

What does it cost you a year to maintain all this stuff? Do you have any idea? What is the total-cost-of-ownership of your high-end servers at work?

At work you are a geek, sort of. You cannot live without e-mail and voice-mail. Nor can you survive without Powerpoint, Microsoft's venerable presentation package. You probably have trouble discussing ideas without Powerpoint (though you are really good at reacting to Powerpoint content). But when anything breaks you have to call the help desk, because you cannot help yourself: we are dependent on technologies we barely understand. This wasn't always the case. The percentage of people whose early driving experiences included the ability to actually fix their cars was much higher than was ever the case with computers, VCRs, or cell phones. Hardly anyone ever bonded with digital technology; instead, a gap appeared almost immediately between digital technology and everyone's ability to keep it humming. Our digital dependency is thus much greater than it was on industrial or analog technology. We are happy when the machines stay up; depressed when they are down. Do you think there is a relationship between this frustration and out interest in outsourcing?

Technology is a fair-weather, multiphrenic friend that simultaneously increases productivity, complexity and confusion. If we take a hard look at our use of technology tools and gadgets the picture's pretty ugly. We spend way too much for way too little, and we are by and large oblivious to the lack of return on our personal or professional technology investments.

There are lots of reasons why this is the case. Most of the obvious ones—like bade business cases, early adoption of unproven technology, and bad management—are easy to understand, but the less obvious ones are what is really driving the technology/productivity disconnect. As suggested above, we still think technology development is evolutionary, still treat its impact on business as only enabling, and still think that technologies (and technologists) are really separate and different from other professionals in our companies. While ROI and TCO models are almost always useful barometers of how much we are spending with what impact, there are those among us completely obsessed with long division. Revenge of the CFOs? Sometimes.

PRIVACY

What about privacy? The linkage of global positioning systems, mass customization, and continuous connectivity means that people who want to sell you stuff know what you like and where you are all the time. While you might think it is cool that your cell phone rings with a message about a sushi opportunity in 15 feet, after it happens 20 or 30 times you will (and your customers will) turn off the phone, and the pager and the PDA, since the sushi message will roll from one device to the other if any of them are dead. Specific knowledge about consumer preferences will be widely shared, sold for huge profits when the consumers are big spenders, and used to develop individual marketing campaigns designed to woo every consumer in the ever-growing database.

A recent survey revealed some truly amazing opinions about privacy. Seems that in spite of rhetoric about the value of privacy, the majority of Americans are willing to sell their consumer preference data—their privacy—for $462 per year. This is clear evidence that those surveyed have absolutely no understanding of the privacy threat, or the survey was populated with people eager to expose all aspects of their personal lives. The sane among us should have serious concerns about the amount of data advertisers and manufacturers are collecting. The annoying telemarketers that call during dinner now have enough data to inform us, during dinner, that we should have clicked on the sale links while browsing online for books at 3:09p.m. The same people also know what we have purchased online, how long we lingered on specific merchandise, and what we are likely to buy at various times during the year. You tell me who is in control?

Or, from another perspective, are you the telemarketer? If you are, or are otherwise committed to customer personalization, then you understand the fragile relationship between privacy and business. (Truth be told, you understand how to exploit our quasicommitment to partial privacy.)

WHO IS TECHNOLOGY?

Technology will play a larger and larger role in our lives. This is inevitable. Much of this technology will become "seamless": it will operate out of sight, like electricity or the way the telephone system used to operate. But make sure that the subtleties of what happened to telephone service do not fall through the cracks of misunderstanding or misrepresentation—or your emerging relationship with digital technology. The government in its infinite wisdom decided that we needed more telecommunications competition, so it broke up Ma Bell's family to protect us from price gouging. Lots and lots of companies were permitted to enter the market. No longer could you make one call to satisfy all of your needs. You had to call one company to arrange for the line to come into your house, another to install the phones that you purchased from a retailer, and two more to service problems that occur, since the company that brought the line to your house refuses to touch anything inside your house, and of course vice versa. The net effect of all this is that we have to actively participate in the planning, implementation and support of our personal and professional telecommunications systems. In order to do this, we have to know something about the technology. So what started as government intervention intended to make telecommunications easier and cheaper ended with a complex, expensive anti-solution. Relatedly, the enterprise application integration (EAI) cottage industry exists today because of all of the disembodied pieces that IT managers have to interconnect. Would someone please think *solutions* before tinkering with regulations or standards?

As the role of digital technology increases, we will all be expected to know more and more about what works, what is expensive and what is obsolete. At home, we will have to know what Nintendo, Sony, and Microsoft have planned for their personal video games and at work what Cisco and IBM have planned for corporate networking. Pretty soon the distinctions between home and work will blur and we will have to track technology integration and interoperability progress as well. The trend toward technology decentralization and democratization will actually require us to continuously upgrade our technology knowledge, just what we need, as we try to manage the collapsing walls of our personal and professional lives. We are all going to learn how it feels to stand in a digital bread line.

What about the World Wide Web (WWW)? The Internet is the most important information technology of the 20th century. When the WWW appeared in the early-1990s unprecedented access to massive amounts of information was granted to anyone with a PC, a modem, and an account at an ISP. Never before had so much information and knowledge been "published." Never before was information and knowledge wrapped in instant, cheap communications: arguably e-mail is the most important communications medium on the planet. But just as certain are search problems on the Web. With millions and millions of Web pages out there organized around all sorts of themes, transactions, and biases, finding what you need when you need it continues to be one of the unpleasant side-effects of universal access. "Search engines" are still pretty dumb, understanding only what you say—not what you mean. So when you ask for information about Indians you will get information about the Sioux, Indians from Madras, and more than you ever wanted to know about the Cleveland Indians baseball team. Redundancy is also a problem: there are numerous sites that do exactly the same thing. You can buy books, CDs, DVDs, and videos at about 100 sites, post your resume on 50, and plan a trip on another 100, each with links to scores of related sites. Skillful Web navigational requires lots of practice. It is been estimated that efficient searching requires between 100 and 300 hours of inefficient searching. Considering that there are 2,080 hours of work in a year, all you need to do is carve out a couple of months of your work year to learn how to find stuff fast on the Web. Of course if you cannot learn while at work you can always take 100 to 300 hours from your personal life. Did I miss something here? There is a thin line between fun and frustration.

The Web provides all kinds of opportunities for technology optimization—and all kinds of opportunity costs. Connectivity, hardware, and software decisions are many and varied, not to mention the most daunting of all: deciding where to go and how to get there. Do you need the newest version of America Online? Should you download plug-ins and players every time you get an on-screen message suggesting you do so? How much time should you devote to all this? People stay up all night, having fun, trying to figure out the Web. Have you? Worse, how many hours a week do your employees surf around looking for what they need to do their jobs (and other stuff)?

NOW WHAT?

This report is about technology optimization through a new relationship between business and technology. It is about cost-effectiveness, return-on-investment, and all that kind of thinking, but it is really about unlocking technology's potential to improve things, not just your image, and especially your business. The arguments here are designed to redistribute control over technology, control that can provide you with opportunities well beyond our current thinking about why we need so much technology in our personal and professional lives.

The larger hypothesis is that we have bet too heavily on technology, that we are still in the early stages of technology optimization, and that we are pretty clueless about how to get the most out all this digital paraphernalia—just as all this stuff is actually beginning to work! Again: "just as all this stuff is actually beginning to work!" The implications are huge. The United States is the first trillion-dollar-per-year-information-technology-spender. Many other countries are closing in on the trillion dollar mark. IBM and Microsoft each spend well over $5 billion a year on research and development. Individual households are spending upwards of 20% of their disposal income on technology in one form or another. It is safe to say that 50% of this personal and professional cash is wasted on misguided investments that do not satisfy serious requirements. The truth is that we are in transition from Stone Age digital processing to bona fide technological ubiquity. But those who market and sell the stuff make us think we cannot live without MP3 wristwatches, Web-enabled cell phones, multiple e-mail accounts, Web services, natural language processing, or storage area networks. Pain killers are always more valuable than vitamins, but technology marketing gurus keep us away from comparisons.

Like a lot of what we think we cannot live without, digital technology is only valuable when it solves specific problems or frees us from difficult, time-consuming, or ridiculous tasks. It is time for a reality check.

Certainly no one believes that technology will not power the new economy. Long term it is the best bet in town. But the timing of all this is tricky—and manipulated. What the Internet stock frenzy really offered was a glimpse of what will happen when the best business models succeed. The exuberance that persisted was a proxy for just how incredible the new economy will be. But if you are like me you are still waiting for reliable, unobtrusive broadband service to your home, still excited more by the promise than the reality of the new economy. Make no mistake: the new economy will be anchored by "pervasive computing" capabilities, with "always on" and continuous computing and communications capabilities. Personal and business processes will be automated and secure. But the new economy's digital infrastructure will not be ready for a while. How long, really? The best we can do until then is buy little pieces of it and coldly assess its meaningfulness to our personal and professional lives—which is precisely where the challenge of technology optimization lies. Buy the wrong pieces and you'll experience all sorts of problems; buy the right ones and the quality of your personal and business life will be enriched. Along the way you'll have to navigate aggressive marketers hawking half-truths about how wonderfully all the pieces fit together.

The emphasis here is on the role technology should play in your personal and professional lives. Technology will not lead your life processes and decisions, it will follow and enhance them—and you will know how to spot and defuse all varieties of hype.

But most importantly this report resets the digital revolution in the context of business and technology progress. When all is said and done, this second major wave of digital technology will earn its place in history by stimulating debates and influencing lasting change. This report launches a debate in an effort to influence the changes digital technology will trigger. We are at a flashpoint: the pace of technology deployment and business velocity has already outstripped our ability to assess its impact on how we live, produce, and distribute. It is now time to think about where all this is going—and how to optimize it. Let us reset the dials:

We have lived through several computing and communications "waves," all the way from mainframe-based computing to business processes that only exist on the Web delivered by distributed servers.

Some of the effects of technology have not been well studied, or studied at all; the life/work balance is out of kilter—and getting worse; life/work disequilibrium is threatening overall productivity and laying the groundwork for future workplace revolutions.

Lots of the early, and current, stuff does not work as advertised, resulting in a cynicism about "business/technology alignment."

The bursting of the Internet bubble resulted in an over-correction, a dubiousness about all varieties of distributed computing equilibrium.

Just when the stuff started to work and just when business models were beginning to morph out of their traditional vertical silos, capital spending in technology collapsed.

In spite of all these reactions and trends, we are now sitting at one of the most important crossroads in the history of business technology, and especially business technology management—the subject of the rest of this report

THE NEW COMPACT, OR "ALIGNMENT" IN THE 21ST CENTURY

Let us assert that the computing and communications technology we have developed and deployed over the past 30 years represents a kind of prototype. Of course it is robust. PCs have gotten cheaper. Serious companies all have access to the Web, and we are now free to think about customer relationship management, Web services, and the emerging semantic Web. But we still struggle with a lack of standards, growing integration, and interoperability problems and chronic disconnects among our back office, front office, and Internet applications. We have gotten good at creating technology pieces, but we are only now beginning to focus on how they all work together.

Let us also assert that business models are evolving, morphing, and accelerating faster than ever before. Let us assert that your company now finds it hard to draw old lines around what it does and how it operates, or around what we used to call it is core competency. In fact, the whole notion of core competencies is now confused since companies began rethinking their supply chains, partnerships and alliances. Here is a thought: is Dell a computer manufacturing company or a supply chain planning and management company? How hard is it to imagine Dell as a company with multiple lines of business including one that sells supply chain management software and services? The other lines of business will focus on manufacturing and distribution using Dell's advanced software platforms. Dell's core competency is this scenario? Supply chain integration.

The last 30 years constitute the first digital revolution. The next 30 will define the second one. Companies that treat the interplay between business and technology as a simple extrapolation from even the most recent past (yes, that includes our initial infatuation with the Web) will be out-maneuvered by companies that see it all differently, that is, as a revolution enabled by the pure business technology convergence. (By the way, this also applies to your personal technology infrastructures and applications, though we will focus more from this point on business technology optimization.)

The convergence of technology and business change is what is different. But unlike a perfect storm when lots of bad things happen at the same time, what we have now is a perfect opportunity fueled by the convergence of technology that is finally ready for prime time, business models that embrace speed and flexibility, and management possibilities that will treat the relationship holistically. Stated a little differently, the stuff now almost works as advertised. Of course, you might argue that "almost" is not good enough, or that you have heard it all before, and that the most prudent approach is to simply wait until the stuff starts really working—as advertised. You can take this approach but the problem is that when convergence hits you'll have to scramble to catch up to those who saw it coming. Remember the Internet: have we forgotten that Microsoft actually missed it, only to respond with the now famous "extend and embrace" initiative? If all this is true then the way we approach business and technology modeling should change dramatically. The approaches we have taken to business/technology "alignment" served us well for a while, but grossly miss the point of holistic modeling.

So how should we proceed? If a picture is worth a thousand words, Figure 1 is one that hopefully communicates the essence of what I am talking about and what this report describes.

What does all this mean? First, it means that questions that belabor business technology "alignment" are obsolete. It means that organizational distinctions

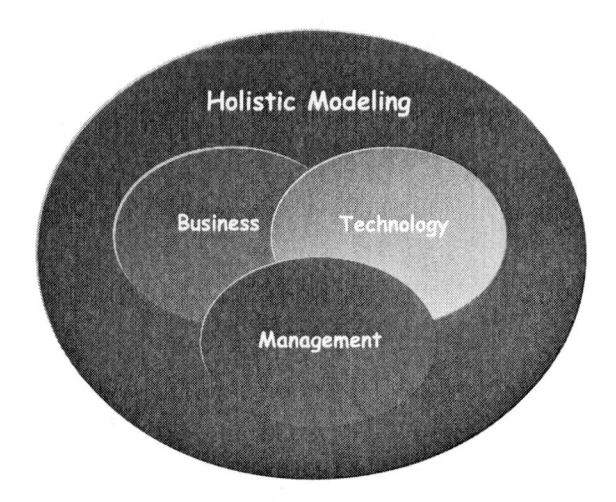

Figure 1. How to think about business technology integration

between business and technology should disappear and be replaced by a seamless interconnection that makes it impossible to address one without the others. It also means that chief generalists (CEOs, CFOs, CIOs, and even CTOs) need to become wider and deeper, redefining the whole notion of generalist practitioners. Can you really lead a company if you do not understand technology architectures and applications? Can you enable a company if all you understand are architectures and applications?

The relationship between technology and business has evolved in a largely productive way that is served each community pretty well over the past 30 years. Companies spend millions and in some cases billions of dollars a year feeding each community that essentially co-exist for the greater good—profit, bonuses, and shareholder value. By and large, the relationship "works," though there is increasingly evidence of dysfunctionality in the trenches where wars continue to erupt among technologists, business managers, and finance professionals, the latter group for the life of them cannot understand how an enterprise resource planning (ERP) implementation project in a Fortune 500 company can cost a 100 million dollars and take 3 years to complete, or why in the same company it takes 3 years to migrate from one desktop/laptop operating system to another.

These wars aside, we now have lots of applications, data bases, devices, communications, virus protection, security, and even the means to resume business if a disaster occurs—and all of this stuff works reasonably well most of the time. Analysts like Paul Strassmann tell us that we are overspending on technology and the relationship between technology investments and productivity is anything but clear. Others, especially in this bearish economy, have resurrected total cost of ownership (TCO) and return-on-investment calculi to derail big technology initiatives.

But things are improving all the time. We whipped the Year 2000 problem, connected just about everyone to the Internet, and have begun to more deeply appreciate the need for integration and interoperability even as proprietary vendors make it tricky to have the pieces all fit together. So life in the trenches is, well, pretty good.

Let us continue with the assertion: we have reached the point where if continue along the same business technology relationship path, we will undermine the very business models and processes we are trying to define and deploy. Worse, the current relationship will eventually collapse under its own weight due to organizational ambiguities, technology complexity, and our inability to satisfy consumer or business-to-business requirements that are appearing and changing faster than they can be supported.

While the evidence tells us that computing and communications technology have made enormous strides over the past 30 years, and we now routinely talk about business/technology "alignment," technology optimization, and how companies can extend their business models through pervasive communications, our current discussions about supply chain planning and management, collaborative forecasting and automation all assume a business/technology relationship that is fundamentally different from our notions of "alignment" or our organizational attempts to get technology to the head table: discussions about whether CIOs and CTOs should report to the CEO or CFO are really very 20th century, since everyone now knows that CIOs and CTOs should breakfast with the big guys (assuming, of course, that they are house-broken).

From another perspective, it is no longer possible for a chief executive or any sane senior management team to conceive of new—or extend an existing—business model without addressing technology requirements, capabilities, and costs. Some of these models are actually created in reverse, where business models extend from what is technologically feasible, not necessarily from solid (read: profitable) business models. Remember the dot.coms?

We have been through a lot over the past few years. We discovered the Internet, successfully managed the Y2K compliance problem, over-hyped Web-based business models, and confused even the most loyal technology investors about what drives capital technology spending.

A longer view, way back to the 1960s, saw the introduction of "data processing" to industries that barely knew what to make of computers, software and data bases.

The 1970s took us to a much higher level where mainframes got a little flexible, minicomputers arrived for the frugal, and PCs began to procreate among parents—like Sinclair and Osborne—long since extinct.

During the 1980s everyone absolutely had to have a PC at work and increasing numbers of us had to have them at home. Even personal software got easier to use, principally through Apple's introduction of the Macintosh, though business applications continued to only slowly evolve.

The 1990s gave us client/server computing, our first real freedom from mainframe architectures, the Internet, the World Wide Web, multitier applications, data warehouses, data mining, applications integration, online exchanges, new security requirements, privacy issues, virtual private networks, application service providers, content management, knowledge management, network services, C++, Java, Perl, Linux, customer relationship management, e-CRM, interactive marketing, Bluetooth, 802.11, and a whole lot more.

Lots of us think we have achieved a new level of "alignment," or the process by which technology supports business. The truth is that the questions that dominate this new alignment consciousness are still the wrong questions. Why? Because technology and business are no longer even "equal" partners—they are an integrated whole.

Here are some questions that no longer make any sense:

- How can we leverage IT onto the "right" business processes in the most cost-effective ways?
- In which technologies should we invest?
- How should we acquire and manage these technologies?

They make no sense because they ignore macro trends in business and technology and they ignore the purposeful aspects of the questions. Why is one technology more important than another? Why is a technology important at all? Technologies without business models are as useless as business models without technologies. This is the key point: technology, business and management can no longer be treated in relative vacuums. They work together or not at all.

The approach we have taken to "aligning" information technology with business models and processes served us well until the year 2000. While most companies never quite got there it is time that they stop trying to achieve the old objective. Not because the goal was wrong, but because it is no longer consistent with the business and technology trends that are upon us: the pace of technology and business change has forever altered the way we should think about how we find and service customers and organize ourselves to compete.

The net effect of all this is that we will continue to spend far more on technology than we should, that we will continue to ask the wrong questions about technology, that we will miss myriad opportunities to leverage IT onto our business strategies, models, and processes, and that we will continue to march to agendas set by consultants and, especially, vendors. We also tend to default to conservative interpretations of what is really outside the box: very few companies are really serious about radically changing, or even challenging, their existing business models. Unfortunately, this perfect storm requires much more than conservative extrapolations of current business models.

While there are certainly organizations that suffer less than others, the lion's share of companies older than 20 are in serious trouble. Really scary is the number of executives unaware of just how serious their condition is. Huge numbers believe that Microsoft, Oracle, or Netscape have the answers to their problems, or management consultants can set them straight, or—even more bizarre—that in-house people that missed seeing the problems can somehow solve them.

All of this occurs as the promise, and reality, of IT is at an all-time high. The producers of IT products and services wax damn near poetic about what they have made possible—and what they plan to do next. Executives crow about how their companies are "upgrading" their "infrastructures" and deploying "state-of-the-art" communications and computing "architectures." The fact is that precious few really understand their own speeches. The consumers of technology are thus at a distinct disadvantage, a disadvantage that is systematically exploited by producers. The facilitators of IT, the consultants, play both sides, offering advice to harried, perplexed consumers and the producers of IT, brokering the relationship often with the finesse of a magician.

Comptrollers are forever writing checks to buy more computers, more telecommunications, more software, and more technology professionals. But Chief Executive Officers want to document the return on their investments. Chief Information Officers find themselves on the defensive far more often than they find themselves in the winner's circle: who wants to be at the head table if you always get served last?

The technology marketplace is one of the largest and fastest growing in the world today. But everywhere one looks, everywhere one goes, and over and over again in the technical journals and trade publications we still see references to the same issues, problems, and challenges: "the software crisis," "the requirements problem," "the return on investment challenge," "process improvement," and "total quality software management," among all sorts of others dopey things. Many of us wrote about requirements problems 2 decades ago, we called for project "dashboards" for managing multiple projects, and committed ourselves to "process improvement." Well, here we are 20 years later asking the same damn questions and, worse, proposing the same damn solutions.

But these are "tactical" problems, problems that are created by—so they can be solved by—the producers of technology products, systems, and services. Is this a conspiracy? You bet (though we could argue about how conscious it really is).

Problems are easy to identify. Solutions are harder to come by. There are legitimate reasons why there are more books about problems than solutions. Perhaps the most obvious is the "moving target syndrome": as business requirements change, technology changes. As technology changes, price/performance ratios change. As price/performance ratios change, corporate cultures change. As corporate cultures change, global competition changes. As global competition changes, profitability changes. As profitability changes, the technology market changes. And so it goes. What do you do first?

Enter the consultants. There are "conventional" consultants, "contrarian" consultants, and consultants who have solutions for problems yet to be invented. And there are vendors—thousands and thousands of vendors. Consultants and vendors seek to reduce problems to their simplest terms, not because problems are by nature cooperative, but because it is the only way they can appear confident enough to convince CIOs, CEOs, and CFOs to spend more money.

This report looks at the intersection of computing and communications technology, people, organizations, requirements, markets and constraints. It documents the problems in an effort to understand and explain them. It challenges "conventional" and "contrarian" wisdom. It takes the prevailing "bunker mentality" and generates prescriptions, predictions, and an adaptive approach to the acquisition, deployment, and management of information technology.

The report assumes that problems and solutions cannot be traced to computers, management, software, people or networks—but to all of above and then some. We are no longer in the age of disembodied solutions to anything; we are in an era of complexity, integration, and synergism. It no longer makes any sense to hire a consultant who knows just about everything there is to know about software but very little about hardware (just as it makes no sense to hire a car mechanic who does not drive).

The report is partly diagnostic but mostly prescriptive. It is about a set of concepts, perspectives and technologies; it is not filled with "silver bullets." It is about how technology can be leveraged onto problems that are difficult to model, certain to change, and often expensive to fix. It acknowledges the complexity of technology-based organizational effectiveness. It stands tall for analysis and discipline—and evidence-based option generation and selection. In other words, if you want to win the second digital revolution, you had better understand and manage the politics swirling around your company.

The report also identifies a suite of principles that define processes that point to methods. Not long ago I received a call from the CFO of a Fortune 500 company about to write a check for $30,000,000 for a network and systems management framework. I asked if his requirements analysis was able to profile his organization's computing assets and network management needs, if his in-house technology professionals had performed trade-off analyses of several alternative frameworks, and if those who would actually be using the framework (to presumably manage their networks better, faster and cheaper than before) had ever used similar tools to help solve network management problems. The CFO asked me to explain what requirements analysis was, the CIO had no idea what network management point solutions were already in the organization, and the network operations center director had not compared the new network management environment about to be acquired with anything else (but liked the vendor's brochures and videotapes). No one had even talked to the network managers who would actually use the application. This short story illustrates how principles, processes and methods can be ignored—and how some relatively simple steps can lead to enhanced productivity and cost-effectiveness.

The analyses and recommendations here are anchored in field and case studies not as evidence or documentation, but as points on a new compass. Over the past 20 years, we have cataloged problems and documented successes. But remember that 20 year old cases are about as relevant today as a single anecdote about a guy down the block who had success with approach A, consultant B, or vendor C. The key lies in the extent to which generalizations hold against the moving target backdrop. For example, who the hell really cares about centralized computing environments today? How can technology investment decisions be made independent of technology forecasts? Who cares about flat data files? The report argues that any rational approach to the technology business of business technology is multidisciplinary, adaptive, cautious, and evidence-based.

So what happens if you read this trends analysis? If you are a CEO you will armed with questions that should be asked of your business and technology professionals—at the same time and in the same room. You will also gain insight into one of the largest, most voracious, yet potentially most significant, "sink holes" in history. A strategy, complete with tactics, will also be developed. If you are a CIO, you will receive some tactical insight on technology acquisition, deployment, and management. You will be cautioned about repeating the mistakes of your competitors. You will think twice before authorizing big technology buys.

Regardless of your role the report will provide you with a new perspective—an analytical compass. The central theme is simple: in spite of all of the hype, all of the serious technology, and all of the rapidly changing business models, we are at a crossroads. It is now time to rethink the business technology relationship and move it from a less-than-equal partner model to an integrated holistic one. The objective of the report is to help you construct a business technology convergence plan that will work for your organization, your people, your corporate culture, and your resources.

HOW TO THINK

True business technology convergence assumes that all discussions about existing or new business models and processes will occur with immediate reference to technology and the best management practices necessary to integrate all of the pieces. With that in mind, let us look at the pieces and how they should be assembled. Let us also keep in mind the earlier perspectives on personal and professional technology, and how the decisions we make from this point forward should be approached.

Business

If a company does not understand its competitive advantages and its current and future business models, it is doomed. Not only will it fail in the marketplace but it will waste tons of cash on technology along the way. We used to ask: "what is our core business?" "What do we do well?" "What markets do we *own*?" The new questions are different:

- "What do we do profitably today?"
- "What should we do tomorrow?"
- "How does technology define and enable these efficiencies?"
- "What business models and processes are underserved by technology?"
- "Which are adequately or over-served by technology?"

Let us look at business models and processes, technology (and management) holistically. Figure 2 offers some ideas and key questions.

It is all about the big questions. Do you know what you do well, poorly, and with whom you compete? Have you thought about what your business will look like in 3 years? Have you segmented what you do according to margins? (One of the more interesting things about the HP/Compaq merger is all of the arguments that broke out over control of a low margin, commodity business.) Larger questions include the long-term survival of your business via partnerships and alliances, the rate at which you can really change, and the interplay among business creativity, technology delivery and management efficiency.

If you are like most companies, those who run "strategic planning" are often one step away from retirement. But holistic modeling requires that business creativity

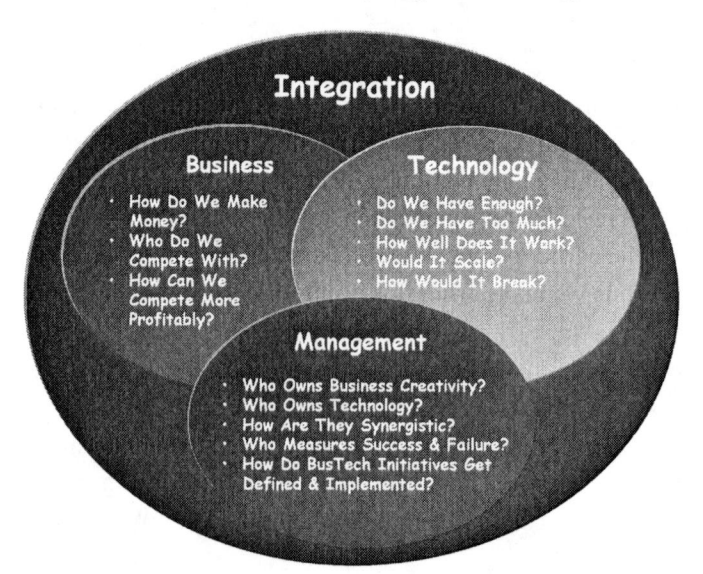

Figure 2. How to think about business, technology, and management

be taken more than seriously, and that those who define and engineer innovation are also good strategic technologists and managers. If they are not, you will miss the historic convergence occurring as we speak.

Do you know if your technology infrastructure, applications, and support all "work?" Do you know if they "match" your business models and processes? Do you know how the pieces might break? And, most importantly, do you know if your technology can grow with your business creativity? If someone asked you if you had too much or too little technology how would you answer them? Would it be easier if they asked if you had the right or the wrong technology?

Who owns creativity? Who owns technology? You have made your first mistake if they live in silos which seldom communicate. (When we talk about organization later you will see just how dangerous business/technology segmentation actually is.) Who manages the integrated process? How is success and failure defined and measured?

Here is the benchmark: if you develop "new" business models (or improve existing ones) and then ask technology if it can support the changes, then you are suboptimizing the business ←→ technology relationship—and you are likely to over- or underspend on the business technology initiative. Why? Because business models cannot exist without enabling technology and technology's only purpose is to support business models and processes. Yes, the implications here are huge. If you revisit the perspectives described at the beginning of this report, you should note that without near perfect synergism, then you will end up with too much, too little, wrong, expensive, and unreliable technology supporting business models that may or may not exceed their potential.

Communications

If we have learned anything over the past few years, it is the importance of pervasive, secure, reliable communications. It not just about the Internet. It is about communications inside and outside of your firewalls and it is about mobile communications. It is about communications among your employees, suppliers and customers—and even your competitors. Have you ever wondered about Dell's (and other) online computer sites that sell Microsoft software and HP's printers? Since all of these vendors need each other to some extent or another, they therefore need to communicate.

It is no exaggeration to say that communications technology will make or break your ability to compete. There are all sorts of issues, problems, and challenges that face your organization as it wrestles with its business strategy, its communications response and its ability to adapt quickly to unpredictable events. Figure 3 identifies some of the synergistic issues and questions. Assessments need to be made first

Figure 3. How to think about communications

Integration

Business
- Who Should We Link?
- How Should We Connect?
- How Should We Define & Deploy "Collaboration?"

Technology
- What Should the Communications Infrastructure Look Like?
- What About Communications Applications?
- Can We Add Partners & Collaborators Seamlessly & Securely?

Management
- Who Defines Communications & Collaboration Policies & Procedures?
- How is Communications Efficiency Defined & Measured?
- How is Infrastructure Migration Managed?

about who will connect to your "network." If your network will be wide—lots of employees, customers, employers, partners—then you may need to re-architect your communications infrastructure. You should simultaneously ask questions about your communications infrastructure, the applications (like e-mail and workflow) that will ride on the infrastructure, and how you'll manage infrastructure migration and measure communications efficiency.

Note that the distinction between business and e-business is gone: all business technology plans should assume full connectivity among a constantly growing number of participants. If you fast-forward to 2009, you will be expected to have the capability to add or delete network nodes and users at a moment's notice. The whole concept of "communications support" is already obsolete, since communications is not a supporting player but an integral part of every aspect of your business process.

Applications

There is a pretty good chance that your applications portfolio is a hodge-podge of applications developed over the past 30 years or so that require some form of life support to exist. You have probably got applications that are mainframe-based, some client-server applications, and some Internet applications that are driving your e-business strategy—and herein lies the problem: for the past 30 years we have been defining applications around silos and fiefdoms. In fairness, we developed

Figure 4. How to think about applications

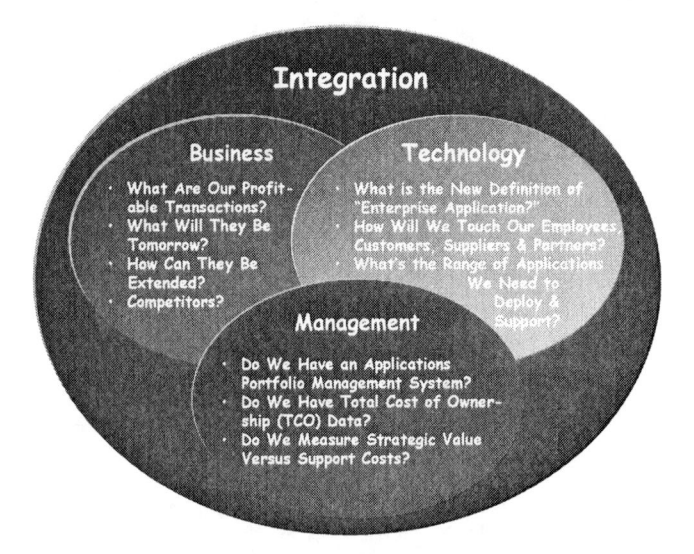

Integration

Business
- What Are Our Profitable Transactions?
- What Will They Be Tomorrow?
- How Can They Be Extended?
- Competitors?

Technology
- What is the New Definition of "Enterprise Application?"
- How Will We Touch Our Employees, Customers, Suppliers & Partners?
- What's the Range of Applications We Need to Deploy & Support?

Management
- Do We Have an Applications Portfolio Management System?
- Do We Have Total Cost of Ownership (TCO) Data?
- Do We Measure Strategic Value Versus Support Costs?

applications around tasks we needed to complete. Initially these tasks were computational; over time, they became transactional, and now they are collaborative. Unfortunately, many of us are still just "computing."

The applications end-game consists of a set of inter-related, interoperable back-office, front-office, virtual-office, desktop, and personal digital assistant (PDA and other thin client) applications that support evolving transactional, collaborative business strategies.

Figure 4 lists some of the questions begging for immediate answers.

A key application question should focus on the relationship between transactions and profit. Do you know which ones yield the greatest profit? Do your applications facilitate the touching of employees, suppliers, customers and partners? How many applications do you have (that run on your desk-tops, lap-tops, PDAs and other access devices)? Do you have an applications portfolio management system that helps you locate and support your applications?

Data

Data is the lifeblood of your applications and your plans to link employees, customers, suppliers, and partners in a virtual world. We now think beyond data base administration and about intelligent decision support, online analytical processing, data warehousing, data mining, metadata, and universal data access. Or at least we should.

Figure 5. How to think about data

Figure 5 lists some of the major questions about data, questions that should get you thinking about the interrelationships among business, data, and management.

The business questions about data are the proverbial ones: can we cross-sell and up-sell? Can we connect everyone? Can we extend our business models through integrated data, content and knowledge bases? The technology questions address variation and integration, and the management questions address administration efficiency and gaps.

Security and Privacy

Security—and its first cousin, privacy—are now household requirements. If you ignore them, you are toast. How did this happen so fast? Blame it on distributed computing, and the distributed steroid known as the Internet. As business models moved into cyberspace we found ourselves facing new threats. We are now surrounded by security and privacy technologies, officers, consultants, and regulators.

Figure 6 lists some key questions.

Trust is critical here. While many consumers have increased their online purchasing, there are still lots that have reservations about making serious purchases over the Web. Business professionals feel the same way about large B2B transactions, and problems with spam and pornography continue to grow.

Denial of service attacks, viruses, sabotage, and full-blown information warfare are all likely to increase as our dependency on digital transaction processing increases.

Figure 6. How to think about security and privacy

[Figure: Venn diagram titled "Integration" with three overlapping circles: Business, Technology, and Management]

Business
- Do Our Employees, Suppliers, Customers & Partners "Trust" Our Transactions?
- Are We Protected from Information Warfare of All Kinds?

Technology
- Do We Have an Adaptive Security & Privacy Policy?
- Do Our Authentication, Authorization, Administration & Recovery Technologies Integrate & Inter-Operate?

Management
- Do We Practice What We Preach?
- Do We Have Security & Privacy Officers With Bold Organizational Lines to Business & Technology?
- Do We Measure Cost-Benefit?
- Do We Share Our Security & Privacy Capabilities?

The technology has to provide trust and protection in cost-effective ways, and all of the trust, protection, and technology pieces have to work together in an environment that is procedural and disciplined. The key point? Trust and protection are business technology goals, not just technology goals.

Standards

The whole area of standards is fraught with emotion. Nearly everyone in your organization will have an opinion about what the company should do about operating systems, applications, hardware, software acquisition, services, and even system development life cycles. Everyone. Even the people who have nothing to do with maintaining your computing and communications environment will have strong opinions about when everyone should move to the next version of Microsoft Office. In fact, discussions about standards often take on epic proportions with otherwise sane professionals threatening to fall on their swords if the organization does not move to the newest version of Windows (or Notes, or Exchange—or whatever).

The other side of standards is technology-based: will the world migrate to Java applications or will extensible mark-up language (XML) obviate the need for common applications architectures? Will fast Ethernet grow dramatically? Will Bluetooth or other wireless standards like 802.11 (a/b) dominate mobile computing?

Why are variation and technology standards so important? Because they determine how much business agility you have, how much business technology efficiency you enjoy, and how much you spend to keep the trains running on time.

It is likely you have heard references to return on investment (ROI) and the total cost of ownership (TCO) every time the subject of standards comes up. Lest there be no misunderstanding here, there is no question that environments with less-rather-then-more variation will save money. Or put another way, you have some choices here. You can aspire to be sane or insane.

What does management really want here? Standards are a second-order business driver. Most businesses do not associate standards-setting with business models, processes, profits or losses. Whether the environment has 1, 5, or 20 word processing systems variation is seldom associated with business performance: it is hard to link homogeneity with sales! But the fact remains that expenses are clearly related to sales, and standards are closely related to expenses. Herein lies the subtlety of standards and 21st century business technology convergence.

What else does business management want? They want flexibility—and here lies the only sometimes-valid argument against standards. If your environment does not support the business computing or communications processes the business feels it needs to compete, there will be loud complaints. Business managers want to compute and communicate competitively. Standards are often perceived as obstacles, not enablers. But almost always, nothing could be farther from the truth.

If we have learned anything over the past few decades, it is that standards are as much about organizational structures, processes, and cultures as they are about technology. The ability to actually control computing and communications environments through thoughtful governance policies and procedures will determine to a great extent how standardized organizations become. We have also learned that the more you succeed the less you will pay.

We have also just recently learned that technology standards yield business flexibility. Not only do standard architectures permit business agility but less-rather-than-more variation in your environment will keep costs manageable. The following figure lists some questions that will help you implement a standards strategy.

Organization

Beyond the endless discussions about death march CIOs who report to CFOs, and Cheshire CIOs who have landed seats at the big table courtesy of their CEO-reporting relationship, are huge issues around how to make IT "work" in your company. Here are a few of them:

Figure 7. How to think about standards

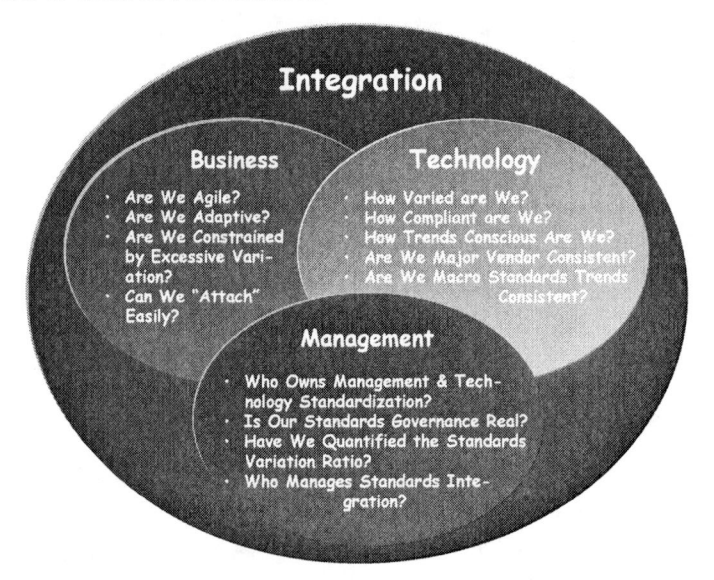

Integration

Business
- Are We Agile?
- Are We Adaptive?
- Are We Constrained by Excessive Variation?
- Can We "Attach" Easily?

Technology
- How Varied are We?
- How Compliant are We?
- How Trends Conscious Are We?
- Are We Major Vendor Consistent?
- Are We Macro Standards Trends Consistent?

Management
- Who Owns Management & Technology Standardization?
- Is Our Standards Governance Real?
- Have We Quantified the Standards Variation Ratio?
- Who Manages Standards Integration?

- The re-engineering of IT organizations will surface as one of the major corporate imperatives of the new millennium: companies will look to IT to (really) integrate with the business and provide competitive advantage; organizations that fail to assume this new role will be ousted in favor of new regimes that "get it."
- Speed and flexibility will become as important as consistency; "good 'ol boy" relationships will be (partially, not completely!) replaced by strategic partnerships that will be judged by performance—not historical inertia.
- As skill sets become obsolete faster and faster, there will be pressure to change IT organizations at a moment's notice. This will dictate against large permanent in-house staffs organized to protect their existence. New applications pressures will kill entrenched bureaucracies and give rise to a new class of results-oriented hired guns.
- The emphasis on business/IT convergence will increasingly focus on business requirements which in turn will lead to business applications and computing and communications infrastructure specifications. Given the pace of technology change, it is essential that your organizational infer requirements and produce specifications quickly and efficiently. This will require companies to tilt toward staff with these kinds of capabilities as they proportionately tilt away from implementation skills. IT organizations will be driven by "architects" and "specifiers"—not programmers.

Companies will find it increasingly difficult—if not impossible—to keep their staff current in the newest business technologies. This means that IT organizations by default will have to outsource certain skills. The approach that may make the most sense is one that recognizes that future core competencies will not consist of in-house implementation expertise but expertise that can abstract, synthesize, integrate, design, plan, and manage IT.

Figure 8 lists the key organization questions. The questions focus essentially on the role that business technology plays in the company as well as the tools necessary to manage business technology assets. As always, incentives should play a pivotal role in business technology optimization.

People

It is naive to believe that behavior will change by redrawing organizational boundaries or by codifying new responsibilities. In order to make 21st century business technology convergence work, several things must be true:

Skillsets must be re-examined: skillsets that supported mainframe-based applications, data center operations, and related activities are less valuable today— and will certainly be less so in the future—than architecture design, systems integration, distributed applications (so-called network centric applications), project management, and program management skillsets.

Figure 8. How to think about organizations

Incentives must be re-examined: we must revisit the reward structure to make certain that the skills, talent, and activities that mean the most to the company are generously rewarded, while those of less importance are rewarded accordingly. It is essential that the "right" message be sent here: employees must believe that (a) there is a clear vision for the business/technology relationship and (b) they will be rewarded for their dedication to this relationship.

A new breed of business/IT professionals must be fielded, professionals with an understanding of broad and specific technology trends, business trends, and how to convert the intersection into system requirements and system specifications. Such professionals will work directly within the businesses to understand how technology can cost-effectively define, enable, and support business models and processes.

Take a look at Figure 9. Many of the serious people questions are there.

BUSINESS TECHNOLOGY TRENDS

Who is in charge of tracking business technology trends in your company? Lots of places have in-house gurus but very few have created formal positions to track the major technology trends that can impact their companies. I must confess that I have always found this amazing given the pace of technology and business change. Maybe it is time for all of us to rethink our technology watch strategies.

So how do you identify the technologies most likely to keep your company grow-

Figure 9. How to think about people

ing and profitable? The explosion in technology has changed the way you buy and apply technology and has forever changed expectations about how technology can and should influence your connectivity to customers, suppliers, and employees.

What you need is a technology investment agenda that helps you identify the technologies in which you should invest more and those that get little or none of your financial attention.

The agenda ultimately must be practical: while blue sky research projects can be lots of fun (especially for those who conduct them), management must find the technologies likely to yield the most growth and profitability, not the coolest write-up in a technical trade publication. But this can be tough especially when there is so much technology to track—and relatively little certainty about what your business models and processes will look like in two or three years.

The trick is to identify the right technologies at the right time and make the most cost-effective investments possible. Or, stated a little differently, it is hard to innovate if you do not track trends in computing and communications technology.

Let us identify some technologies that should appear on everyone's list, technologies that will impact a wide range of horizontal and vertical industries:

- Objects
- Wireless
- Peer-to-Peer
- Optimization
- Web services
- Artificial intelligence
- Customization & personalization
- Data integration
- Applications integration
- Security solutions

No doubt there are others. This is an almost generic list; the key is to distill it down to those likely to most impact your business by addressing each of the technologies with reference to trends in business models and processes and the management policies and procedures necessary to optimize performance.

Simplicity is important here. A long list of cool technologies does not help many companies with their business/technology alignment. The key is to reduce the number to those that can be monitored and piloted. This is the proverbial technology hit list—as famous for what is on it as for what is not.

The purpose of technology monitoring and assessment is to develop lists of technologies likely to impact your business. Hit lists are excellent devices for rank-ordering and screening technologies. They also focus attention on specific technology opportunities. But remember that the holistic approach requires that you identify technologies with reference to current and future business models and processes.

Pilot projects should be real projects. They should have project managers, schedules, milestones, and budgets. They also need dedicated professionals to objectively determine where the promise really lies.

Pilot projects should not last too long: a pilot project that requires 6 or more months to yield the classic go/no go result is much less likely to succeed than one that yields an answer in 60 days. In fact, if you institutionalize the piloting process, your ability to attract funds to conduct technology pilots will correlate to how quickly you have delivered results in the past.

Investments you make in new technologies (and in the pilots that justify these investments) should be measured over time to determine if the technology is delivering on the promise you expected. Metrics should be developed that address the rationale for the technology's deployment, metrics such as cost, speed, effectiveness, and the like, as well as business value metrics, like customer satisfaction, market share, and profitability.

Again, and as argued throughout this report, technology trends assessment in a vacuum is useless: technology's value is only calculated through business success. So we also need to track the major trends in business models and processes, such as:

- Transaction processing
- Execution speed
- "Agility"
- Adaptation
- Collaboration
- Supply chain integration

All of this is summarized in Figure 10.

10 TAKE-AWAYS

This trends report is about the relationship among business, technology, and management. Ideally, after reading the report you would think a little differently, know some things you did not know before, and be able to do some new things.

Here is a summary of some of the key ideas in the report that address the think/know/do possibilities:

Figure 10. How to think about business technology trends

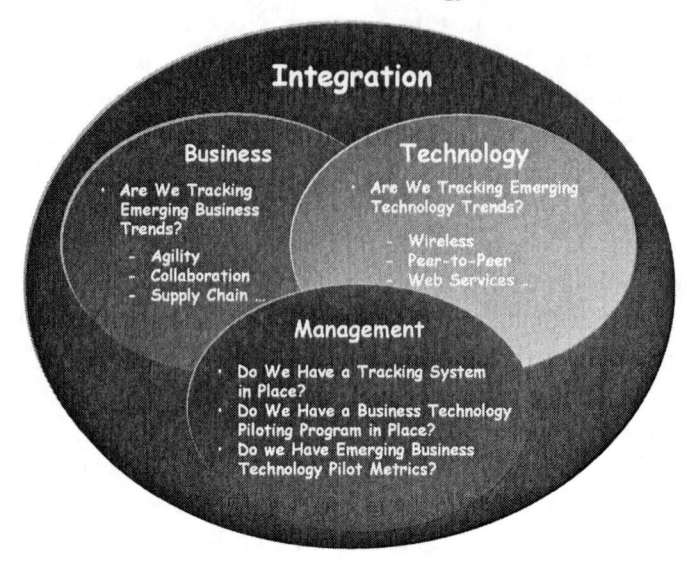

Integration

- **Business**
  - Are We Tracking Emerging Business Trends?
    - Agility
    - Collaboration
    - Supply Chain

- **Technology**
  - Are We Tracking Emerging Technology Trends?
    - Wireless
    - Peer-to-Peer
    - Web Services

- **Management**
  - Do We Have a Tracking System in Place?
  - Do We Have a Business Technology Piloting Program in Place?
  - Do we Have Emerging Business Technology Pilot Metrics?

1. Technology decisions cannot be made in a business or management vacuum; all technology decisions touch internal (employees) and external (customers, suppliers, partners) players and all technology investments should be driven by holistic strategies.
2. There are "levels" of technology, infrastructure technology, enabling technology, and applications technology, that work inside and outside of their corporate firewalls.
3. There are management processes that can make technology investments more cost-effective, processes such as business case development, assessments of total cost of ownership (TCO) and return-on-investment, performance metrics management, and due diligence4. .
4. The relationship between business, technology, and management is inseparable—no matter how hard we try to treat them independently.
5. The range of mainstream technologies necessary to define and support successful business models, such as communications, applications, data, and security distributed across the infrastructure/enabling/application technologies landscape.
6. The range of emerging technologies, like wireless, Web services, natural language understanding, and automation (including the "semantic" Web) likely to impact business the most.
7. The range of business technology management tools and techniques necessary to deploy the right technologies in the right way, tools, and techniques

like ROI, TCO, metrics, EVA, due diligence, business cases, benchmarking, requirements modeling, and systems analysis.
8. The range of organizational and people (political) strategies necessary to make all of the pieces work together including strategies like decentralization, standards-setting, project management, and e-learning.
9. Develop a strategic business technology management plan with all of the necessary components, such as business cases for specific technology projects (like ERP or CRM projects), the financials around TCO, ROI, and strategic EVA), organizational strategies, people strategies, and a plan to monitor mainstream and emerging technologies.
10. Communicate—"sell"—the plan to a skeptical audience.

WHERE IS IT ALL GOING?

The business technology relationship is changing. Let us look at the drivers that are defining the new business technology relationship.

Transformation Trend # 1: Operational vs. Strategic Technology

Discussions about the commoditization of technology are accurate to a point. There is no question that PCs, laptops, and routers are commodities; even some services—like legacy systems maintenance, data center management, and certainly programming—have become commoditized. But the real story here is not the commoditization per se but the bifurcation of business technology into operational and strategic layers. Operational technology is what is becoming commoditized; strategic technology is alive and well—and still a competitive differentiator, in spite of what some recent articles have been arguing.[1] What is the difference? Operational technology supports current and emerging business models and processes in well-defined ways with equipment and services whose costs have become stable and predictable and have generally declined significantly over the past decade. Hardware price/performance ratios are perhaps the most obvious example of this trend but there are others as well including what we are willing to pay for programming. Strategic technology, on the other hand, is the result of creative business technology convergence where, for example, a Wal-Mart streamlines its supply chain, a Starbucks offers wireless access to the Web, and a Vanguard leverages its Web site to dramatically reduce it is costs. There is no limit to how creative business technology convergence can become; there is no limit to how strategic the business technology relationship can be.

Implications

Figure 11 draws the important line. Above the line is where management lives. It is also where big front office applications—like CRM applications—live. Below the line are the back office applications and the infrastructure that enables digital contact with customers, suppliers, partners, and employees.

Companies should segment their technologies into operational and strategic layers and adjust their acquisition, deployment, and support procedures accordingly. They should also segment the professionals who tend to these tasks. Beyond this, consideration should be given about the need for Infrastructure Officers, Business Technology Officers, and Business Technology Strategists. Today's Chief Information Officers and Chief Technology Officers do not map well onto these new requirements. The segmentation suggested here would better support the management of the transformed organization.

Transformation Trend #2: Consolidation

The technology industry is consolidating, not in absolute numbers, but in terms of the number of companies from which companies buy most of their technology. There is much more of a tendency to buy from those with the largest market share than from smaller vendors, even if these vendors are spectacularly hungry for business (and therefore more willing to deal). Some of this is because of risk aversion, and some because we are now comfortable with a relatively small number of vendors as suppliers of our computing and communications technology and services. Requests

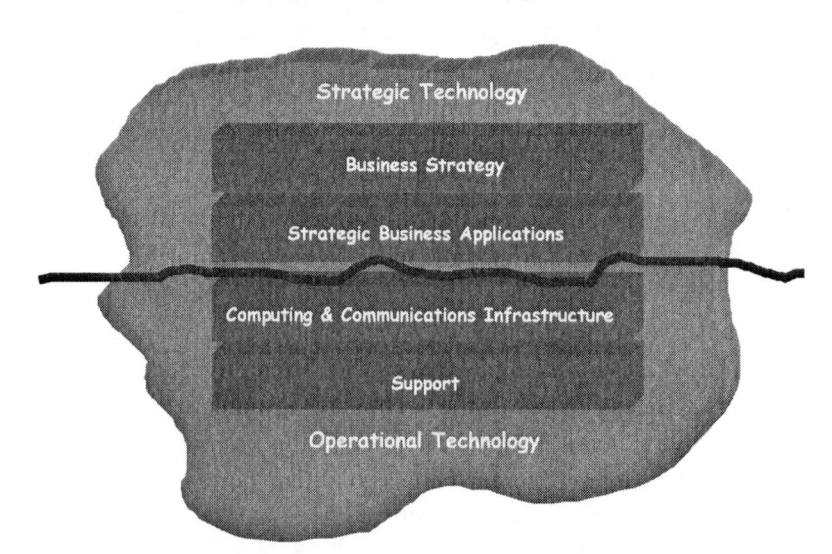

Figure 11. Strategic versus operational technology

for Proposals (RFPs) that used to be sent to ten PC/laptop manufacturers or VARs are now sent to three or four; requests for data base (DB) management platforms now only go to two or three DB vendors.

Implications

The good news is that consolidation reduces complexity. It is easier, and safer, to select among 5 rather than 50 alternatives. The industry has matured to the point where "best of breed" and "single source" decisions have all but disappeared. Often—though certainly not always—the best of breed is the dominant (single) source.

Transformation Trend #3: Discipline

There is also a general increase in discipline used to acquire, deploy, and support technology. We are getting much more disciplined about the use of business cases, total cost of ownership models, return on investment calculations, and project management best practices than we were a decade ago. Some of this is attributable to the general bear technology market over the past few years but some from our discovery of key relationships, like technology variation and support costs or formal vs. "informal" project management and project success. We are just better at making technology investment decisions.

Implications?

Since we no longer have to beg for logic in the technology investment process, we are more free to think about high impact strategic technology and cost-effective operational technology. We can also institutionalize policies and procedures for technology acquisition, deployment and support so we do not have to fight about these things nearly as often as we have in the past.

Transformation Trend #4: Integration & Interoperability

A significant additional transformation trend is our increasing ability to make disparate stuff—data bases, hardware systems, applications—work with one another. We have seen the field move from application programming interfaces (APIs) to data extraction, translation, and loading (ETL) to enterprise application integration (EAI) to Web services—all in about a decade. Good progress to an even better end game: ubiquitous transaction processing through seamless technology integration. Integration is now a core competency in many companies and the industry has responded with a barrage of new tools to make things cooperate. This trend will continue and threatens to dramatically alter software architectures and technology infrastructures.

Implications

We are now able to wrap older legacy systems in newer standards-based technologies. We are able to integrate supply chains with Web services, and we are able to think holistically about application integration and the transactions it supports—like up-selling and cross-selling.

Transformation Trend #5: Sourcing

Another major trend is our willingness to optimize sourcing. As more and more technology gets commoditized, we will see more and more hybrid sourcing models. Some companies outsource lots of processes while others have adopted a cosourcing model where their own people work closely with the outsourcer. The trend, however, is clear: companies are re-evaluating their sourcing strategies and have lengthened the list of potential candidates for full or partial outsourcing. Some of these include help desk support, production programming and application maintenance. If we extend this trend it is likely that we will see a lot more hosting of large applications that companies will increasingly rent instead of wrestling with implementation and support challenges.

Implications

We are faced with an interesting optimization challenge that is ultimately tied to core competency assessments. Figure 12 shows the decision matrix we all need to master:

Transformation Trend #6: Thinfrastructure

Network access is now almost ubiquitous today: we use desktops, laptops, personal digital assistants (PDAs), thin clients, and a host of multifunctional converged devices, such as integrated pagers, cell phones, and PDAs to access local area networks, wide area networks, virtual private networks, the Internet, hosted applications on these networks, as well as applications that run locally on these devices. These networks work.

Many companies provide multiple access devices to their employees, and employees often ask their companies to make personal devices (like PDAs) compatible

Figure 12. Sourcing optimization matrix

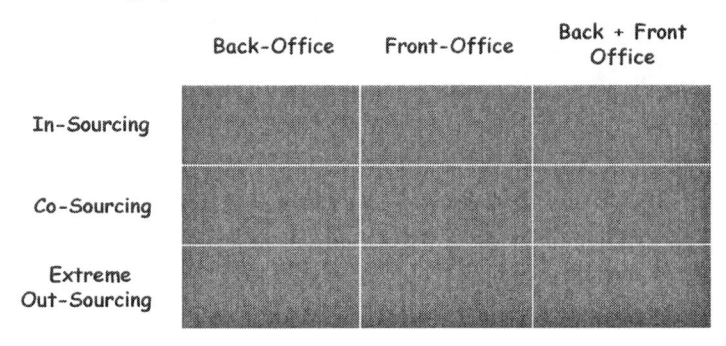

	Back-Office	Front-Office	Back + Front Office
In-Sourcing			
Co-Sourcing			
Extreme Out-Sourcing			

with their networks. All of this is made even more complicated with the introduction of wireless networks which make employees more independent and mobile.

Small, cheap, reliable devices that rely on always-on networks make sense. Shifting computing power from desktops and laptops to professionally managed servers makes sense. Moving storage from local drives to remote storage area networks makes sense. Fat clients should lose some weight—as we bulk up our already able servers. The total-cost-of-ownership (TCO)—not to mention the return-on-investment (ROI)—of skinny client/fat server architectures is compelling (to put it mildly).

Implications

Anyone updating their infrastructures should think about thinfrastructure. Migrations should be filtered with thinfrastructure opportunities. Pilots should be launched to collect TCO and ROI data, and you should begin the education process in your organization to raise everyone's consciousness about the potential for thinfrastructure in your company.

Transformation Trend #7: Collaboration

As data bases become more integrated, as shopping becomes more digital, and as "always on" access devices become more pervasive, we can expect to be treated to all sorts of offers. I already get e-mail (and snail mail) from companies that have profiled me. The have analyzed data about where I live, what I earn, and what I buy to determine what I like and what I would pay for what they are selling. This is first generation mass customization, child's play compared to what is coming.

Built on much the same data that mass marketing assumes, mass customization infers beyond the simpler correlations—like age, wealth, time of year—to specific ideas about what you and me would really like to buy, based on inferences about us as part of a larger group *and* as individual consumers

Contact can be "personalized" with customers, suppliers, partners, and employees through all varieties of messages including sales, marketing, service, and distribution. Over time, given how low digital transaction costs are compared to other ways companies touch the members of their value and supply chains, and how ubiquitous digital communication is becoming, companies will reassess their advertising and marketing budgets. They will go increasingly personal.

Supply chains will become more and more integrated over the next few years, and though we are still probably ten years away from real-time analysis and optimization, all of the business models and technologies assume that real-time is the end-game.

Figure 13 tries to map these trends. Companies will strive toward the top of the cube.

Implications

Think holistically and then make your technology integrate and interoperate. Cross-selling, up-selling, and supply chain optimization are only a few of the outcomes we should target. It is also important to let all of the collaborators in: who will be your suppliers, distributors, customers, partners?

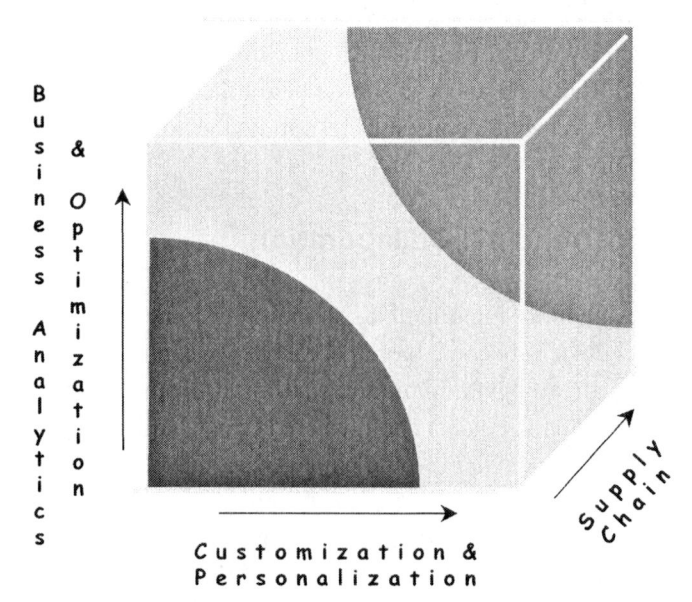

Figure 13. Collaboration trends

Transformation Trend #8: Renting

Five years ago application service providers (ASPs) were all the rage. Lots of companies emerged to host dot-com Web sites and facilitate business-to-business (B2B) and business-to-consumer (B2C) transactions. There was talk about extending the idea to all kinds of hosting, but before they could launch the next wave the dot.com bubble burst. The large enterprise software vendors breathed a sigh of relief: they were safe for a while at least. But little by little the ASPs—like zombies—came back from the dead. Worse, the software vendors themselves broke ranks and began to host their own applications for customers that were too small or scared to implement the software in-house. Up until now, this trend has been relatively slow to mature. But now even Siebel Systems has decided, again, to host its CRM software. SAP's are also doing it. What is going on?

Research in the field tells us that the probability of a successful enterprise software implementation is somewhere around .25. We know that unless you have got Houdini working for you, it is almost impossible to decipher enterprise software licenses. Unless you are blessed with internal "competency groups" (and the deep pockets necessary to keep them happy) you will need consultants to help you implement and support the application (and even deeper pockets to keep *them* happy). But much more importantly, unless your company adopts the processes that underlie the application—that is, unless your company has, for example, a customer centric culture with supporting models and processes—you are unlikely to ever recoup your technology investment in, for example, a customer relationship management (CRM) application. Need more logic? Compute the (real) total-cost-of-ownership and return-on-investment (ROI) of a major enterprise application and see if the numbers make sense over some reasonable period of time. There are also some risk calculations we could walk through.

There is a trend toward renting. Sometime this year Salesforce.com, the premier CRM ASP, is likely to go public—and lots of people think that it is likely to be a very successful IPO. Siebel Systems has responded with its own hosting services. Others have followed suit with more on the way. In fact, when we step back and look at a variety of trends in the industry, we see even more evidence that hosting will expand with a vengeance. Web services, utility computing, thin client architectures, and even the Semantic Web are all conceptually consistent with the hosting trend.

Are we sliding toward to new implementation standard? Are large scale enterprise application projects dinosaurs? Will paying-by-the-drink become the new revenue model for enterprise software vendors?

Implications

Given your odds of implementation success, given the uncertainties around TCO and ROI, and given other supporting trends, should you buy or rent? Without question, it makes sense to pilot some hosted applications. While you may not want to give up the enterprise applications and its supporting infrastructure, you might be surprised by the results of a well-conceived pilot.

The industry itself will also have to reconfigure its software licensing and pricing models—something it is been unwilling embrace. Many companies have already experienced the pain of shelfware; hosted applications open the door to customer expectations about "paying by the drink."

Transformation Trend #9: Clustering

Technologies have limited impact until full clusters develop around them consisting of all of the things necessary for technologies to grow, all of the applications, data, support, standards, and developers that keep technologies alive and well over long periods of time. It is really about wide and deep acceptance. Some technologies—like business rule servers—have crossed the prototype-to-cluster chasm, but the cluster is relatively small (and so is the technology's impact). Other technologies, like large enterprise resource planning platforms, have enormous supporting clusters and have had huge business impact. It is too early to tell if many of what the technology trade publications declare as "technologies-to-watch" will become high impact technologies. Real-time synchronization, business process modeling, grid computing and utility computing, among others, may or may not yield successful prototypes, which may or may not evolve into full-blown clusters. It is our job to objectively segment technology concepts, emerging technology prototypes, and technology clusters—before pulling out our checkbooks.

So what do you think should go on each list?

Some concepts would include:

- The Semantic Web
- Realtime synchronization
- Dynamic business processing modeling

Some emerging technologies would include:

- Grid computing
- Nanotechnology

- Web services
- Personalization
- Customization
- Voice recognition
- Thin clients
- The segway
- RFID

Some technology clusters include:

- ERP
- CRM
- B2B transaction processing
- Business intelligence
- Wireless communications
- Application servers
- Security services
- Technology outsourcing

Implications

The trick is to correctly categorize technologies before making major financial commitments to them. The prize goes to whoever can predict which technologies will cross the emerging technologies/technology clusters chasm. If you can do this accurately you can get a jump on the competition—but if you do it poorly you will waste a lot of money. Marching orders? Mostly buy clusters, occasionally invest in prototypes, and enjoy (but do not buy any) concepts. Or put another way, unless you are in the technology business, do not be an early adopter, a pioneer, or live on the bleeding edge.

Transformation Trend #10: 2007

Seems like only yesterday that technology companies were going public at an incredible pace and stock prices reached inexplicable highs. But then April of 2000 happened and everyone sobered up on exactly the same day. So what can we expect now that everyone is "cured"? We have weathered some tough times since the bubble burst. Does 2007 look good? Bad? What are the trends that matter?

2007 will see the return of solid, though not exuberant, technology spending. Stuff is wearing out, but do not expect those ten aging servers to be replaced by ten new ones. Capacity and architecture will reduce the need for additional boxes and in many cases enable technology managers to reduce the number of boxes they deploy (and the cost per box). PC sales will improve but instead of desktops and laptops, we will see more laptops, high-end PDAs, and thin clients dotting the network access and computational landscape. Total spending will increase slightly for access devices but do not expect PCs sales to skyrocket.

Enterprise software acquisition is going to get interesting in 2007 to 2009. More and more companies will invest in their existing enterprise ERP and CRM platforms and some will decide to rent them. The net effect of this acquisition/optimization shift will be modest increases in mega enterprise applications platforms and larger increases in optimization software and enterprise application hosting through vendor-specific and third party ASPs.

The data base world will continue its movement toward a triumvirate of Oracle, IBM, and Microsoft, but the real action will be in business intelligence and the tools that make it possible to better understand internal business processes and external customers, suppliers, and partners. Look for a lot of new products and investments here consistent with the optimization theme.

Lots more companies will get out of the technology business in 2007. Outsourcing will continue to increase as companies export more and more of their technology infrastructures to third parties in the US and abroad. It is not just about saving money; it is much more about core competencies, technology complexity, and flexibility. Companies will outsource their technology infrastructures as aggressively as they protect their business technology strategies. The big story in 2007 will be the increase in front-office outsourcing built on top of the technology infrastructure outsourcing that is been growing for some time now.

Technology *managers* will become even more practical than they have been over the past few years because more and more of what they buy, deploy, and support has been commoditized. There is also a solid appreciation for basic best practices, like standardization. Technology *strategists,* on the other hand, will see things holistically, blurring distinctions between business models and the technologies that enable them. Organizationally, we can expect to see technology organizations bifurcate into "operational" and "strategic" units, with the latter ideally moving to Mahogany Row. Lots of CIOs will think about changing their titles and job descriptions opting to spend more time with business managers than technologists. Companies will begin to reorganize around the operational/strategic distinction.

We will also see some important technologies mature in 2007. On the top of the list is wireless technology, followed by Web services and thin client architectures. Other technologies—like the Semantic Web and grid computing—will have more hype than substance in 2007 (though 2008 could be their breakout year).

The structure of the technology industry will continue to morph through acquisitions and mergers. As we come out of the technology capital expenditures slump, and companies begin to see the value of their currency, their stock prices and revenue prospects, rise, we can expect companies to try to maximize their relative strengths in specific sectors like software, communications, networking, and devices.

The really big story in 2007 will be psychological: for the first time in several years more technology professionals will see the glass as half full—not half empty.

The key is to recognize the crossroads we are at now and begin to take steps to think about which new direction to take. Start with a business technology management health check framed by the questions posed in Figures 2 through 10. The answers to these questions might help you better understand where you are now and where you should be going.

ENDNOTE

[1] Perhaps the most famous of these treatises was Nickolas G. Carr's "IT Doesn't Matter," published in the *Harvard Business Review*, May 1, 2003.

About the Author

Stephen J. Andriole's career has focused on the development, application, and management of information technology and analytical methodology to complex business problems. These problems have been in government and industry; Dr. Andriole has addressed them from academia, government, his own consulting company, a global insurance and financial services company, and from the unique perspective of a venture capitalist.

Dr. Andriole was the Director of the Cybernetics Technology Office of the Defense Advanced Research Projects Agency (DARPA) where he managed a $25 million program of research and development that led to a number of important scientific and technological advances in the broad-based information, decision, and computing sciences.

Dr. Andriole served as the Chief Technology Officer and Senior Vice President of Safeguard Scientifics, Inc., where he was responsible for identifying technology trends, translating that insight into the Safeguard investment strategy, and leveraging trends analyses with the Safeguard partner companies to help them develop business and marketing strategies.

Dr. Andriole was the Chief Technology Officer and Senior Vice President for Technology Strategy at CIGNA Corporation where he was responsible for the enterprise information architecture, computing standards, the technology research and development program, and data security, as well as the overall alignment of enterprise information technology investments with CIGNA's multiple lines of business.

As an entrepreneur, Dr. Andriole founded International Information Systems (IIS), Inc., which designed interactive systems for a variety of corporate and government clients. He is also the founder of TechVestCo, a new economy consulting consortium that identifies and leverages technology trends to help clients optimize business technology investments.

Dr. Andriole is currently the Thomas G. Labrecque Professor of Business Technology at Villanova University where he teaches and directs applied research in business/technology alignment and pervasive computing. He is also a Fellow at the Cutter Consortium.

He is formerly a Professor of Information Systems & Electrical & Computer Engineering at Drexel University in Philadelphia, Pennsylvania, and a member of the faculty of George Mason University as a Professor and Chairman of the Department of Information Systems & Systems Engineering; he was awarded an endowed chair from the university becoming the university's first George Mason Institute Professor of Information Technology.

Some of Dr. Andriole's books include *Interactive Computer-Based Systems Design and Development* (Petrocelli Books, Inc., 1983), *Microcomputer Decision Support Systems* (QED Information Sciences, Inc., 1985), *Applications in Artificial Intelligence* (Petrocelli Books, Inc., 1986), *Information System Design Principles for the 90s* (AFCEA International Press, 1990), the *Sourcebook of Applied Artificial Intelligence* (McGraw-Hill, 1992), a (co-authored with Len Adelman) book on user interface technology for Lawrence Erlbaum Associates, Inc., entitled *Cognitive Systems Engineering* (1995), and a book for McGraw-Hill entitled *Managing Systems Requirements: Methods, Tools, & Cases* (1996). He has recently published articles in *Communications of the AIS,* the *Communications of the ACM* and the *Cutter IT Journal.* IGI/CyberTech published Dr. Andriole's *The 2nd Digital Revolution* in 2005.

Dr. Andriole received his BA from LaSalle University in 1971 and his Masters and Doctorate degrees from the University of Maryland in 1973 and 1974. His masters and doctoral work was supported by a National Defense Education Act fellowship. His PhD dissertation was funded by DARPA.

Details about Dr. Andriole's career can be found at www.andriole.com.

Index

Symbols

802.11a standard 271

A

Accenture's research program 111
access devices 128, 134
acquisition targets 47
adaptive, intelligent interface technology 89
advanced technology investment guidelines 91
advanced technology targets 86
advanced technology trends 109
agents 343
agile enterprise 221
AI tools and techniques 338
analytical hierarchy process (AHP) 27
applications 342, 370
application services and support 55
applications management 54
applications targets 47
application targets, range of 49
authentication 81
automation 304

B

bandwidth management 320
bandwidth optimization 320
blogs 117
Broadband 321
budget cycle alignment 10
business 367
business-to-business integration 234
business case drivers 40
business collaboration drivers 101
business intelligence 120, 302, 308
business models, current and future 108
business process management (BPM) 62
business technology cases 36
business technology convergence 101
business technology integration 361
business technology integration, trends in 349
business technology partnership 42
business technology trends 377
business technology trends analysis 99
business technology trends analysis methodology 100

C

carrier-grade ThinAir server 140
certificate process 85
Cisco Aironet 271
Cisco Aironet 1200 Series Access Point 271
citrix 166, 266

Citrix technology 167
Citrix technology, investment drivers 167
client-side software 116
clustering 388
collaboration 307, 385
collaborative business questions 37
collaborative communications 163
commercial-off-the-shelf (COTS) software 261
communications 25, 369
communications access 59
communications activity-based investment targets 61
communications targets 58
communications technology investment guidelines 64
communications trends 317
competition 243
competitive advantages 242
computable problems 337
connectivity technology investment targets 59
consolidation 382
content sources 282
convergence CRM 125, 134
corporate performance management (CPM) 62
cost/benefit questions 38
criteria weighting 27, 29
CRM scenarios 127
crowdsourcing 116
customer relationship management (CRM) 9, 211
customer relationship management (CRM), 161
customization 302

D

data 371
data, information & knowledge requirements matrix 67
data investment guidelines 72
data products 72
data targets 66
design aids 343
development services 316

differentiation 22
digital signal processors (DSPs) 308
discipline 383
document content management 306
due diligence 1
due diligence, LiquidHub 222
due diligence, LiquidHub, "packaging" and communications 228
due diligence, LiquidHub, "politically correct" products and services 227
due diligence, LiquidHub, "right" technology 223
due diligence, LiquidHub, budget cycle alignment 224
due diligence, LiquidHub, changes to processes and culture 226
due diligence, LiquidHub, differentiation 227
due diligence, LiquidHub, experienced management 228
due diligence, LiquidHub, horizontal and vertical strength 226
due diligence, LiquidHub, industry awareness 227
due diligence, LiquidHub, infrastructure requirements 223
due diligence, LiquidHub, multiple exits 226
due diligence, LiquidHub, partners and allies 227
due diligence, LiquidHub, quantitative impact 225
due diligence, LiquidHub, recruitment and retention 227
due diligence, LiquidHub, solutions 226
due diligence, NexTone 185
due diligence, NexTone, "right" technology 185
due diligence, NexTone, changes to process or culture 187
due diligence, NexTone, compelling differentiation story 187
due diligence, NexTone, end-to-end solutions 187
due diligence, NexTone, experienced management 188

due diligence, NexTone, horizontal and vertical strength 186
due diligence, NexTone, industry awareness 188
due diligence, NexTone, infrastructure requirements 185
due diligence, NexTone, multiple defaults 187
due diligence, NexTone, quantitative impact 186
due diligence, Oracle 202
due diligence, PFRT, "packaging" and communications 173
due diligence, PFRT, "politically correct" products and services 170
due diligence, PFRT, "right" technology 165
due diligence, PFRT, budget cycle alignment 167
due diligence, PFRT, changes to processes and culture 168
due diligence, PFRT, differentiation 173
due diligence, PFRT, experienced management 173
due diligence, PFRT, horizontal and vertical strength 169
due diligence, PFRT, industry awareness 170
due diligence, PFRT, infrastructure requirements 166
due diligence, PFRT, multiple exits 169
due diligence, PFRT, partners and allies 170
due diligence, PFRT, quantitative impact 168
due diligence, PFRT, recruitment and retention 172
due diligence, PFRT, solutions 169
due diligence, Postiva 251
due diligence, Postiva, "packaging" and communications 256
due diligence, Postiva, "politically correct" products and services 255
due diligence, Postiva, budget cycle alignment 252
due diligence, Postiva, changes to processes and culture 253
due diligence, Postiva, differentiation 255
due diligence, Postiva, experienced management 255
due diligence, Postiva, horizontal and vertical strength 254
due diligence, Postiva, industry awareness 254
due diligence, Postiva, infrastructure requirements 252
due diligence, Postiva, multiple exits 254
due diligence, Postiva, partners and allies 255
due diligence, Postiva, quantitative impact 253
due diligence, Postiva, recruitment and retention 255
due diligence, Postiva, solutions 254
due diligence, Prudential Fox Roach/Trident (PFRT) 165
due diligence, radio frequency identification (RFID) 191
due diligence, remote access technology 157
due diligence, RFID 202
due diligence, RFID, "packaging" and communications 208
due diligence, RFID, "politically correct" products and services 207
due diligence, RFID, "right" technology 202
due diligence, RFID, budget cycle alignment 203
due diligence, RFID, changes to processes and culture 205
due diligence, RFID, differentiation 208
due diligence, RFID, experienced management 208
due diligence, RFID, horizontal and vertical strength 206
due diligence, RFID, industry awareness 206
due diligence, RFID, infrastructure requirements 202
due diligence, RFID, multiple exits 206
due diligence, RFID, partners and allies 206
due diligence, RFID, quantitative impact

204
due diligence, RFID, recruitment and retention 207
due diligence, RFID, solutions 205
due diligence, TAA, "right" technology 148
due diligence, TAA, changes to process and culture 148
due diligence, TAA, experienced management 151
due diligence, TAA, horizontal and vertical strength 150
due diligence, TAA, industry awareness 151
due diligence, TAA, infrastructure requirements 148
due diligence, TAA, multiple defaults 150
due diligence, TAA, outcome 153
due diligence, TAA, packaging and communications 151
due diligence, TAA, partners and allies 152
due diligence, TAA, quantitative impact 150
due diligence, TAA, solutions 150
due diligence, TechVestCo 263
due diligence, TechVestCo, "packaging" and communications 268
due diligence, TechVestCo, "politically correct" products and services 267
due diligence, TechVestCo, "right" technology 264
due diligence, TechVestCo, budget cycle alignment 265
due diligence, TechVestCo, changes to processes and culture 266
due diligence, TechVestCo, differentiation 268
due diligence, TechVestCo, experienced management 268
due diligence, TechVestCo, horizontal and vertical strength 267
due diligence, TechVestCo, industry awareness 267
due diligence, TechVestCo, infrastructure requirements 265
due diligence, TechVestCo, multiple exits 266
due diligence, TechVestCo, partners and allies 267
due diligence, TechVestCo, quantitative impact 265
due diligence, TechVestCo, recruitment and retention 268
due diligence, TechVestCo, solutions 266
due diligence, ThinAirApps (TAA) 148
due diligence, Villanova University 273
due diligence, Villanova University, "packaging" and communications 279
due diligence, Villanova University, "politically correct" products and services 278
due diligence, Villanova University, "right" technology 273
due diligence, Villanova University, budget cycle alignment 275
due diligence, Villanova University, changes to processes and culture 276
due diligence, Villanova University, differentiation 278
due diligence, Villanova University, experienced management 279
due diligence, Villanova University, horizontal and vertical strength 277
due diligence, Villanova University, industry awareness 278
due diligence, Villanova University, infrastructure requirements 274
due diligence, Villanova University, multiple exits 277
due diligence, Villanova University, partners and allies 278
due diligence, Villanova University, quantitative impact 275
due diligence, Villanova University, recruitment and retention 278
due diligence, Villanova University, solutions 277
due diligence, voice-over-IP (VOIP) 175
due diligence, wireless communications technology 136
due diligence, wireless technology 270
due diligence analysis, off-the-shelf tools 285

398 Index

due diligence case studies 135
due diligence criteria 2
due diligence process 1
due diligence project management 285
due diligence template 287, 290
due diligence tools and techniques 280

E

e-mail trust services and products 238
e-mail trust solutions 229
EAI 212, 216
EAI/IAI/exchange integration 298
EA reference model 212
easier to use software applications 258
electronic publishing 306
EM-agents 230
EM-Agents.tm 230
embedded applications 308
embedded systems 343
emerging natural interface technology 90
enterprise applications 52
enterprise architecture (EA) 212
enterprise architecture path 213
enterprise architecture planning (EAP) 212
enterprise architecture reference model 215
enterprise architecture service capability 210
enterprise information portals (EIPs) 315
enterprise investing 157, 191, 270
enterprise investing in radio frequency identification (RFID) 191
enterprise investing in remote access technology 157
enterprise investing in wireless technology 270
enterprise resource planning (ERP) 161
enterprise resource planning (ERP) applications 52, 264
enterprise services architectures 219
enterprise services path 216
enterprise services transformation 213
enterprise services transformation roadmapSM (ESTR) 212
ERP, 211

ERP portals 307
ESA IT budget allocation comparison 220
EST technology, investment drivers 225
evaluation process 262
execution 40
experienced management 23

F

"food chain" costs and margins 96
first generation (1G) distributed computing 5
folksonomies 117
full service TSPs 314

G

general business trends 100

H

"hertical" service capabilities 93
HCI routine identification process 261
HCI workbench 261
horizontal and vertical market strategy 201
horizontal strength 17

I

IBM's research program 110
industry awareness 19
infrastructure engineering services 317
infrastructure integration 76
infrastructure investment framework 77
infrastructure investment guidelines 76
infrastructure requirements 9
infrastructure support services 74
infrastructure targets 73
integrated security requirements 80, 83
integration 197, 383
intellectual property (IP) 185
intelligent agents 343
intelligent decision-making 330
intelligent military planning 344
intelligent systems 342, 343
intelligent systems technology 337
intelligent systems technology, trends in 327

interactive knowledge-based workbench 260
interoperability 383
intersecting trends 133
investing in knowledge-based user-computer interaction 258
investment drivers 11, 204, 225
investment opportunities 80, 83
investment opportunity 159

K

knowledge-based user-computer interaction 258
knowledge base structure 263

L

legal compliance 231
LiquidHub, approach 212
LiquidHub, background 210
Liquid Hub, challenge 211
LiquidHub, EAI 216
LiquidHub, enterprise architecture path 213
LiquidHub, enterprise services architectures 219
LiquidHub, enterprise services path 216
LiquidHub, enterprise services transformation roadmap 221
LiquidHub, enterprise services transformation roadmap (ESTR) 223
LiquidHub, Inc. 210
LiquidHub, service oriented architectures (SOAs) 217
LiquidHub case 210

M

manageability 197
market analysis 180
market evolution 182
marketing strategy 184
master data management 120
master data management for business intelligence 120, 133
McCarran (Las Vegas) Airport 194
messaging 230
messaging optimization 234, 240

Microsoft's R&D agenda 109
military planning processes 344

N

NASA Dryden Flight Research Center 194
natural interfaces 308
network applications 322
network security solutions 319
network services 322
NexTone, background 177
NexTone, competition 179
NexTone, intellectual property 184
NexTone, management team 178
NexTone, market 179
NexTone, market analysis 180
NexTone, market evolution 182
NexTone, marketing strategy 184
NexTone, prior financing 178
NexTone, product strategy 184
NexTone, strategic partnerships 180
NexTone, technology drivers 181
NexTone Communications (NexTone) 175
NexTone Communications case 175

O

off-the-shelf tools for due diligence analysis 285
office productivity applications 161
online transaction processing (OLAP) 68
operational technology 381
optical networking 323
Oracle, horizontal and vertical market strategy 201
Oracle, integration 197
Oracle, manageability 197
Oracle, RFID hardware considerations 196
Oracle, RFID standards 196
Oracle, scalability 197
Oracle, security 197
Oracle's RFID development roadmap 195
Oracle's RFID investment strategy 193
Oracle case 191
Oracle context 207
Oracle RFID partners 199
Oracle USA 191

organization 374
outcomes 40
outsourced service providers 312

P

"politically correct" products 20
"politically correct" services 20
packaging 25
peer-to-peer computing 308
people 376
personal computers (PCs) 5
personal digital assistant (PDA) 48, 350
personal digital assistants (PDAs) 5
personalization 302
pervasive computing, major computing eras 295
pervasive computing, trends in 293
pervasive computing action plan 324
pervasive computing technology trends 298
PFRT, investment opportunity 159
podcasts 117
portals, personal and professional 307
Postiva 229
Postiva, "right" technology 252
Postiva, business-to-business integration 234
Postiva, CFO 249
Postiva, competition 243
Postiva, competitive advantages 242
Postiva, competitive map 244
Postiva, CPO 249
Postiva, CTO 250
Postiva, executive vice president 248
Postiva, Inc., background 229
Postiva, legal compliance 231
Postiva, management 246
Postiva, market 231
Postiva, marketing 236
Postiva, messaging optimization 234
Postiva, opportunity 231
Postiva, positioning 236
Postiva, president & CEO 246
Postiva, privacy 231
Postiva, privacy consulting 238
Postiva, productivity 233
Postiva, products 238
Postiva, projected financial performance 246
Postiva, revenue and business model 238
Postiva, RightPathtm compliance system 239
Postiva, senior architect 251
Postiva, strategy 236
Postiva, technology 245
Postiva, vice president 250
Postiva "trusted message" seal program 238
Postiva case 229
Postiva development plan 237
POSTiva Platform 230
Postiva solution 236
powerful business cases 35
privacy 231, 356
product strategy 184
prototyping process 262
Prudential Fox & Roach/Trident /Trident context 172
Prudential Fox Roach/Trident (PFRT) realtors 157
Prudential Fox Roach/Trident case 157

Q

quantitative impact 13

R

"right" technology 4, 148
radio frequency (RF) waves 191
radio frequency identification (RFID) 191
real estate trends 163
recruitment 22
relative criteria weightings 291
relative importance 25, 27
remote access technology 157
renting 387
requirements modeling process 261
research and development (R&D) 43
retention 22
return-on-investment (ROI) 12, 129
RFID development roadmap 195

RFID hardware considerations 196
RFID investment strategy 193
RFID partners 199
RFID standards 196
RFID system 191
RFID tags, types of 193
RFID technology, investment drivers 204, 253
rich content aggregation/management 305
rich Internet applications 116
RightPathtm compliance system 239
right team, organizing 32
risk questions 39
RSS 116

S

sales force automation (SFA) 161
savvy consultants 33
scalability 197
scheduling 34
security 197
security and privacy 372
security investment guidelines 85
security targets 79
server-side software 116
service-oriented architectures 311
service oriented architectures (SOAs) 217
services investment guidelines 97
services investment target map 72
services targets 92
services trends 312
software applications, easier to use 258
software applications investment guidelines 56
software configuration management 306
software development and delivery 112, 132
software trends 298
sourcing 384
sourcing optimization matrix 88, 385
standards 373
strategic partnerships 180
strategic technology 381
supply chain connectivity 302
supply chain management (SCM) 6, 211
supply chain planning (SCP) 6

T

"trusted message" seal program 238
TACPLAN aid 345
talent profiler 24
technology 356
technology drivers 181
technology investments 41, 43, 44
technology product and service development 210
technology questions 38
technology services map 96
technology timeline 5
technology trends 111
technology trends analysis 293, 327, 349
technology trends analysis methodology 281
TechVestCo, interactive knowledge-based workbench 260
TechVestCo, software applications 258
TechVestCo, user interface workbench 259
TechVestCo case 258
Telecom 321
text-to-speech (TTS) technology 319
ThinAirApps ("TAA") 136
ThinAirApps, accomplishments 143
ThinAirApps, additional funds 146
ThinAirApps, CEO 144
ThinAirApps, company description 138
ThinAirApps, CTO 144
ThinAirApps, customers 142
ThinAirApps, executive vice president 145
ThinAirApps, financials 146
ThinAirApps, forecast 146
ThinAirApps, history 143
ThinAirApps, investment highlights 147
ThinAirApps, management team 143
ThinAirApps, market 138
ThinAirApps, marketing 143
ThinAirApps, risks 145
ThinAirApps, sales 142
ThinAirApps, strategy 138
ThinAirApps case 136
ThinAirApps opportunity 137
ThinAirApps solutions 140
ThinAir for the mobile professional 140

ThinAir Mail™ ("TAM") 138
ThinAir Server, pricing 141
ThinAir Server™ ("TAS") 138
ThinAir Server™, product line 141
ThinAir server for the enterprise 140
Thinfrastructure 384
thinfrastructure 384
total-cost-of-ownership (TCO) 12, 129
touch technologies 323
transaction platform development 301
transaction support 330
transformation trend 381, 382, 383, 384, 385, 387, 388, 389
trends in pervasive computing 293

U

user-computer interaction 258
user interface workbench 259

V

"vorizontal" service capabilities 93
venture capitalists (VCs) 8
venture investing 136, 175, 229
venture investing in e-mail trust solutions 229
venture investing in voice-over-IP (VOIP) 175

venture investing in wireless communications technology 136
vertical industry processes 108
vertical knowledge integration 90
vertical strength 17
vertical xSPs (VSPs) 315
Villanova University, background 270
Villanova University, challenge 271
Villanova University, results 273
Villanova University, solution 271
Villanova University case 270
voice-over-IP (VOIP) 175
voice recognition 308

W

Web-based applications and desktops 115
Web 2.0 115, 132
Web content management 306
Web protocols 116
Web services 311
Wikis 117
wireless applications 317
wireless communications technology 136
wireless technology 270
wireless technology, investment drivers 276
workbench's master menu 264